Advances in Intelligent Systems and Computing

Volume 1162

AF869645

Series Editor

Janusz Kacprzyk, Systems Research Institute, Polish Academy of Sciences, Warsaw, Poland

Advisory Editors

Nikhil R. Pal, Indian Statistical Institute, Kolkata, India

Rafael Bello Perez, Faculty of Mathematics, Physics and Computing, Universidad Central de Las Villas, Santa Clara, Cuba

Emilio S. Corchado, University of Salamanca, Salamanca, Spain

Hani Hagras, School of Computer Science and Electronic Engineering, University of Essex, Colchester, UK

László T. Kóczy, Department of Automation, Széchenyi István University, Gyor, Hungary

Vladik Kreinovich, Department of Computer Science, University of Texas at El Paso, El Paso, TX, USA

Chin-Teng Lin, Department of Electrical Engineering, National Chiao Tung University, Hsinchu, Taiwan

Jie Lu, Faculty of Engineering and Information Technology, University of Technology Sydney, Sydney, NSW, Australia

Patricia Melin, Graduate Program of Computer Science, Tijuana Institute of Technology, Tijuana, Mexico

Nadia Nedjah, Department of Electronics Engineering, University of Rio de Janeiro, Rio de Janeiro, Brazil

Ngoc Thanh Nguyen, Faculty of Computer Science and Management, Wrocław University of Technology, Wrocław, Poland

Jun Wang, Department of Mechanical and Automation Engineering, The Chinese University of Hong Kong, Shatin, Hong Kong

The series "Advances in Intelligent Systems and Computing" contains publications on theory, applications, and design methods of Intelligent Systems and Intelligent Computing. Virtually all disciplines such as engineering, natural sciences, computer and information science, ICT, economics, business, e-commerce, environment, healthcare, life science are covered. The list of topics spans all the areas of modern intelligent systems and computing such as: computational intelligence, soft computing including neural networks, fuzzy systems, evolutionary computing and the fusion of these paradigms, social intelligence, ambient intelligence, computational neuroscience, artificial life, virtual worlds and society, cognitive science and systems, Perception and Vision, DNA and immune based systems, self-organizing and adaptive systems, e-Learning and teaching, human-centered and human-centric computing, recommender systems, intelligent control, robotics and mechatronics including human-machine teaming, knowledge-based paradigms, learning paradigms, machine ethics, intelligent data analysis, knowledge management, intelligent agents, intelligent decision making and support, intelligent network security, trust management, interactive entertainment, Web intelligence and multimedia.

The publications within "Advances in Intelligent Systems and Computing" are primarily proceedings of important conferences, symposia and congresses. They cover significant recent developments in the field, both of a foundational and applicable character. An important characteristic feature of the series is the short publication time and world-wide distribution. This permits a rapid and broad dissemination of research results.

**** Indexing: The books of this series are submitted to ISI Proceedings, EI-Compendex, DBLP, SCOPUS, Google Scholar and Springerlink ****

More information about this series at http://www.springer.com/series/11156

Prachi Deshpande · Ajith Abraham ·
Brijesh Iyer · Kun Ma
Editors

Next Generation Information Processing System

Proceedings of ICCET 2020, Volume 2

Springer

Editors
Prachi Deshpande
Department of Computer Engineering
Dr. Babasaheb Ambedkar Technological University
Lonere, Maharashtra, India

Brijesh Iyer
Department of Electronics and Telecommunication Engineering
Dr. Babasaheb Ambedkar Technological University
Lonere, Maharashtra, India

Ajith Abraham
Machine Intelligence Research Labs (MIR Labs)
Auburn, WA, USA

Kun Ma
School of Information Science and Engineering
University of Jinan
Jinan, Shandong, China

ISSN 2194-5357 ISSN 2194-5365 (electronic)
Advances in Intelligent Systems and Computing
ISBN 978-981-15-4850-5 ISBN 978-981-15-4851-2 (eBook)
https://doi.org/10.1007/978-981-15-4851-2

© Springer Nature Singapore Pte Ltd. 2021
This work is subject to copyright. All rights are reserved by the Publisher, whether the whole or part of the material is concerned, specifically the rights of translation, reprinting, reuse of illustrations, recitation, broadcasting, reproduction on microfilms or in any other physical way, and transmission or information storage and retrieval, electronic adaptation, computer software, or by similar or dissimilar methodology now known or hereafter developed.
The use of general descriptive names, registered names, trademarks, service marks, etc. in this publication does not imply, even in the absence of a specific statement, that such names are exempt from the relevant protective laws and regulations and therefore free for general use.
The publisher, the authors and the editors are safe to assume that the advice and information in this book are believed to be true and accurate at the date of publication. Neither the publisher nor the authors or the editors give a warranty, expressed or implied, with respect to the material contained herein or for any errors or omissions that may have been made. The publisher remains neutral with regard to jurisdictional claims in published maps and institutional affiliations.

This Springer imprint is published by the registered company Springer Nature Singapore Pte Ltd.
The registered company address is: 152 Beach Road, #21-01/04 Gateway East, Singapore 189721, Singapore

Preface

Dr. Babasaheb Ambedkar Technological University, Lonere, is a State Technological University of the State of Maharashtra in India. Over the years, the Department of Electronics and Telecommunication Engineering of this University has been organizing faculty and staff development and continuing education programmes.

In the year 2013, the department had taken a new initiative to organize international conferences in the areas of next-generation computing technologies. The ICCET series (earlier ICCASP) is an outcome of this initiative. The 5th ICCET 2020 is jointly organized by the Department of E&TC Engineering of the University and Department of Computer Science and Engineering, MGM College of Engineering, Nanded, India. Keynote lectures, invited talks by eminent professors and panel discussion of the delegates with the academicians and industry personnel are the key features of 5th ICCET 2020.

This volume aims at a collection of scholarly articles in the area of next-generation computing technologies. We have received a great response in terms of the quantity and quality of individual research contributions for consideration. The conference had adopted a "double-blind peer review" process to select the papers with a strict plagiarism verification policy. Hence, the selected papers are the true record of research work in their edict.

We are thankful to the reviewers, session chairs and rapporteurs for their support. We also thank the authors and the delegates for their contributions and presence. We are extremely grateful to the university officials and organizing committee of MGMs College of Engineering, Nanded, for the support for this activity.

We are pledged to take this conference series to greater heights in the years to come with the aim to put forward the need based research and innovation.

Thank you one and all.

Lonere, India — Dr. Prachi Deshpande
Auburn, USA — Prof. Dr. Ajith Abraham
Lonere, India — Dr. Brijesh Iyer
Jinan, China — Dr. Kun Ma

Contents

About the Editors

Dr. Mrs. Prachi Deshpande earned her Ph.D. degree in Computer Engineering from Indian Institute of Technology, Roorkee, India in 2016. She is working as an Associate Professor in the University Department of Computer Engineering at Dr. Babasaheb Ambedkar Technological University, Lonere (A State Technological University). She is a recipient of MHRD fellowship during her doctoral degree. She has authored and coauthored over 30 research publications in peer-reviewed reputed journals and conference proceedings, special issue journals on the topics like IoT, big data, and next-generation technologies. She is having an experience of 14+ years in the academics at various positions and had authored 02 books in the area of cloud computing and big data. She has worked as the program committee member of various international conferences and reviewer for various international journals. Her research interests include cloud computing, big data analytics, IoT, data science and cognitive science.

Prof. Dr. Ajith Abraham works as the Director of Machine Intelligence Research Labs (MIR Labs), a Not-for-Profit Scientific Network for Innovation and Research Excellence connecting Industry and Academia. As an Investigator/Co-Investigator, he has won research grants worth over 100+ Million US$ from Australia, USA, EU, Italy, Czech Republic, France, Malaysia, and China. Dr. Abraham works in a multi-disciplinary environment involving machine intelligence, cyber-physical systems, Internet of things, network security, sensor networks, Web intelligence, Web services, and data mining and applies them to various real-world problems. In these areas, he has authored/coauthored more than 1,300+ research publications out of which there are 100+ books covering various aspects of Computer Science. About 1000+ publications are indexed by Scopus, and over 800 are indexed by Thomson ISI Web of Science. Some of the articles are available in the ScienceDirect Top 25 hottest articles. He has 700+ co-authors originating from 40+ countries. Dr. Abraham has more than 32,000+ academic citations (h-index of 83+ as per Google Scholar. He has given more than 100 plenary lectures and conference tutorials (in 20+ countries). For his research, he has won seven best paper awards at prestigious international conferences held in Belgium, Canada, Bahrain,

Czech Republic, China, and India. Dr. Abraham received Ph.D. degree in Computer Science from Monash University, Melbourne, Australia (2001), and a Master of Science Degree from Nanyang Technological University, Singapore (1998).

Dr. Brijesh Iyer received his Ph.D. degree in Electronics and Telecommunication Engineering from Indian Institute of Technology, Roorkee, India in 2015. He is an Associate Professor in the University Department of E&TC Engineering at Dr. Babasaheb Ambedkar Technological University, Lonere (A State Technological University). He is a recipient of INAE research fellowship in the field of Engineering. He had 02 patent to his credit and authored over 40 research publications in peer-reviewed reputed journals and conference proceedings. He had authored 05 five books on curricula as well as cutting-edge technologies like sensor technology. He has served as the program committee member of various international conferences and reviewer for various international journals. His research interests include RF front-end design for 5G and beyond, IoT, biomedical image/signal processing.

Dr. Kun Ma received his Ph.D. degree in Computer Software and Theory from Shandong University, Jinan, Shandong, China, in 2011. He is an Associate Professor (University Professor) in Shandong Provincial Key Laboratory for Network-based Intelligent Computing and School of Information Science and Engineering, University of Jinan, China. He has authored and coauthored over 60 research publications in peer-reviewed reputed journals and conference proceedings. His entire publications have been cited over 330 times (Google Scholar). He has served as the program committee member of various international conferences and reviewer for various international journals. He is the Co-Editor-in-Chief of International Journal of Computer Information Systems and Industrial Management Applications (IJCISIM) and International Journal of Autonomic Computing (IJAC). He is the managing editor of Journal of Information Assurance and Security (JIAS). He is the editorial board member of International Journal of Grid and Utility Computing and International Journal of Intelligent Systems Design and Computing. He is the guest editor of International Journal of Grid and Utility Computing, Informatica, International Journal of Communication Networks and Distributed Systems, International Journal of High Performance Computing and Networking, Recent Patents on Computer Science, and International Journal of Services Technology and Management. His research interests include model-driven engineering (MDE), data-intensive computing, and big data management. He has obtained 7 patents for inventions.

UAV Communication in FANETs with Metaheuristic Techniques

Meghna Goswami(✉), Rajeev Arya, and Prateek

National Institute of Technology Patna, Patna, Bihar, India
{meghna.ec18, rajeev.arya, prateek.ec18}@nitp.ac.in

Abstract. The recent advancement in communication technologies has paved the way for the flying ad hoc networks (FANETs) by enabling the deployment of small UAVs unmanned aerial vehicles (UAVs). Owing to its potential features, FANETs have a wide variety of applications. However, the UAVs face certain restrictions in terms of their battery power and mobility, resulting in short lifetime and unreliable routing. In this paper, we try to primarily address the issue plaguing to the short lifetime of the FANETs, namely the limited power availability of the UAVs. The paper aims to minimize the energy consumption with the aid of a clustering scheme and observe its impact on the lifetime of the network. Two different clustering methods are employed to achieve this, and a comparative analysis based on its performances is presented. The first scheme implemented is the combination of the k-means clustering algorithm and the firefly algorithm while the second method is based on the glowworm swarm optimization (GSO) and the firefly algorithm. The primary clusters are formed with the k-means clustering and the GSO, respectively, while the firefly algorithm elects the cluster heads and derives the optimal positions of the UAVs in the cluster. The performance of the schemes is further compared in terms of the cluster building time and energy consumption.

Keywords: FANETs · Clustering · Cluster head selection · GSO with firefly · Minimum energy consumption

1 Introduction

With the introduction of small UAVs, a new dimension of ad hoc network has been introduced under the banner of flying ad hoc networks (FANETs) [1]. In virtue of its unique characteristics such as scalability, adaptability, cost and installation efficiency, FANETs have found its operation in a variety of military and civil application domains. [2] The efficiency of FANETs is restrained primarily due to its highly mobile nature, sparse deployment, and limited battery energy availability [3]. In general, the speed of the UAVs ranges from 30 to 460 km/h with a flight time of 30 min [1, 3].

The work presented is motivated primarily by the limitations observed in the FANETs and its simultaneous consequences. They are:

- The continuously changing position of the nodes due to the highly mobile nature of FANETs calls for apt broadcast of the relevant information to the other nodes for proper and uninterrupted communication.

© Springer Nature Singapore Pte Ltd. 2021
P. Deshpande et al. (eds.), *Next Generation Information Processing System*,
Advances in Intelligent Systems and Computing 1162,
https://doi.org/10.1007/978-981-15-4851-2_1

- The scarce deployment scenario of the UAV nodes calls for higher power requirements for communication. This is because the distance among the UAVs is very high. In addition to this, UAVs have limitations in terms of computational power and bandwidth requirements.

Communication in FANETs through proactive routing and reactive routing has very high routing overheads, resulting in adverse effects on the lifetime of the network [4]. An efficient solution would be clustering, wherein a cluster typically consists of cluster members and a cluster head, which is elected among the nodes present in a cluster based of certain specified criteria. The cluster members communicate to the cluster heads both for inter-cluster and intra-cluster communications and finally to the destination. The primary advantage of this approach is its scalability along with reduced overheads.

While designing a cluster, the salient aspects required to be considered are the proper selection of cluster heads, measures to reduce the energy consumption at each node, maintaining a stable topology and simultaneous improvement of overall performance. Overheads in clustering technique are primarily in terms of the control messages required for the forming and maintaining the clusters [3]. In the recent years, swarm intelligence has found its importance in devising clustering techniques.

Based on these observations, in this paper, we presented two methods of clustering using bio-inspired metaheuristic techniques for designing a stable communication network with the limited energy resources in order to improve the lifetime of the network.

2 Related Work

The existing clustering techniques implemented for the various other ad hoc networks are reviewed for possible application in FANETs. With certain addition, in accordance with the distinctive features of FANETs, the proposed techniques may be considered for application. In this section, a brief overview of the proposed techniques is discussed.

An optimal cluster head selection method was proposed by Farhan et al. for FANETs by using a variant of k-means clustering algorithm [3]. The basic goal of the algorithm was to keep the computational and the communication overhead to the minimum. A modified k-means algorithm was proposed, wherein the cluster formation criteria included both the distance and the energy factor. Different transmission power levels were defined for the cluster members to select as per their requirement for communication. This approach reduced the energy consumption of the individual UAVs. A modified algorithm of GSO was proposed in [5] which provided different solutions for the optimal position of the centroid. They aimed to maximize the inter-cluster distance and minimize the intra-cluster distance. Also, they proposed three fitness function based on the above criteria. In [6], a synchronous firefly algorithm was proposed, in which the potential fireflies are selected by the tournament selection process after implementing the firefly algorithm. This was followed by the reproduction of those fireflies through crossover and mutation which are then together considered for the next iteration process. The algorithm aimed to improve the network performance in terms of packet loss ratio and energy consumption. A variation of the krill herd algorithm was proposed in [7] for finding the cluster center. The paper introduced two different mutation schemes depending upon the fitness of the individual krill. They focused on avoiding the best individual in terms of fitness to be trapped in the local optimum. The proposed algorithm was found to perform better than the fundamental krill herd algorithm in terms of accuracy of data and overall efficiency.

Two outlooks for clustering using particle swarm optimization (PSO) was enlightened in [8]. Considering the pros and cons of the k-means algorithm, a hybrid of the k-means and the PSO algorithm was proposed. The algorithm focuses on improving the distinction of the clusters formed by the k-means clustering by the PSO. The second technique proposed was an extension of PSO, wherein clustering is based on the user-specified number of clusters. A hybrid algorithm based on the concept of GSO and fruitfly optimization algorithm was proposed in [9]. The advantages and disadvantages of both the algorithm were considered, and accordingly, an amalgamation of the two was devised. The algorithm primarily focused on achieving its objective of minimizing the delay and maximizing the lifetime of the nodes. Again, a hybrid clustering technique based on the GSO and krill herd algorithm was presented in [4]. The algorithm aims to cut down the energy consumption and improve the routing of the FANETs through a three-step process. The first step was the cluster formation which involved the GSO algorithm that clustered the UAVs in accordance with its energy and position. It was followed by the cluster management based on the behavior of the krill herd. The routing was further carried out using a path detection function accompanied by a cluster maintenance phase. In order to improve the lifetime of the network, minimize the energy consumption and to find a optimize path for routing, a hybrid technique was presented in [10]. The technique addresses both the clustering and the routing aspect of the network. It conducted the clustering by a hybrid method of k-means and PSO, where k-means formed the clusters and PSO is implemented for finding the optimal cluster head. This was followed by the implementation of the GSO for selecting the optimum path which was dependent on the energy of the nodes and the distance to the cluster head.

2.1 Implemented Technique

In this paper, two different techniques for clustering are applied for minimizing the energy consumption in the network. The initial step involved the deployment of the UAV nodes over the network. The deployed nodes are further divided into distinct clusters. This is depicted in Figs. 1 and 2, respectively. Here two methodologies are being applied. The first technique forms the clusters by the k-means clustering technique. The nodes are distinguished on the basis of the Euclidean distance of each sensor node. The second technique employed is the glowworm swarm optimization technique [11], where the clusters are formed based on the sensing range and the energy value of the sensor nodes. The nodes within the sensing range of each other form the cluster. This is followed by the detection of the centroid based on a probability function value. Thus, with the formation of the clusters, the selection of the optimal cluster head from each of the clusters is carried out using Firefly Algorithm [12]. Finally, the position of the cluster members is updated with respect to the cluster head.

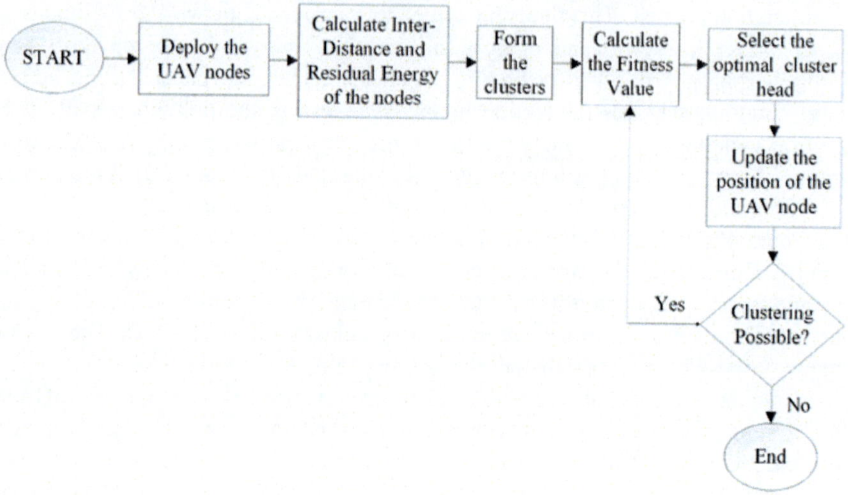

Fig. 1. Flowchart representation of the approach taken for implementation of both the algorithms

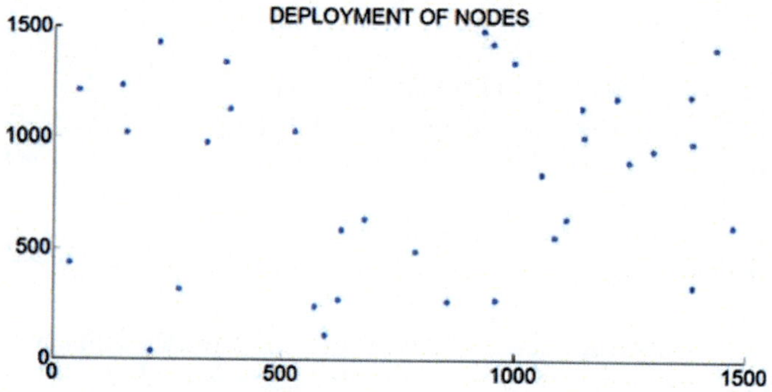

Fig. 2. Deployment of UAV nodes over an area

The result of the above technique is the formation of distinct clusters and the selection of the optimal cluster head in each cluster, as shown in Fig. 3. The communication among the inter-cluster and intra-cluster members is being considered. Periodically, the position and the energy of the UAV nodes are being broadcasted, and accordingly, optimal cluster head is being selected and dead nodes are declared. Communication continues till there are sufficient nodes for the clustering-based process that can take place.

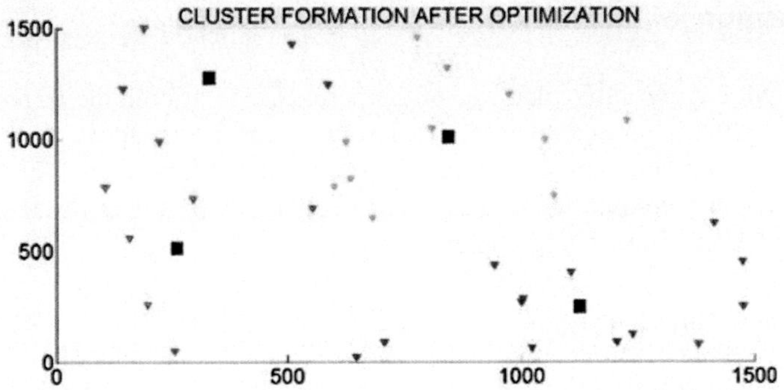

Fig. 3. Formation of the cluster after implementing the presented technique

2.2 Fitness Function

The technique implemented primarily addresses the limited energy capacity of a UAV. The energy consumption of the UAV nodes during its transmission and reception is high enough as the UAVs are very scarcely deployed. Also, owing to the high mobility of the UAVs, overhead requirement in terms of control message will be present. Thus, the main aim of the work was to minimize energy consumption and maximize the residual energy or the lifetime of the UAVs. Thus, the condition to be optimized is given by Eq. (1):

$$\mathrm{Max}(\arg(\mathrm{Node\_Lifetime})) = \left(w_1 \times \frac{1}{\mathrm{Internodal\_distance}}\right) + (w_2 \times \mathrm{Residual\_nodal\_energy}) \tag{1}$$

where w_1, w_2 are weight coefficients w_1, w_2 € [0, 1].

Algorithm	
1.	Start Input ← UAV nodes Output → Clustered UAV nodes
2.	Randomly deployment of the UAV nodes
3.	For all the nodes in the network
4.	– Calculate ← distance between each node and residual energy
5.	– Initial cluster ← GSO and K-Means
6.	– Calculate ← fitness function at each node
7.	For all the clusters,
8.	– highest fitness node → optimal cluster head
9.	– Update → the position of the cluster members
10.	End for
11.	End for

3 Performance Evaluation

In this section, a comparative analysis of the performance of both the techniques, i.e., GSO with firefly and k-means with firefly, is carried out. The analysis is carried out in terms of cluster building time and energy consumption of the network. The techniques are simulated considering different sizes of the network, where the UAVs are deployed, and the number of UAVs being deployed is varied.

3.1 Cluster Building Time

The cluster building time defines the computational complexity involved in a network to build a cluster based on the given constraints and the goals to be achieved. Lower the computational complexity of an algorithm implemented, lower will be its time required for the optimal cluster formation.

Performance based on the cluster formation time is carried out for three different areas of 1000, 2000, and 3000 m^2 for different numbers of UAV nodes deployed ranging from 25 to 40. The time required for all the scenarios is shown in Figs. 4, 5, and 6, respectively. It is evident from the graphs that, GSO with firefly technique performed better with respect to the k-means with firefly technique. The computational complexity of the network is also increased with the number of nodes deployed. Table 1 shows the quantitative analysis of the same. M1 in the table is abbreviated for the k-means with firefly technique, while the M2 is GSO with firefly.

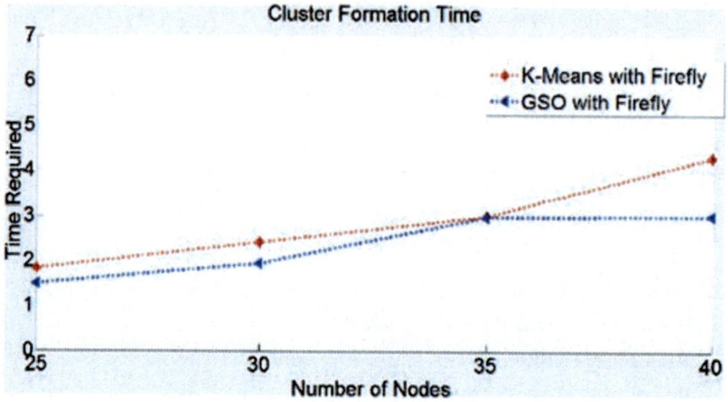

Fig. 4. A comparative analysis of the cluster building time required by both the techniques over an area of 1000×1000 m^2 for different number of node deployment

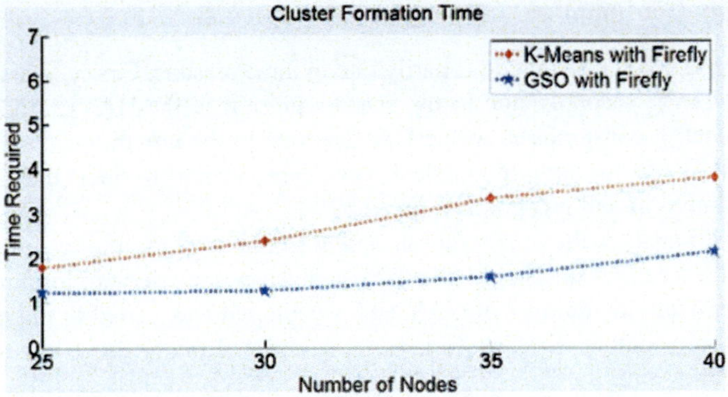

Fig. 5. A comparative analysis of the cluster building time required by both the techniques over an area of 2000 × 2000 m^2 for different number of node deployment

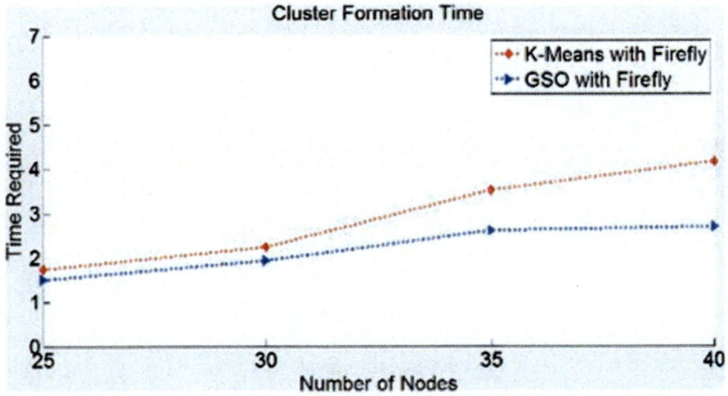

Fig. 6. A comparative analysis of the cluster building time required by both the techniques over an area of 3000 × 3000 m^2 for different number of node deployment

Table 1. Quantitative analysis of the energy consumption of the network and the cluster building time for different network area and different size of the nodes

Nodes	Energy consumption in the network						Cluster formation time of the network					
	Network area = 1000 m^2		Network area = 2000 m^2		Network area = 3000 m^2		Network area = 1000 m^2		Network area = 2000 m^2		Network area = 3000 m^2	
	M1	M2	M1	M2	M1	M2	M1	M2	M1	M2	M1	M2
25	3.76	2.08	3.87	2.80	3.8	2.47	1.83	1.50	1.80	1.23	1.75	1.51
30	3.87	2.93	4.3	3.32	4.46	3.57	2.41	1.95	2.40	1.27	2.24	2.24
35	4.39	3.29	3.52	3.04	4.61	2.71	3.01	2.98	3.34	1.58	3.51	2.61
40	4.38	3.80	4.54	4.13	5.65	5.79	4.31	3.02	3.81	2.16	4.61	2.70

3.2 Energy Consumption

Energy in an UAV is consumed primarily due to three reasons. Firstly, due to the flight time of the UAV, secondly due to the sensors present in the UAVs. And lastly, the maximum energy consumption in the UAVs is due to the communication among the UAV nodes. The first-order radio model is considered which considers the energy spent during transmission and reception of data.

In order to analyze the energy consumption in the network, the communication is carried out for a certain simulation time. The analysis was carried out for three different network areas, and the number of UAV nodes deployed was varied in these networks. The performance output is shown in Figs. 7, 8, and 9. Table 1 shows the quantitative analysis for the same.

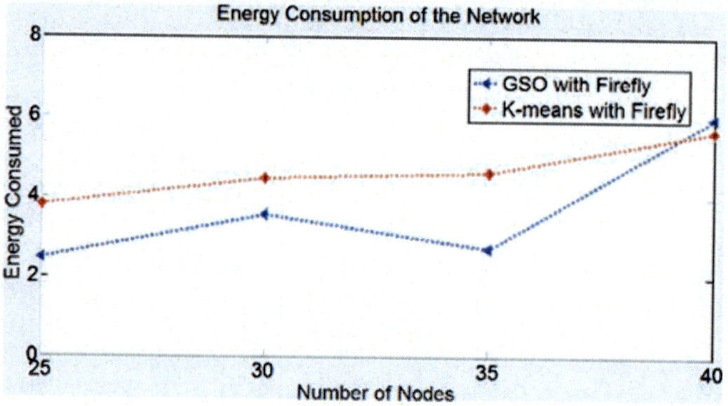

Fig. 7. A comparative analysis of the energy consumed by the network with both the techniques over an area of 3000×3000 m^2 for different number of node deployment

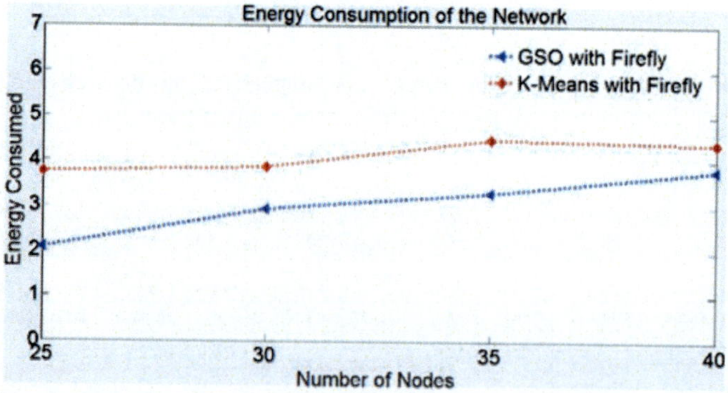

Fig. 8. A comparative analysis of the energy consumed by the network with both the techniques over an area of 1000×1000 m^2 for different number of node deployment

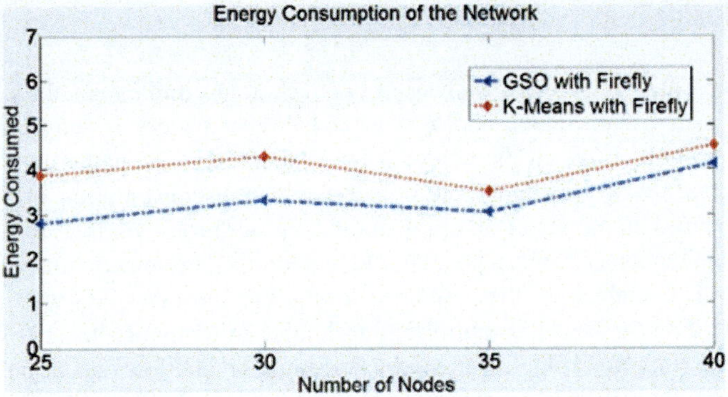

Fig. 9. A comparative analysis of the energy consumed by the network with both the techniques over an area of 2000 × 2000 m^2 for different number of node deployment

From the performance results obtained, it can be inferred that the energy consumption in case of GSO with firefly is less as compared to the other method. The less energy consumption of the network may be due to the selection of the cluster heads in the network. The k-means method tends to trap under the local optimum due to the initial random selection of the cluster centers. But GSO selects the cluster heads based on the probability function.

3.3 Performance Evaluation with Other Techniques

One of the limitations of GSO is the reduction of the convergence rate as the node size increases. However, the firefly algorithm has the innate search capability due to which its convergence rate is relatively immune to variances in the node size. Due to sub-optimal random initial cluster head selection when using only k-means model, we may obtain a result which may not perform competitively. To combat this shortfall, we go for firefly algorithm, which immediately takes care of the suboptimal CH selection by first considering their fitness values followed by the selection of the most suitable CHs. In some of the existing literature, PSO has been incorporated by using predefined dataset for various simulation parameters. In contrast, the current work focuses on random node deployment followed by the implementation of partition-based clustering and metaheuristic methods. In a similar work, implementation of gray wolf optimization (GWO) and ant colony optimization (ACO) was used to cluster the sensor nodes, and the performance results included cluster building time and the energy consumption the network. The proposed work in this paper trounces the preceding results, indicating the usefulness of k-means plus firefly algorithm and GSO with firefly. In another related work, fruitfly algorithm was used in which the position and velocity are changed in accordance with the odor of fruits due to auxin hormones which cause ripening of fruits. This fruitfly algorithm was incorporated with GSO for optimal cluster head selection, whereupon smaller node sizes employed GSO and larger node sizes employed fruitfly algorithm. The proposed work differs from this literature in terms of taking the help of luciferin levels for all node sizes, thereby simplifying the optimization procedure greatly.

4 Conclusion

The paper presents comparative analysis of two techniques implemented for improving the lifetime of a UAV. Scarce deployment and limited battery resources along with high mobility put a check on the potential UAV operation. The paper addresses these issues and presents two techniques to reduce the energy consumption in each nodes. The paper focused on the selection of the optimal cluster heads. Further, to stabilize the topology, simultaneous optimum position of the nodes is determined with respect to the cluster head. The analysis of the simulated results showed that GSO with firefly performed better than the k-means with firefly technique in terms of cluster building time and energy consumption. The lower energy consumption and low cluster building time may be attributed to the randomization property of both the GSO and firefly algorithm, preventing it to fall under local optimum. Future work may include the implementation of a hybrid algorithm to further improve the overall efficiency. In order to curb the limitation of the high mobility, efficient routing techniques may be implemented. To further optimize the energy consumption aspect, fitness function may include certain other constraints that can further aid in improving the lifetime of a UAV.

References

1. Khan, M.A., Safi, A., Qureshi, I.M., Khan, I.U.: Flying Ad-hoc Networks (FANETs): a review of communication architectures, and routing protocols. In: 2017 First International Conference on Latest trends in Electrical Engineering and Computing Technologies (INTELLECT), Karachi, pp. 1–9 (2017). https://doi.org/10.1109/intellect.2017.8277614
2. Bekmezci, I., Sahingoz, O.K., Temel, S.: Flying Ad-Hoc Networks (FANETs): a survey. Ad Hoc Netw. **11**, 1254–1270 (2013)
3. Aadil, F., Raza, A., Khan, M.F., Maqsood, M., Mehmood, I., Rho, S.: Energy aware cluster-based routing in Flying Ad-Hoc Networks. Sensors **18**, 1413 (2018). https://doi.org/10.3390/s18051413
4. Khan, A., Aftab, F., Zhang, Z.: BISCF: bio-inspired clustering scheme for FANETs. IEEE Access (2019). https://doi.org/10.1109/ACCESS.20192902940
5. Alijarah, I., Ludwig, S.A.: A new clustering approach based on glowworm swarm optimization. In: IEEE Congress on Evolutionary Computation, 20–23 June 2013, Cancun, Mexico
6. Baskaran, M., Sadagopan, C.: Synchronous firefly algorithm for cluster head selection in WSN. Sci. World J. Article ID 78087 (2015)
7. Li, Q., Liu, B.: Clustering using an improved krill herd algorithm. Algorithms **10**, 56 (2017). https://doi.org/10.3390/a10020056
8. Merwe, D., Engelbrecht, A.P.: Data clustering using particle swarm optimization
9. Dattatraya, K.N., Rao, K.R.: Hybrid based cluster head selection for maximizing network lifetime and energy efficiency in WSN. J. King Saud Univ. Comput. Inf. Sci.
10. Asha G.R., Gowrishankar: An energy aware routing mechanisms in WSNs using PSO and GSO algorithm. In: 5th International Conference on Signal Processing and Integrated Network (SPIN) (2018)

11. Kalaiselvi T., Nagaraja P., Basith A.: A review on glowworm swarm optimization. Int. J. Inf. Technol. (IJIT) **3**(2) (2017)
12. Fister, I., Fister Jr., I., Yang, X., Brest, J.: A comprehensive review of firefly algorithms. Swarm Evol. Comput. **13**, 34–46 (2013)

On-Demand Joint Predictive Flooding in Vehicular Ad Hoc Networks

H. N. Jithendra(✉) and D. Rekha

Department of SCSE, VIT University, Chennai, India
jithendra.hn2013@vit.ac.in

Abstract. Vehicular ad hoc networks (VANET) are flavor of mobile ad hoc networks. With the increase in the development of mobile technology in the last decade or so, we see increased interest in the development in VANET protocols as well. In this paper, we propose a routing protocol that uses predictive position information of neighboring nodes for data dissemination. The highly dynamic topology of VANET results in route discovered by a routing protocol to be intermittent. Hence, in this work we propose a predictive flooding in VANET (PFV) which has a flavor of flooding protocol for data transfer. Our results show that the PFV considerably reduces the network load in terms of forwards with an increase in the performance of data dissemination as well.

Keywords: VANET · Kalman filter · MANET · PFV

1 Introduction

With the continuous development in the automotive industry, vehicles today come with highly equipped sensors. Improvement in mobile communication research and development has enabled almost any two devices to communicate anywhere irrespective of the difference in technology stacks. One such advancement is in the direction of vehicular ad hoc networks (VANET). VANET is a flavor of ad hoc network created among the moving vehicles to share useful information.

The network created is often temporary and for a specific purpose. Given that the movement of the vehicles is very ad hoc; VANET comes in handy when specific information is to be communicated across the network and for small stipulated period of time.

VANET is also a mobile ad hoc network (MANET). MANET does not have any central network with multiple nodes communicating with each other in a distributed manner. In the contrary, the possibility of a node joining and leaving a network in VANET is very high. This property of VANET makes most of MANET routing protocols reusable in VANET. VANET is used in multiple applications like traffic prediction and control, collision detection and avoidance, getting through emergency traffic situations.

The characteristics of VANET are as follows:

- Dynamic topology: Varying direction and speed of vehicles result in having a dynamic topology.

© Springer Nature Singapore Pte Ltd. 2021
P. Deshpande et al. (eds.), *Next Generation Information Processing System*,
Advances in Intelligent Systems and Computing 1162,
https://doi.org/10.1007/978-981-15-4851-2_2

- Intermittent connectivity: Devices are paired in a network for a very short duration. With the dynamic topology, every node in the network is available for communication in short bursts and should be used effectively.
- Mobility patterns: Within the area of consideration, a large section of vehicles follows a specific pattern in the movement that is guided by signals, speed limits, streets, road conditions, etc. These patterns are used to create routing protocol for VANET.
- Unlimited power and storage: In contrast to MANET, VANET is assumed to have unlimited storage and power leading to building protocols that are not constrained by battery lifetime and storage wastage.
- In-built sensors: VANET comes up with multiple in-built sensors that could be used for data gathering.

VANET also forms a very important part in intelligent transport systems as insights are produced from the information being exchanged by the vehicles and other devices in the VANET.

The vehicular ad hoc networks (VANETs) have been emerged from the advances in ad hoc wireless technology and share most of the MANET common characteristics. Many forms of wireless communication technologies are available for deploying VANETs, and the most commonly used technology is Wi-Fi (IEEE 802.11 based). The rapid movement of nodes and frequent change of network topology in VANETs limits the communication time and link connection between the vehicles. The two communicating nodes which are not in each other's communication range could use multi-hop connection, and the communication goes through number of intermediate nodes. The infrastructure or environment characteristics like highly dense distribution of vehicles, high-speed mobile vehicles, and narrow roads would affect directly the route discovery and transmission performance of network information. The traditional MANET routing protocol may not give the best functional results and better performance in VANET.

One of the major limitations of VANET is in terms of higher network load. Highly dynamic topology causes the discovered routes to be temporary. To overcome this limitation, the literature suggests routing protocols with very limited network load that has been beneficiary. Recent works [1–3] also indicate that the link retention or link stability is one of the open-ended problems in VANET.

2 Related Work

In this section, we review the literature in research related to our work. Many protocols for routing have been proposed in the literature for use in MANETs. They are mainly classified as proactive, reactive, and hybrid protocols. This classification is done depending upon the maintenance of routing information at every node in the network.

Multiple performance measures have been studied for evaluating and to compare works in the literature related to routing protocols in MANET in different scenarios and problems [4–12].

Breakthrough work related to MANET's misbehavior was first taken up by Marti [13]. This work aims at mitigating misbehavior of the nodes in the network.

Subsequently, many other works followed the similar objective. Predominantly, the objective was to detect and isolate misbehaving nodes, and other schemes to encourage cooperation were also proposed, either by credit-based [14], or by reputation-based schemes [15, 16].

There are few other literatures in the past which addresses the issue of nodes that behave differently at different point in time. These nodes are most often than not treated as a malicious node. This work addresses them as selfish nodes as they do not intend to harm the network in any means.

Active route timeout established ad hoc on-demand distance vector (E-AODV) routing protocol 1 [16] and applied some mechanisms by changing parameter values that involve active route timeout and hello interval to enhance the AODV protocol performance and better routing and quality of service.

The adaptive hybrid routing protocol (AHR) [14] integrates the features from both geographic and reactive routing protocols along with proactive routing schemes to reduce routing overhead and enhanced the routing performance.

The hybrid routing mechanism is best suitable for VANET's environments which usually comes with diversified infrastructure consisting of different kinds of mobile nodes, roadside units (RSU), control centers, stationary nodes, etc.

The characteristics of vehicular network that makes the data dissemination are a challenging task in VANET applications. The reliable and efficient data dissemination is crucial, and most of the existing algorithms generate redundant data and saturate the network. In [17], the authors proposed the new algorithm efficient encounter-based event dissemination protocol (E-BED) which performs efficient data dissemination in both urban and highway scenarios based on the distance and the encountering probability of the vehicles to the event.

VANET has been used in multiple applications like traffic control, health care, natural disaster management, and so on. The work in [18] deals with efficient TDMA-based broadcast scheduling in wireless mesh networks for disaster management. This work maximizes the number of transmissions in a stipulated time slot. The genetic solution to the same problem is provided in [19]. The later is designed to work on ad hoc networks, whereas the former is designed for wireless mesh networks. One of the challenges in VANET to cater to the above-mentioned applications is routing. The work in [20] addresses the routing problem in VANET by proposing EAMDF algorithm that uses information about the node which is situated at the edge of radio range of the sender node, and then the packet is transmitted by using the trustworthy greedy position-based routing approach through that node. The work establishes higher PDR as compared to state-of-the-art approaches.

3 Our Algorithm

In this section, we discuss the holistic purpose of this work by setting objectives and aims to be achieved. The concept of routing through the vehicles in the road with minimal beacon exchanges is considered as a part of this paper.

The objective of this paper is to address the problem of attending the medical emergency with the help of vehicular network. An example use case for this problem is attending the medical emergency in time by ambulances. This requires the gathering of traffic information in terms of signals, traffic density, shortest route, etc. Also, it is important to have a communication established between hospital and the ambulance to give instructions on medical emergency.

We approach the solution to this problem as a flavor of flooding to discover the route for data transfer. It is quite often noted that the vehicles move at different speeds and in different directions leading to highly dynamic network topology. With high level of dynamic nature effort spent on route discovery need to be as minimal as possible, any route discovered is intermittent and route discovery needs to be re-initiated in short periods.

To reiterate, the choice of flooding algorithm is because of very nature of VANETs where the link stability is very less. Flooding is one of stronger baselines in VANETs because of high dynamic nature of the network. Our objective in this work is to reduce the network load while maintaining the benefits of flooding in the network.

The route discovery process is proposed in this paper that predicts the node movement and proactively estimates the presence of a node in the network to decide on participating in the route discovery process. The main contribution of our approach is in the route discovery of the VANET. The following assumptions are considered in building the algorithm for route discovery.

Assumptions

The following are the assumptions that made in our approach.

- We are being conservative and biased against every node participating in the route discovery process, and hence, the assumption is that the network is dense, i.e., considering the scenarios for VANET in town/cities where the traffic is dense.
- Every vehicle is associated with determined traffic route that it would follow.
- Each vehicle is assumed to participate in the route discovery algorithm (however, not all vehicles would be actively involved).

Predictive Flooding in VANET (PFV)

In this paper, we propose a predictive flooding VANET (PFV) algorithm which is a variant of classical flooding. Flooding is a very successful way of forwarding the information in a highly dynamic topology network. Even though the number of packets in the network at any given point in time is high, we will be able to cater the need of data transmission in a highly volatile network. The problem of dangling packets (packets getting struck in the network without reaching the destination) can be resolved by the use of time-to-live (TTL).

PFV algorithm takes all the positive aspects of the classical flooding and builds upon it to take an intelligent decision making at every node to flood the 135 packets further or not. Every node in the network upon reception of a data packet forwards the packet to all its neighboring nodes. In PFV, every node stores the information about the distance to its neighbors and also predicts the speed and direction of every neighbor.

We present our main algorithm in Algorithm [21]. This is a distributed algorithm that is run by every node in the network to decide whether to participate in the information exchange or not for that time instance. Every node i has a list of neighbors Ni, and their position information (for last 10 instances) stored in its neighbor table. Each neighbors next 10 predictive positions are calculated using kalman filter predictions as explained in Algorithm [22]. With this predicted information, node i calculates how long the neighboring nodes will be in the communication range. With this, the forward probability is set for every node. If the forwarding probability Fi is greater than the threshold, then the node participates in the data dissemination process.

Algorithm 1 Predictive Flooding in VANET

procedure PFV(i) ▷ Decision of the node i forwarding the data packet
    $\forall F_i \leftarrow 0$
    $\forall N_i$; $pred(N_i) \leftarrow kalman\_prediction(N_i, t_{1...10})$
    $pred(i) \leftarrow kalman\_prediction(i, t_{1...10})$
    for every N_i **do**
        $pdist(N_i) \leftarrow euclidian\_distance(pred(i), pred(N_i))$
        if $\forall pdist(N_i) < R_c$ **then**
            $F_i \leftarrow F_i + 1/count(N_i)$
    if $F_i > thresh$ **then**
        return $True$
    else
        return $False$

Algorithm 2 Kalman Filter Prediction

procedure KALMAN_PREDICTION(i,$t_{1..10}$) ▷ Kalman filter based position prediction for multiple time units
    for every t_i **do**
        $new\_pos \leftarrow kalman\_prediction(cur_pos, t_i)$
        $pred(i)$.**append**(new_pos)
    return $pred(i)$

4 Implementation

- Simulating a wireless network environment uses vehicular nodes during route discovery process.
- Each node should be intelligent enough to choose their respective forward probability based on the predictive movement.
- Instead of considering the exact position of the node, we consider the node movement probability for transmission.
- We need to run the simulation for different network scenarios (varying nodes, traffic, and mobility) to estimate the difference in the route discovery considering forward probability in one case and without forward probability as another case.
- With variation in every parameter, we need to count the number of forwarders during route discovery in both the cases for every source and destination pairs.
- A state-of-the-art flooding routing algorithm is used as a baseline and compared with modified PFV algorithm to show the effectiveness of our algorithm. With the above-mentioned objectives, we design our system to evaluate the route discovery performance of the network.

Following are the parameters that are involved as a part of this work. We divide the parameter involved in observation parameters and tweaking parameters. Observation parameters are those parameters that are evaluated after every simulation settings. In other words, these are the output of our simulations.

Tweaking parameters are those which we vary one or more of them in order to simulate different wireless network scenarios.

The observation parameters used are:

- Number of forwarders
- Packet delivery ratio
- Route discovered for every pair of nodes during simulation along with the route metric value.

The tweaking parameters used are:

- Network traffic
- Node placement
- Mobility
- Number of nodes
- Packet size
- Queue length
- Data rate.

With these parameters in place, we design our algorithm to capture the minimal flooding in VANET during route discovery process. We have simulated all our experiments in NS-2 with varying number of nodes and recording the packet delivery ratio as a performance measure. For each scenario considered, we have run multiple experiments in the order (8–10) and the averaged packet delivery ratio is obtained. With every scenario experimented, we calculate the number of forwarders who were actively involved in forwarding the data packet.

5 Results

The following are some of the observations of the PFV algorithm as compared to that of state-of-the-art flooding. The number of forwarders is an estimate of how well the algorithm judges the node movement and picks up the steady route to forward the beacon. This reduces the chance of message losses and in turn the number of re-transmissions. The results show that the number of forwarders is reduced by a large extent as compared to the state-of-the-art flooding.

As evident in the results shown in Fig. 1, PFV has considerably less number of forwards as compared to classical flooding algorithm. Also, we see that as the number of nodes increase, rate at which the number of forwards increase in PFV is less than that of the number of forwards in the classical flooding algorithm.

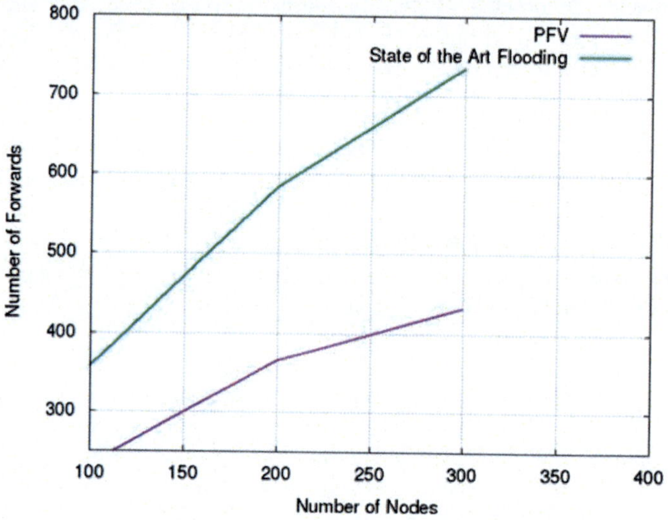

Fig. 1. Number of forwards in the network

This indicates that the network density is relatively reduced in the proposed algorithm leading to lesser packet collisions.

With reduced number of forwards, it is important to see the impact in terms of performance as compared to flooding algorithm. Figure 2 shows the packet delivery ratio achieved by our algorithm as against the classical flooding algorithm. The results give a clear indication of PFV outperforming the classical flooding algorithm with different number of nodes.

With the decrease in number of forwarders, the other metric to estimate is the packet delivery ratio. The results in the graph give clear indication of increase in PDR leading to the performance stability of our algorithm.

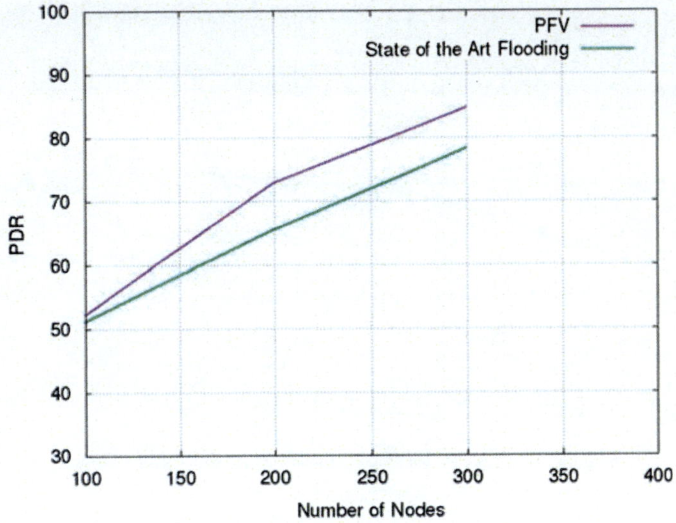

Fig. 2. Packet delivery ratio

The conservative nature of our PFV algorithm ensures that the network load is considerably reduced without affecting the PDR. Table 1 showcases the number of forwards in the network with varying threshold values. It is evident from the table that the increased threshold value makes the algorithm n = more conservative. The threshold value is globally set, and Fig. 1 represents the threshold of 0.5.

Table 1. Number of forwards done with varying threshold

Number of nodes	Threshold 0.5	Threshold 0.4	Threshold 0.2
100	220	282	350
150	300	364	434
200	380	414	543
250	400	498	610
300	430	598	675

With the variations in threshold, the number of forwards would change as shown in the table. Also, it is important to see are these disjoint curves or overlapping curves. We demonstrate that in Fig. 3. The disjoint curve here indicates that the increase in the threshold is definitely better to remove the overhead in the network. However, the trade-off is that the PDR starts reducing once we start increasing the threshold. As the number of forwards reduces, there is a chance of losing the packet completely in the transit. We demonstrate that in Fig. 4.

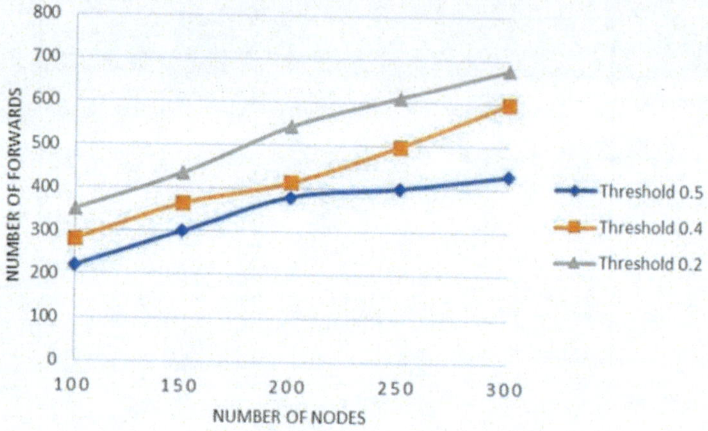

Fig. 3. Number of forwards for different threshold values

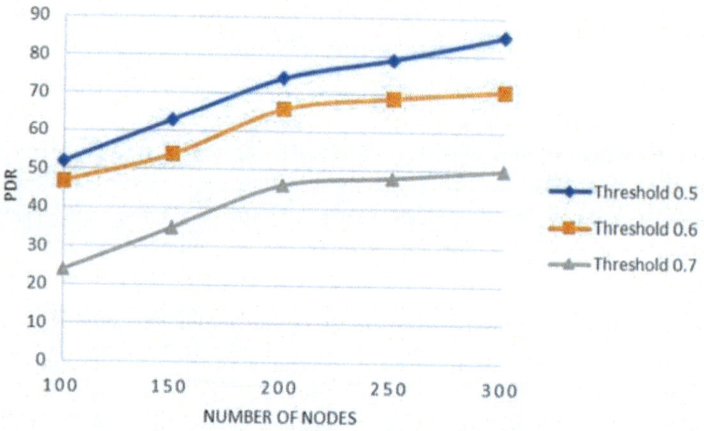

Fig. 4. PDR for different threshold values

We also run our experiments with varying mobility to see the improvements in the performance as we vary the number of nodes moving in the network. Figure 5 shows the PDR for different scenarios with varying mobility. It is evident from the graph that with high mobility the PDR drops due to the unreachable neighboring nodes.

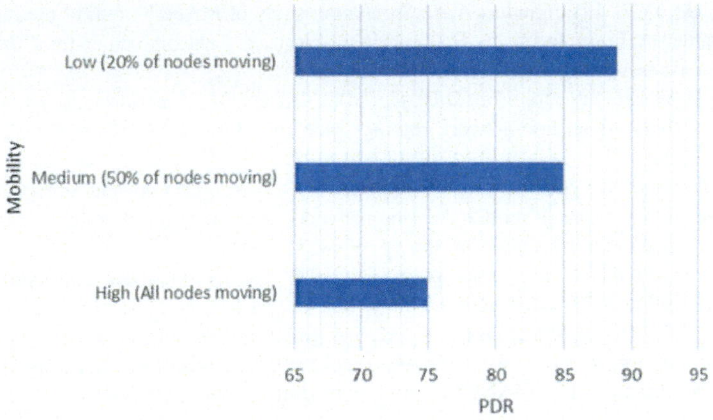

Fig. 5. PDR for varying mobility

6 Conclusions

In this paper, we proposed a PFV algorithm that takes into consideration the predictive movement of neighboring nodes to decide on the active participation in data dissemination. PFV consistently outperforms flooding both in terms of reduced network load and increased packet delivery ratio. The experimental results show that the proposed algorithm can also be fine-tuned with the threshold based on the node density in the network. We managed to improve the performance of state-of-the-art flooding protocol for VANETs by utilizing information of node movements. Without significantly increasing routing overhead, we are able to choose a better path using forwarding probability of the nodes in the network. We compared our work with the state-of-the-art flooding where the paths were chosen based on the node position only.

One of the advantages of multiple receptions of the same data is data validation, reduced number of forwarders indicate decrease in the number of receptions of the same data at every node making it difficult to validate the data using traditional voting algorithms. Current work can be extended to ensure the data validation in highly dynamic network infrastructure.

References

1. Zhang, D., Zhang, T., Liu, X.: Novel self-adaptive routing service algorithm for application in VANET. Appl. Intell. **49**(5), 1866–1879 (2019)
2. Wang, H., et al.: A improved routing scheme based on link stability for VANET. In: 2019 14th IEEE Conference on Industrial Electronics and Applications (ICIEA). IEEE (2019)
3. Li, N., et al.: Probability prediction-based reliable and efficient opportunistic routing algorithm for VANETs. IEEE/ACM Trans. Netw. (TON) **26**(4), 1933–1947 (2018)
4. Samara, C., Karapistoli, E., Economides, A.A.: Performance comparison of MANET routing protocols based on real-life scenarios. Ultra modern telecommunications and control systems and workshops

5. Kampitaki, D.G., Economides, A.A.: Simulation study of MANET routing protocols under FTP traffic. In: Presented at the Conference on Electrical Telecommunication Computation, CETC, Lisbon, Portugal (2013)
6. Broch, J., Maltz, D.A., Johnson, D.B., Hu, Y.C., Jetcheva, J.: A performance comparison of multi-hop wireless ad hoc network routing" protocols. Proc. ACM/IEEE MobiCom **1998**, 85–97 (1998)
7. Mbarushimana, C., Shahrabi, A.: Comparative study of reactive and proactive routing protocols performance in mobile Ad Hoc networks. Advanced Information on Networking Application Workshops, AINAW '07, pp. 679–684 (2007)
8. Bilandi, N., Harsh, K.V.: Comparative analysis of reactive, proactive and hybrid routing protocols in MANET. Int. J. Electr. Comp. Sci. Eng. **1**(03), 1660–1667 (2012)
9. Perkins, C.E., Royer, E.M., Das, S.R., Marina, M.K.: Performance comparison of two on-demand routing protocols for ad hoc networks. IEEE Pers. Commun. **8**(1), 16–28 (2001)
10. Maan, F., Mazhar, N.: MANET routing protocols vs mobility models: a performance evaluation. In: Third International Conference on Ubiquitous and Future Networks (ICUFN), pp. 179–184 (2011)
11. Fan, B., Sadagopan, N., Helmy, A.: IMPORTANT: a framework to systematically analyze the impact of mobility on performance of routing protocols for Adhoc networks. In: Proceedings of INFOCOM, pp. 825–835 (2003)
12. Ho, Y.H., Ho, A.H., Hua, K.A.: Routing protocols for intervehicular networks: A comparative study in high-mobility and large obstacles environments. Comp. Commun. **31** (12), 2767–2780 (2008)
13. Marti, S., Giuli, T.J., Lai, K., Baker, M.: Mitigating routing misbehavior in mobile ad hoc networks. In: Proceedings of the 6th Annual International Conference on Mobile Computing and Networking (MobiCom '00), pp. 255–265. ACM, New York, NY, USA (2000)
14. Divyalakshmi, D., Manjusha, D.: Adaptive hybrid routing protocol for VANETs. Int. J. Recent Innov. Trends Comput. Commun. **5**(5), 1085–1091 (2017)
15. Sonja, B., Le Boudec, J.-Y.: Performance analysis of the CONFIDANT protocol. In: Proceedings of the 3rd ACM International Symposium on Mobile ad hoc Networking and Computing. ACM (2002)
16. Mittala, P., Yadav, B.: Improving AODV routing in vehicular ad hoc networks. Int. J. Innov. Res. Comput. Commun. Eng. **4**(9) (2016)
17. Nikolovski, T., Pazzi, R.W., Akabane, A.T., Villas, L.A.: Efficient encounter-based event dissemination protocol (E-BED) for urban and highway vehicular ad hoc networks. In: 2017 IEEE Symposium on Computers and Communications (ISCC), Heraklion, pp. 1085–1090 (2017)
18. Arivudainambi, D., Rekha, D.: Heuristic approach for broadcast scheduling problem in wireless mesh networks. Elsevier Int. J. Electron. Commun. **68**(6), 489–495 (2014)
19. Arivudainambi, D., Rekha, D.: Broadcast scheduling problem in TDMA ad hoc networks using immune genetic algorithm. Int. J. Comput. Commun. Control **8**(1), 18–29 (2012)
20. Narasappa, J.H., Rekha, D.: Energy aware methodical data forwarding (EAMDF) mechanism in VANET. J. Intell. Fuzzy Syst. 1–11 (2019). https://doi.org/10.3233/JIFS-169986
21. Siva Ram Murthy, C., Manoj, B.S.: Ad hoc Wireless Networks: Architectures and Protocols. Pearson Education (2004)
22. Abolhasan, M., et al.: A review of routing protocols for mobile ad hoc networks. Ad Hoc Netw. **2**(1), 1–22 (2004)

Entity-Centric Combined Trust (ECT) Algorithm to Detect Packet Dropping Attack in Vehicular Ad Hoc Networks (VANETs)

Kuldeep Narayan Tripathi(✉), Gourav Jain, Ashish Mohan Yadav, and S. C. Sharma

Electronics and Computer Discipline, DPT, Indian Institute of Technology Roorkee, Roorkee, Uttarakhand 247667, India
ktripathi@pp.iitr.ac.in

Abstract. Road safety and security are the most challenging topics in recent times. Vehicular ad hoc network, an active part of the intelligent transportation system, provides the advanced technology to tackle the various security and safety issues of vehicles. Vehicle is an integral part of the VANET that often gets compromised with numerous types of attackers and starts to perform malicious activities. To provide secure and smooth communication among the vehicles, we are required to detect those infected nodes and remove them from the network. In this paper, we proposed a trust-based security algorithm based on cooperation among the vehicles. The proposed algorithm detects the malicious activities, mainly centered on packet dropping, with the help of the trust-based security measurements. Our proposed model primarily based on the two things one is cooperation among the vehicles, and the other is monitoring network traffic of moving vehicles. Simulation analysis and results demonstrate the incredible growth in the performance of the AODV routing under the proposed algorithm. We present the simulation of the proposed algorithm in network simulator NS-2, and result evaluation by using network parameters like successful data delivery fraction and end-to-end delay. Simulation and result analysis show that our algorithm works better to detect malicious entities (vehicles) in the network and enhances the performance of the network of moving vehicles.

Keywords: Data dissemination · Trust-based security · Vehicle cooperation · Trust value · Malicious nodes · Packet drop attack · VANETs

1 Introduction

Vehicular ad hoc networks, a temporary network of fast-moving vehicles, are intended to enhance the user driving experience with safety and security in the road. Vehicular networks are designed to advance the traffic efficiency with the help of the information transfer among the devices in the network. Mainly two types of communication take place in the vehicular network: First one is vehicle-to-vehicle (V2V) and another one is vehicle-to-infrastructure (V2I) [1]. Communication among the various vehicles takes

© Springer Nature Singapore Pte Ltd. 2021
P. Deshpande et al. (eds.), *Next Generation Information Processing System*,
Advances in Intelligent Systems and Computing 1162,
https://doi.org/10.1007/978-981-15-4851-2_3

place with the help of the wireless communication protocols known as dedicated short-range communication (DSRC) standard protocol [2]. The vehicular network is an open and dynamic network without a central administrator and fixed infrastructure and suffers from many security threats. Information dissemination in the original form under some hard time limit and frequent disconnections due to highly dynamic topology is the main challenges that the vehicular network often faces [3]. If during data transmission any link is failed, then the current node stores the data and starts the route repairing process. If the route repairing process is successful, then the current node sends the data to the next node. Secure routing is one of the most challenging issues in VANET due to the emergence of numerous types of routing attacks. Ad hoc on-demand distance vector routing (AODV) is the most widely used reactive routing protocol for wireless multi-hop communication. Proactive routing protocols are not suitable for highly dynamic networks such as a vehicular network. Another reactive routing protocol such as AOMDV creates extra overhead on the network resources. So, AODV is the most efficient and suitable routing protocol for the vehicular environment.

Vehicular technology is the backbone of the intelligent transportation system (ITS) and also comes with various networking challenges [4]. The vehicular network is a special subclass of the mobile ad hoc network with some extra features such as sufficient energy, fast-moving vehicles, dynamic topology, and distributed architecture [5]. In a vehicular network, various types of selfish or malicious vehicles try to affect the safety and security of the network for its own benefits [6]. Security is a primary concern in a vehicular ad hoc network. Compromising nodes and delayed or altered information can lead to serious consequences such as road accidents [7]. To maintain the security of the vehicular network, we have to satisfy the various security attributes such as authentication, confidentiality, integrity, availability, and privacy [4]. Our main aim is to provide safety to the user in the network. Vehicular networks are constantly suffering from some serious types of attacks like denial-of-service (DOS) attack, distributed DOS attack, and packet dropping attacks [8]. Packet dropping attacks such as black-hole attack, gray-hole attack, and warm-hole attack are most common in the vehicular network and used to disrupt the regular functioning of the network [4]. To identify these kinds of malicious nodes, we calculate the direct trust (based on direct observations) and indirect trust (based on recommendations from others) of the nodes by analyzing the behavior of the nodes. To distinguish the malicious behavior of the particular entity in the network, trust is a vital consideration. Trust of the node in the network can be explained as the degree of the subjective as well as cumulative belief about the particular node [9]. By using the trust value of the specific node, we can predict the behavior of the node to some extent. In the proposed algorithm, first we estimate the satisfaction value of the node based on the three parameters: reliability of transitions, position closeness, and contact frequency. After the calculation of satisfaction value, we evaluate the direct trust of the node which is based on the weight of the data transmission and satisfaction value. With the help of the direct trust of various nodes, we can calculate the indirect/recommendation trust of the particular node. And with the help of combined trust, we can identify the malicious nodes in the network. After identifying the malicious node, we can easily isolate the particular infected node from the network.

The remaining of this paper is systematized as: Sect. 2 presents the literature survey which consists of a discussion of various trust-based security algorithms as well as details of various network attacks in the vehicular ad hoc network. Further, this section also contains the limitations of the various previous security models in VANETs and the motivation behind the study. Section 3 introduces the proposed model based on an entity-centric combined trust-based model in VANET. Section 4 covers the simulation and performance evaluation of the proposed algorithm, and in the last Sect. 5 contains conclusion and future work.

2 Literature Survey

There are various types of trust management models defined in the literature and used by different authors to protect the vehicular network from unauthorized and malicious access. In this section, we discussed the various types of trust algorithms used to provide security to the vehicular network. In VANET, inter-vehicle communication and communication with the roadside unit attract many intruders to fetch the information or alter the information shared. There are numerous types of attacks possible in the vehicular environment. Dynamic network topology and fast-moving speed of nodes further increase the challenge to secure the network. An attacker exploits the vulnerabilities of vehicular ad hoc environment to launch the attack in the network [10]. There are gigantic varieties of attacks that can be performed by selfish nodes from the inside of the network or outside the network. A variety of solutions have been proposed to deliver security to the communication process in a vehicular network. Trust-based security models are the prominent and widely used models to secure the vehicular network and identify the malicious activity in the network. Based on the trust calculation, we can categorize the trust-based models into three categories: (i) entity-oriented trust algorithms, (ii) data-oriented trust algorithms, and (ii) hybrid trust algorithms [11]. In entity-centric trust algorithms, the main concentration of the trust algorithm is on the entity or node. Data-oriented trust models give emphasize on the trust of data transferred among the vehicles. And the hybrid trust model considers the trust of the node as well as data to consolidate the trust of the transmission process in the vehicular ad hoc network. An author can select the trust model based on the application and nature of the problem.

Djahel et al. [12] presented an acknowledgment-based security mechanism to defend the network against the packet dropping attacks. The acknowledgment-based algorithm enhances the security of the network and isolates the malicious node from the network. In [13], author proposed a model that detects the malicious nodes based on the trust value of the node and calculated by monitoring the behavior of the node. This proposed model converts the trust values into binary values and transfers the trust information in the form of binary values to reduce the complexity of the presented model. Shabut et al. [9] proposed a recommendation trust-oriented algorithm to identify the attack in the network. The proposed scheme dynamically filters the dishonest recommendations based on the clustering method. Chen et al. [11] proposed a hybrid trust model based on global trust value to identify the trustworthy nodes in the network. Zhang et al. [14] proposed a reputation-based trust management scheme to identify the malicious attack and optimize the efficiency of the network. With the help of subjective trust and recommendation trust with cosine similarity measure, they evaluated the reputation value of the node. Liu et al. [15] presented an algorithm to detect anomalies in VANET based on trust calculation and affinity propagation. The inclusion of trustworthy cluster heads enhances the performance of the proposed model. In [16], the author presented an algorithm to provide the security against the denial-of-service (DoS) attack. They have introduced the P-secure protocol which calculates the *p*-value for the RSU vehicle based on the vehicle speed.

In Table 1, we presented the comparative analysis of some of the existing trust-based algorithms based on various parameter measures such as the approach used, simulation tool, parameters for performance evaluation, and advantages of various methods. Based on the literature survey, we identify the advantages and limitations of various existing security algorithms.

Table 1. Comparison of various security mechanisms in VANET

IDs	Objective	Approach/algorithm	Tool	Parameters	Remark
[5]	Detection of black-hole attack	Route backtracking and statistics-based approach	NS-3	PDR, Throughput	Present algorithm effective for a single malicious node
[6]	Identification of malicious nodes to secure the n/w	Watchdog and Bayesian theory-based approach	Not mentioned	False positive, detection ratio	Lightweight algorithm with low overhead
[8]	Prevention from the DDoS and Masquerade attacks	Genetic model-based approach using Gaussian modeling	MATLAB	Tracking time, processing delay, network recovery time	Genetic model improves the detection rate
[9]	To filter out the misbehaving nodes	Recommendation-based trust model	NS2	Throughput, packet loss	Recommendation manager criteria not defined
[12]	Security against the packet dropping attack	Acknowledgment-based mechanism to enhance the security of VANETs	OPNET 11.5	PDR, routing overhead, detection rate	Excluded the malicious nodes from the forwarding process
[13]	Detect rogue nodes in the network	Combined trust-based model with binary trust values	NS-2	PDR, delay, throughput, overhead	Proposed model can detect limited malicious activities
[14]	Detect the malicious attack and improve the n/w efficiency	Subjective and recommendation trust-based reputation model	NetLogo	Reputation, misreporting nodes	Reputation calculated based on subjective and recommendation trust
[17]	To detect the malicious activity in network	Reputation and plausibility check-based approach	Qualnet 5.0	Avg. time, number of packets dropped	Algorithm suited for real-time applications

3 Proposed Entity-Centric Combined Trust (ECT) Algorithm

In this work, we proposed a reliable and secure routing for efficient data dissemination in vehicular ad hoc networks. This model can detect the malicious nodes, which perform the packet dropping attacks and packet alteration/modification attack. We used a trust model based on direct trust and indirect trust, which can calculate the trust and satisfaction value of a particular node without additional control overheads. Direct trust can be calculated based on the satisfaction value of the node and the weight factor of that node. The satisfaction value is calculated based on the three parameters observed during the data transmission process for a particular time. Many times only direct trust is not the sufficient measure of the reliability of the particular node. After direct trust, the proposed model asks about the recommendations from the other neighboring nodes, corresponding to the observed node. Based on recommendations, we calculate the indirect trust or recommendation trust of the observed node (Fig. 1), and based on the direct and indirect trust, we form the opinion about the observed node. The detailed discussion of our proposed algorithm and the various nomenclatures used is given below:

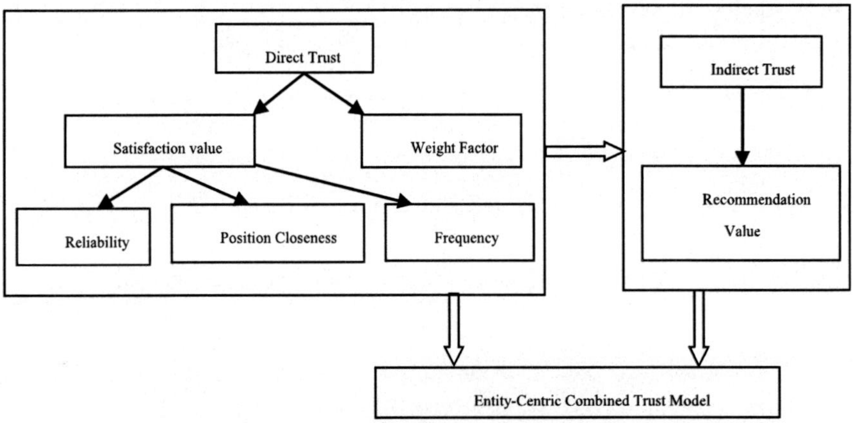

Fig. 1. Framework for proposed secure and optimized routing

3.1 Assumptions for the Proposed Algorithm

Some assumptions have been followed in the proposed algorithm that is listed below:

- All vehicles cooperate to detect the malicious node or packet dropping attack in the network.
- All nodes send true recommendations when asked to send.
- After recognition of malicious node, information is broadcasted to all other nodes in the network regarding the current status of the malicious node.

- The routing table has one extra entry as the status of the neighboring nodes as malicious or trustworthy.
- There are three categories of vehicles in the network: (i) malicious nodes, (ii) normal nodes, (iii) fixed nodes known as RSUs.
- All nodes are in promiscuous mode so that they can also monitor the traffic of the neighboring nodes.

3.2 Proposed Algorithm

The proposed algorithm first calculates the satisfaction value of the particular node over the period of time '*t*' and '*k*' transmissions. After calculating the satisfaction value, we identify the direct trust value of that node with the help of the weight of the transmitted data. After estimation of direct trust for node *u*, the verifier node also computes the indirect trust or recommendation trust for node *u* based on the recommendations from other nodes present in the range at that time. A combination of direct and indirect trust (also known as combined trust) gives an exact idea about the behavior of the node *u*. Here node *v* is used as a verifier node that verifies the trust of node *u* known as observed node in the proposed algorithm.

$\mathrm{Sat}_k^t(u,v)$ = Total satisfaction value of node *u* observed by node *v* over the period of

time *t*, based on the total *k* transmissions

$wt_{(j)}$ = Weight of *j*th transmission between node *u* and node *v*

Total satisfaction value for the *k* transmission can be calculated as:

$$\mathrm{Sat}_k^t(u,v) = \omega * \mathrm{Sat}_{\mathrm{curr}} + (1-\omega) * \mathrm{Sat}_{k-1}^t(u,v) \tag{1}$$

where $\mathrm{Sat}_{\mathrm{curr}}$ is the satisfaction value for the current transmission and ω is the weight coefficient.

Satisfaction value for the current transmission can be defined as below:

$$\mathrm{Sat}_{\mathrm{curr}} = \begin{cases} 1, & \text{if the transmission is satisfactory} \\ 0, & \text{if the transmission is unsatisfactory} \\ (0,1), & \text{otherwise} \end{cases} \tag{2}$$

Calculation of satisfaction value for the node *u* can be calculated based on a multi-functional criterion, because single criteria cannot give an exact realization of the trust value for the vehicle. We calculate the satisfaction values of the nodes based on the following three parameters:

- Reliability of delivery $(\mathrm{Rd}_{(u,v)})$: Reliability of the delivery can be calculated based on the forwarding rate of packets by the observed node *u* over the period of time *t* during *k* transmissions. Node with a higher rate of successful packet forwarding can get higher reliability of delivery for the successive transmissions. A node with malicious behavior will drop the packet after receiving it.

$$\mathrm{Rd}_{(u,v)} = P_\mathrm{f}/P_\mathrm{r} \tag{3}$$

where $\mathrm{Rd}_{(u,v)}$ = Reliability of delivery of node u observed by node v, P_f = Number of packets forwarded by node u, P_r = Number of packets received by node u.

- Position closeness $\left(\mathrm{Pc}_{(u,v)}\right)$: Position closeness is the parameter that defines the closeness of the location of the observed node to that verifier node. When the verifier node is closer to the observed node, then the probability of data transmission will be higher using the observed node.

$$\mathrm{Pc}_{(u,v)} = 1 - (d/R) \tag{4}$$

where d = distance between the observed node and verifier node, R = circumference of the communication range.

- Contact frequency $\left(\mathrm{Cf}_{(u,v)}\right)$: Contact frequency can be defined as the interaction of two nodes for a particular time period. It can be calculated based on the number of times a node transferred the information via another one. If any node has a higher contact frequency with another node, then the probability of transmitting data using this node is higher.

$$\mathrm{Cf}_{(u,v)} = \mathrm{Cf}_\mathrm{next}/\mathrm{Cf}_\mathrm{other} \tag{5}$$

where $\mathrm{Cf}_\mathrm{next}$ = contact frequency with the next node and $\mathrm{Cf}_\mathrm{other}$ = contact frequency with other nodes.

Satisfaction value for the node u can be calculated as:

$$\mathrm{Sat}_k(u, v) = \alpha_1 * \mathrm{Rd}_{(u,v)} + \alpha_2 * \mathrm{Pc}_{(u,v)} + \alpha_3 * \mathrm{Cf}_{(u,v)} \tag{6}$$

where $\alpha_1, \alpha_2, \alpha_3$ are the coefficients and $Sat_{(u,v)} \in [0, 1]$.

Direct trust for the vehicle u can be defined as the:

$$\mathrm{Dt}_{(u)} = \frac{\sum_{j=1}^{k} \mathrm{Sat}_j(u, v) * \mathrm{Wt}_{(j)}}{\sum_{j=1}^{k} \mathrm{Wt}_{(j)}} \tag{7}$$

where $\mathrm{Dt}_{(u)}$ = Direct trust of node u, $\mathrm{Sat}_j(u, v)$ = Satisfaction of node u for the jth transmission, and $\mathrm{Wt}_{(j)}$ = Weight of the jth transmission.

After the estimation of direct trust for node u, the verifier node invites the recommendations from other nodes that are presented in the vicinity. Suppose in a particular time, there are 'n' nodes that are present in the range of the node u and node v. All the nodes send the recommendation about the trust of node u to the node v. Indirect trust of the node u based on the recommendations from the other nodes can be calculated as:

$$\mathrm{Rt}_{(u)} = \frac{\sum_{i=1}^{n} \mathrm{Dt}_{(u)}^{i} * \mathrm{Dt}_{(i)}}{\sum_{i=1}^{n} \mathrm{Dt}_{(i)}} \tag{8}$$

where $\mathrm{Rt}_{(u)}$ = Recommendation trust of node u, $\mathrm{Dt}_{(u)}^{i}$ = Direct trust of node u relative to vehicle i, and $\mathrm{Dt}_{(i)}$ = Direct trust of node i.

The combined trust of the vehicle u can be calculated with the help of the direct and indirect trust as below:

$$\mathrm{Ct}_{(u)} = \gamma \mathrm{Dt}_{(u)} + \mu \mathrm{Rt}_{(u)} \tag{9}$$

Based on the combined trust value, a malicious node identified in the network.

4 Simulation and Performance Analysis

To evaluate the performance of our proposed ECT algorithm, we used network simulator NS-2 [18], a discrete event simulator used to simulate the various types of the communication network. To generate the real traffic environment, we used the VANET-MobiSim as a traffic generator. To simulate the proposed algorithm, we used the Ubuntu 16.04 platform. A comparative analysis of our proposed algorithm has been performed with a successful packet delivery ratio (SPDR) and end-to-end delay (e2ed). Various simulation parameters used in the simulation process are defined in Table 2.

Table 2. Parameters for simulation in NS-2

Parameters for simulation	Values used in simulation
Simulator	Network simulator NS-2
Network area	3000×3000 m^2
Number of vehicles	100, 200, 300, 400, 500, 600
Number of malicious nodes	10, 20, 30, 40, 50, 60
% of malicious nodes	10%
Simulation time	500 s
Routing protocol	AODV, AODV-BH
Mobility	Highway
Maximum speed	0–30 m/s
Attack analysis	Black-hole attack
Traffic	Constant Bit Rate (CBR)
Performance metrics	Successful packet delivery fraction, end-to-end delay

4.1 Results and Discussion

We analyzed the performance of the network with varying number of nodes (with 10% malicious vehicles of the total vehicles) using network simulator NS-2. We compared the performance of our proposed algorithm with the AODV routing protocol with and without the presence of the black-hole attack in the network. In this paper, we performed the performance comparison of AODV [10] algorithm, AODV-BH, and our proposed work ECT with varying numbers of rogue nodes in the network. Initially, we run the simulation process with a basic AODV algorithm without the presence of malicious activity. Further, we inserted some dishonest nodes in the network to perform a black-hole attack with AODV routing in vehicular networks. In the last, we simulated our proposed model with the presence of dishonest nodes and compared the results of all three scenarios. We also used the network traffic generator VANET-MobiSim to generate the simulation traffic close to the real traffic. There are various types of nodes used in our simulation to perform the different duties: (1) attacker node, (2) normal node, (3) RSU nodes (fixed nodes).

Figure 2 shows that the proposed algorithm performs better in terms of successful packet delivery ratio in comparison with AODV and AODV-BH algorithms. The number of malicious nodes increases, the number of hop counts in the trusted path used to transfer data. So the presence of malicious nodes significantly drops the successful delivery fraction. The AODV-BH algorithm without any security implementation suffers more in comparison with other security algorithms. So SPDR decreased significantly in AODV-BH while other shows less decrement due to some security measures. ECT performs better than all other algorithms in terms of packet delivery fraction, due to the selection of only trusted vehicles in the path between sources to the destination to relay the data packets.

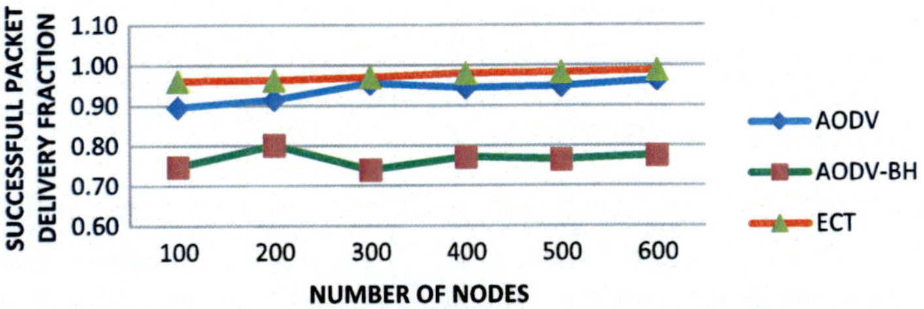

Fig. 2. Comparison of successful packet delivery fraction

In Fig. 3, we can see the impact of the increasing number of nodes on end-to-end delay. For better network performance, the end-to-end delay should be as minimized. In the case of VANET, timely delivery of information packet is prime attention. The delay is slowly increasing in AODV, AODV-BH, and ECT with the increasing number of selfish nodes in the vehicular network. The proposed algorithm ECT displayed the lesser end-to-end delay compare to the other algorithms.

From the above results, we can summarize that the proposed model ECT performs better with respect to AODV and AODV-BH. We also presented the qualitative analysis of the proposed model with respect to AODV and AODV-BH (Table 3). ECT

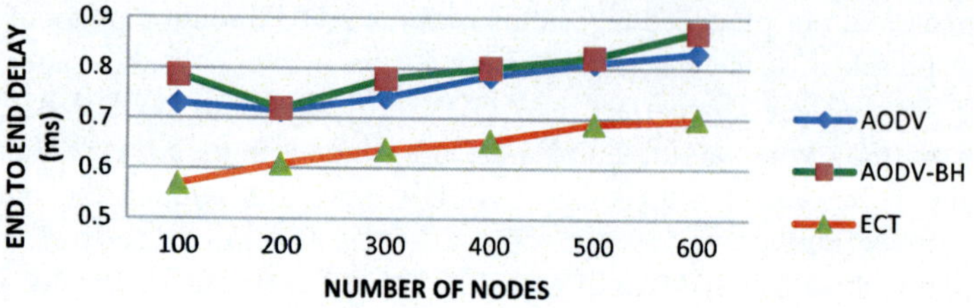

Fig. 3. Comparison of end-to-end delay

Table 3. Performance comparison

Performance parameters	AODV algorithm	AODV-BH algorithm	ECT algorithm
SPDF	Medium	Low	High
E2ed	Medium	High	Low
Malicious nodes	Not present	Present	Present
Overhead	Low	High	Medium
Path breakage	Medium	High	Low
Packet loss %	Medium	High	Low
Throughput	Medium	Low	High
Trusted nodes %	All	Low	Medium

algorithm performs better without any extra overhead. ECT algorithm successfully isolates the malicious nodes from the network.

5 Conclusion and Future Scope

Traditional and basic routing strategies in the vehicular network are incapable to detect attack behavior and malicious activities. So traditional routing strategies cannot fulfill the current data transfer requirements. The presence of a malicious node degrades the network performance. To overcome these limitations and to enhance the performance of the vehicular network, we proposed a trust-based algorithm based on the cooperation among the vehicles. The proposed trust model is entity-centric and based on the direct and indirect trust calculation. The proposed model shows better network performance of AODV algorithm in terms of successful packet delivery fraction and end-to-end delay. The proposed model mainly identifies the packet dropping attack in the network and isolates the attacker node from the network. In the future, we can also add a data-centric trust model to provide better accuracy to the trust estimation. We can also use the best available similarity measure techniques to identify the false recommendation values provided by the selfish nodes in the network. The proposed work can also be extended to detect some other variety of attacks such as a gray-hole attack and denial-of-service attack, and Sybil attack.

References

1. Qu, F., Wu, Z., Wang, F.Y., Cho, W.: A security and privacy review of VANETs. IEEE Trans. Intell. Transp. Syst. **16**(6), 2985–2996 (2015)
2. Lim, K., Manivannan, D.: An efficient protocol for authenticated and secure message delivery in vehicular ad hoc networks. Veh. Commun. **4**, 30–37 (2016)
3. Arya, K.V., Tripathi, K.N.: Power aware and secure routing in mobile and ad-hoc networks. In: 2013 IEEE 8th International Conference on Industrial and Information Systems, pp. 477–482. IEEE (2013)
4. Malathi, A., Sreenath, N.: Black hole attack prevention and detection in VANET using modified DSR protocol. Int. J. Comput. Appl. **168**(7) (2017)
5. Tobin, J., Thorpe, C., Murphy, L.: An approach to mitigate black hole attacks on vehicular wireless networks. In: 2017 IEEE 85th Vehicular Technology Conference (VTC Spring), pp. 1–7. IEEE (2017)
6. Rupareliya, J., Vithlani, S., Gohel, C.: Securing VANET by preventing attacker node using watchdog and Bayesian network theory. Procedia Comput. Sci. **79**, 649–656 (2016)
7. Abumansoor, O., Boukerche, A.: A secure cooperative approach for nonline-of-sight location verification in VANET. IEEE Trans. Veh. Technol. **61**(1), 275–285 (2011)
8. Malhi, A.K., Batra, S.: Genetic-based framework for prevention of masquerade and DDoS attacks in vehicular ad-hoc networks. Secur. Commun. Netw. **9**(15), 2612–2626 (2016)
9. Shabut, A.M., Dahal, K.P., Bista, S.K., Awan, I.U.: Recommendation based trust model with an effective defence scheme for MANETs. IEEE Trans. Mob. Comput. **14**(10), 2101–2115 (2014)
10. Mokhtar, B., Azab, M.: Survey on security issues in vehicular ad hoc networks. Alexandria Eng. J. **54**(4), 1115–1126 (2015)
11. Chen, J.M., Li, T.T., Panneerselvam, J.: TMEC: a trust management based on evidence combination on attack-resistant and collaborative internet of vehicles. IEEE Access (2018)
12. Djahel, S., Naït-Abdesselam, F., Zhang, Z., Khokhar, A.: Defending against packet dropping attack in vehicular ad hoc networks. Secur. Commun. Netw. **1**(3), 245–258 (2008)
13. Tripathi, K.N., Sharma, S.C.: A trust based model (TBM) to detect rogue nodes in vehicular ad-hoc networks (VANETS). Int. J. Syst. Assur. Eng. Manag. 1–15 (2019)
14. Zhang, S.S., Wang, S.W., Xia, H., Cheng, X.G.: An attack-resistant reputation management system for mobile ad hoc networks. Procedia Comput. Sci. **147**, 473–479 (2019)
15. Yang, S., Liu, Z., Li, J., Wang, S., Yang, F.: Anomaly detection for internet of vehicles: a trust management scheme with affinity propagation. Mob. Inf. Syst. (2016)
16. Fotohi, R., Ebazadeh, Y., Geshlag, M.S.: A new approach for improvement security against DoS attacks in vehicular ad-hoc network. Int. J. Adv. Comput. Sci. Appl. **7**(7), 10–16 (2016)
17. Dhurandher, S.K., Obaidat, M.S., Jaiswal, A., Tiwari, A., Tyagi, A.: Vehicular security through reputation and plausibility checks. IEEE Syst. J. **8**(2), 384–394 (2013)
18. Pandey, K., Raina, S.K., Rao, R.S.: Performance analysis of routing protocols for vehicular adhoc networks using NS2/SUMO. In: 2015 IEEE international advance computing conference (IACC), pp. 844–848. IEEE (2015)

Optimized Low-Energy Adaptive Clustering Hierarchy in Wireless Sensor Network

Sumit Pundir(✉), Mohammad Wazid, Ayan Bakshi, and Devesh Pratap Singh

Department of Computer Science and Engineering, Graphic Era Deemed to be University, Dehradun, India
sumitpundir1983@gmail.com

Abstract. Energy which drives the whole world is also required for the propitious technology called wireless sensor network (WSN), which is the driving force of many helpful applications in many areas like health, environment, and industry monitoring, mainly for detecting the threats. The ease of portability, availability, accuracy, and advancement in wireless communication has made it popular among the masses. Sensing nodes are set up in a geographical area of interest, and then, these nodes sense the environmental physical phenomenon and accumulate the information and transfer it to the sink (base station) for application specific action. One of the major concerns with WSN is the network lifetime because of the battery-powered sensing devices. It is one of the most challenging problems which is related with all the issues present in the WSN. The recharge of the sensing devices and prolonging the network lifetime can be achieved through energy harvesting techniques. The primary goal is to enhance the lifetime of the battery-powered sensors and prolong the network lifetime. Numerous of routing protocols have been proposed by the researchers like many versions of LEACH protocols, RPL, 6LowPAN, etc. Here, we propose optimized form of LEACH routing protocol which improves the lifetime of the network, and the simulation results show that the proposed work has improved the network lifetime and stability of the protocol compared with the other existing techniques.

Keywords: Network lifetime · Optimized LEACH · Wireless sensor networks

1 Introduction

Amelioration of wireless technology leads to the better development of WSN. It has been almost three decades for persistent research that has been carried out for the advancement of WSN [1]. Applications of WSN are in different domains such as military, medical, boarder surveillance, geographical area, industrial, agriculture abetment, real-time applications, and various other scientific and human assisting applications. Designing an energy-efficient and reliable routing protocol which prolongs the network lifetime is one of the biggest challenges, since the sensor nodes are battery powered or dependent on the energy harvesting techniques. Most of the energy consumed during the data transmission phase which reduces the lifetime [2].

© Springer Nature Singapore Pte Ltd. 2021
P. Deshpande et al. (eds.), *Next Generation Information Processing System*,
Advances in Intelligent Systems and Computing 1162,
https://doi.org/10.1007/978-981-15-4851-2_4

WSN is an assemblage of thousand wireless sensor devices covering a field of interest, to gather the environmental physical phenomenon and transfer the information to the sink (base station) for further application specific action. Communication among the network nodes is done using radio frequency signal links of scientific, medicine, and industrial (SMI) radio bands (902–929 MHz and 2.4–2.5 GHz) [3]. Base station is one of the critical points of WSN which collects the information from sensing devices present in the network and process it for the remote users. It connects the outside world with the sensor network. Therefore, most of the time, it is resource-rich node equipped with multiple communicating facilities like radio frequency (RF), general packet radio service (GPRS), global system for mobile communication (GSM), or Wi-fi features. Most of the time, sensor network is required at the hazardous or unattended locations which require appropriate fault tolerance to maintain the proper coverage of the desired area and connectivity of the sensor nodes. Resource-constrained sensor nodes which are very cheap and easily available can cover a large area with very less cost. However, with such a resource-constrained environment, we need a very efficient routing algorithm which maintains the quality, reliability, security, and accuracy of the network [4–6].

1.1 Routing Protocol Classification

There are two broad categories of routing protocols [1, 5]:

- Proactive approach—Applications which require persistent information from the desired location. Sensor nodes present in this type of environment persistently sense the required information and keep transmitting it to the sink (base station) for further processing as required by the application. However, in eminently dynamic environment, the proactive approach requires resource-rich devices to keep routing information up-to-date and reliable.
- Reactive approach—Applications where the information is not required on the regular basis, information required for sudden changes or crossing certain limits or threshold values. Sensor nodes present in this environment continuously sense the data but transmits the data only when it is required. These protocols are used for real-time applications and may also work for time-critical applications

We will discuss our work optimized LEACH in the coming section and organization of rest of the paper is as follows: Motivation of work is in Sect. 2, Sect. 3 consists of related work, and Sect. 4 is all about proposed work. Section 5 is result simulations and comparisons. Section 6 is conclusion and future directions, and the last section is references.

2 Motivation

Wireless communication technology has taken the whole world with a sweep with its key applications in the field of health care, military and defense, geographical monitoring, seismic activities, security, and the agriculture. India is a developing nation and primal user of wireless sensors. Taking the example of agriculture, in India, agriculture forms the backbone of the nation. With increase of population, demand for food is also increasing at an alarming rate. Recent survey reports warn that growth of food production is far below the growth of population [4]. Moreover, India is the one of the largest exporters of food pulses and grains in global market. Thus, better methods need to be implemented to increase the production to an incredible level and also make the works of farmers and the end users easy. This is where WSN can be put into good use. Wireless sensing devices can also be used to monitor the soil temperature, volumetric water content, soil composition (sand, clay, nutrients), fertilizer levels, and controlling the irrigation system. This will not only save much of time and effort, but also increase the production. For that, the WSN-based solutions in agriculture should be available at low cost, efficient, and manageable at a low cost. So, is goes for all the other fields. Each requires the low-cost implementation of the wireless sensor network. There have been numerous researches conducted for an efficient, low energy-consuming sensor network routing protocols. Effective routing protocol can lead to lower power consumption of the sensors and thus lower the cost of implementation.

The prime motive of the researches is to provide the users with such wireless sensor systems and thus providing an optimal routing solution. We have worked to get out the most from the available protocols and to provide with an optimized solution to solve the problem [1, 5, 6].

3 Related Work

Continuous research has been going on for last couple of decades for designing an energy-efficient routing protocol for the energy-constrained WSN environment. Researchers have been proposing variety of protocols to prolong WSN life span for long period of time. However, when it comes to provide effective routing solution, we have to compromise with some of the network parameters such as latency, throughput, loss rate, etc. Thus, we are still in a need to provide a better solution for the energy-efficient routing solution.

Heinzelman et al. [7] proposed an energy-efficient routing protocol which distributes the load of a single node called as cluster head to multiple nodes which are present in the network called "low-energy adaptive clustering hierarchy protocol (LEACH)." It prolongs the network life based on rotation-based cluster head selection using random number. Nodes which participates in the process of selection of cluster head generate a random number between 0 to 1, and if the random number is less, then the value calculated from Eq. 1 becomes cluster head for the given round, and it is a proactive protocol.

Manjeshwar et al. [1] proposed a reactive protocol named "threshold-sensitive energy-efficient sensor network protocol (TEEN)" which is based on two threshold values, where at every cluster change time, the two threshold values are broadcasted by the cluster head. Using these two values, the sensing device saves energy and sends the required information to the cluster head when it is required unlike proactive WSN protocols which keep on sending the data. Reactive protocols are well suited for time-critical data sensing applications.

4 Proposed Work

In the current section, we will discuss the proposed optimized low-energy adaptive clustering hierarchy protocol. It has two key features:

- It is proactive.
- It is optimized LEACH protocol which increases the number of clusters per round, thus reducing per cluster size and energy required by each node to send the sensed data to its neighboring cluster head.

To describe the proposed protocol, first we discuss the LEACH algorithm. LEACH is typically a self organized and adaptive protocol. Selection of cluster head is based on the predetermined threshold probabilistic value. Nodes generate a random value r between 0 and 1. If the randomly generated values by the node is less, then the value calculated using the formula below in Eq. 1 is eligible to become cluster head for the current round [1, 7].

$$\left\{ T_n = \frac{p}{1 - p\left[r \cdot \bmod\left(\frac{1}{p}\right)\right]} \right. \tag{1}$$

All cluster heads broadcast their TDMA schedule for the nodes associated with their cluster.

The proposed work uses the same formula but after modifying it. The modified formula is mentioned below as in Eq. 2,

$$T_n = \left[\frac{p * S(i) \cdot E}{\left(1 - p\left[r \cdot \bmod\left(\frac{1}{p}\right)\right]\right) \cdot \left(\text{total}_{\text{energy}} + a_{\text{alpha}}\right)}\right] \tag{2}$$

The LEACH formula has been modified and addition of extra factors increases the selection of number of cluster head for each cluster. Here, "$S(i) \cdot E$" is the initial energy of the particular node which is up for selection, $\text{total}_{\text{energy}}$ is the sum of initial energies of all the nodes, and alpha is a value taken whose value is 1 for stability of equation while introducing maximum output. In (2), the process of cluster head election is almost same except the probability to choose the cluster head becomes more, and as a result, there will be more than one cluster heads in a cluster in a particular round. However, earlier it is used to be only one. This results in forming smaller virtual clusters within a single cluster, each again having its own cluster head for that cluster. Thus, the distance between the neighboring sensor nodes and the corresponding CHs decreases as there is more than one option of nearest cluster head for each node. Thus, they have to transmit sensed data to a more nearest CH, and the energy needed for transmitting is also reduced. This in turn increases the life of each node by reducing the energy utilization level, thus increasing the network lifetime, respectively [1, 8].

During the testing, it has been recorded that it has been optimized by more than 300% of the original one, thus making it an energy-efficient routing protocol. Due to low energy utilization for transmission, each node has a longer life span and remains operational for a longer period. The nodes elected as cluster heads including some of the sensor nodes are the ones among the dead nodes. This increases the lifetime of the WSN infrastructure as well as decreases the cost as the nodes do not need replacement and charging (if possible) for a long period of time. Thus, the proposed work ensures longer exposure of the wireless nodes to physical environment with minimal energy usage.

Figure 1 shows the cluster formation in OPTLEACH and the process of sensing data. Note that, more than one cluster heads are formed within the previous-formed single cluster, each forming a cluster of nodes, virtually. That is how we get more than one cluster heads CHs within a single cluster.

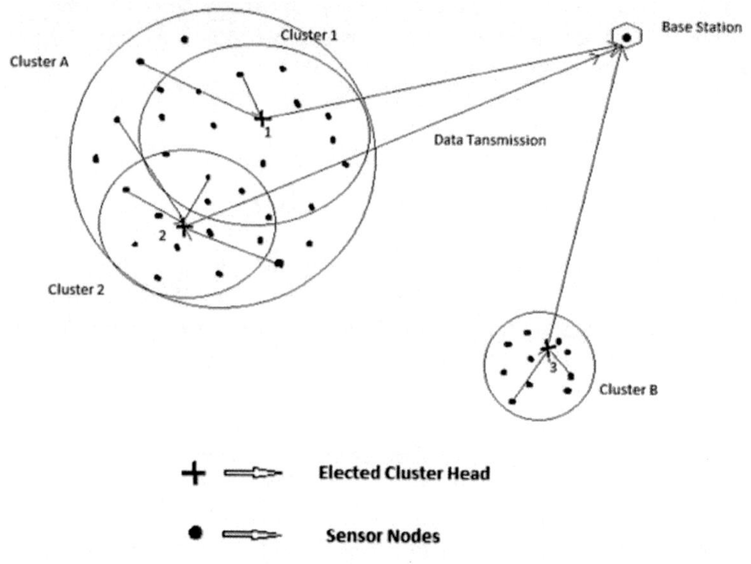

Fig. 1. OPTLEACH hierarchy

As shown in the above figure, two CHs are formed within a single Cluster A, each virtually forming its own cluster. The nodes nearest to the particular cluster head send its data to it and thus saving the transmission energy which would otherwise been higher for sending to the other cluster head if it was the only one. This result demonstrates significant decrease in the count of dead nodes and increase in the total number of alive nodes, thus increasing network lifetime.

5 Result Simulation

In this section, results of optimized LEACH algorithm are discussed with the help of some diagrams which are obtained during the simulation process of optimized LEACH on MATLAB software. Comparing the results of optimized LEACH with traditional

LEACH protocols which is one of the best, traditional, and authentic protocols provides us the details of hierarchal clustering.

Figure 3 shows the comparison of optimized LEACH with standard LEACH routing protocol, which shows a significant decrease in the count of dead nodes in each round. Thus, number of rounds has also increased, which is very efficient (Fig. 2).

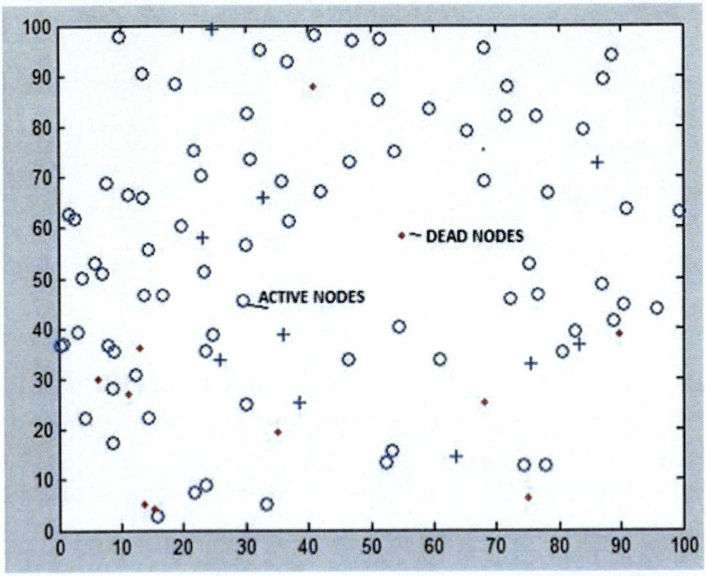

Fig. 2. Active and dead node

The red line shows the fall in alive nodes with consecutive rounds in LEACH while blue line shows the fall in consecutive rounds of the optimized LEACH. Almost zero nodes are dead in 300 rounds in optimized LEACH compared to the LEACH in which maximum nodes are dead in first 100 rounds (Tables 1 and 2).

Here, the total number of dead nodes is less than compared to total number of alive nodes in Fig. 2.

Fig. 4 shows the comparison of the basic average energy of each node for optimized LEACH algorithm and LEACH algorithm. It can be observed that the fall of energy of the nodes in LEACH is faster in given rounds, with a rapid downfall compared to optimized LEACH where the average energy of the nodes is sustaining longer. This in turn leads to longer lifetime of the nodes which prolongs network life.

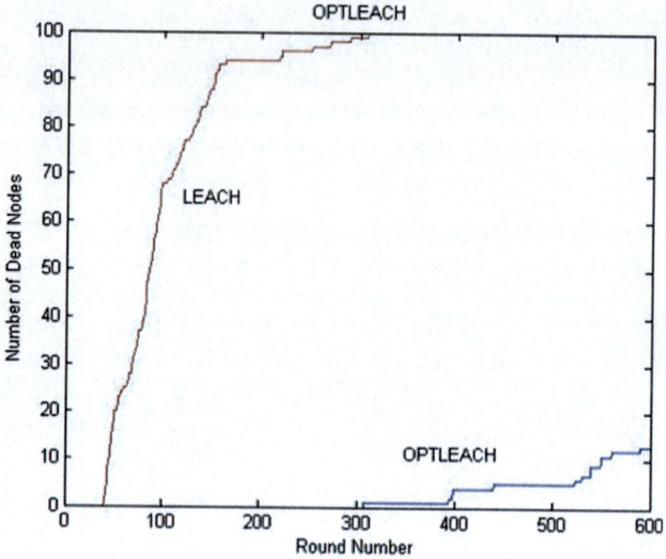

Fig. 3. Comparison between optimized LEACH and LEACH

Table 1. Simulation parameters

Parameters	Value
Dimension of field taken	100 × 100
Percent (%) of advanced nodes, *m*	0.1
Probability of cluster head selection, *p*	0.1
Amplification energy (short distance), E_{fs}	10 pJ/bit/m^2
Amplification energy (long distance), E_{mp}	13 pJ/bit/m^2
Energy for data aggregation, E_{DA}	5 nJ/bit/signal
Initial energy, E_0	0.1 J

Table 2. Comparison table

Parameters	Dead nodes in 300 rounds	Dead nodes in 600 rounds	Average energy remaining after 50 rounds	Average energy in remaining after 300 rounds	Average energy in remaining after 600 rounds
LEACH	100	>100	0.04	0.00	0.00
OptLEACH	0	12	0.11	0.07	0.035

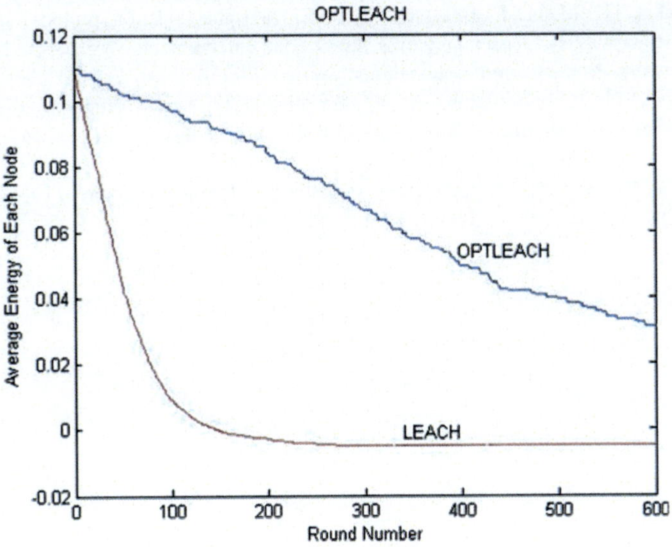

Fig. 4. Comparison of average energy of each node

6 Conclusion

This paper focuses on the lifetime of WSN and proposes a routing algorithm optimized LEACH which is optimized by a great margin compared to traditional LEACH. It also proved its improved efficiency compared to the other routing protocol. Various comparison graphs and models are presented, and each node is sustaining for a longer time period. For future, the proposed algorithm can be improved for more than two hierarchy levels for heterogeneous nodes. Protocol can be improved for various other fields such as precision monitoring in agriculture.

References

1. Manjeshwar, A., Agarwal, D.P.: TEEN: a routing protocol for enhanced efficiency in wireless sensor networks. In: Presented in 1st International Workshop on Parallel and Distributed Computing Issues in Wireless Networks and Mobile Computing (2001)
2. Riga, N., Bestavros, A., Matta, I.: DIP: Density Inference Protocol for wireless sensor networks and its applications to density unbiased statistics. Technical Report BUCS-TR-2004-023 (2004)
3. Maurya, S., Jain, V.K.: Fuzzy Based Energy Efficient Sensor Network Protocol for Precision Agriculture (2006)
4. Wu, H., Zhao, C., Zhu, L.: Study on an energy-aware routing algorithm for agriculture WSN. Indonesian J Electr. Comput. Sci. **11**(7) (2013)
5. Lu, H., Liao,Y.M.: Multipurpose watermarking for image authentication and protection. IEEE Trans. Image Process. **10**, 1579–1592 (2001)

6. Smaragdakis, G., Matta, I., Bestavros, A.: SEP: a stable election protocol for clustered heterogeneous wireless sensor networks. In: Second International Workshop in Sensor and Actor Network Protocols and Applications (SNPA) (2004)
7. Heinzelman, W., Chandrakasan, A., Balakrishnan, H.: Energy-efficient communication protocols for wireless microsensor networks. In: Proceedings of Hawaiian international conference on systems science (2000)
8. Fu, C., Wei, W., Wei, A., Jiang, Z. (2013). An energy balanced algorithm of LEACH protocol in WSN. Int. J. Comput. Sci. Issues (IJCSI) **10**(1) (2013)

QoS in Vertical Handoff for Wireless Network Using Hungarian Model and Data Parceling Technique

G. U. Mali(✉) and D. K. Gautam

Department of Electronics Engineering and Technology,
North Maharashtra University Jalgaon, Jalgaon, Maharashtra, India
ganeshumali@gmail.com, dkgautam_nmu@hotmail.com

Abstract. In a heterogeneous wireless network, the server–client architecture plays a vital role. As the clients request the data from the servers, they analyze the requests and stream the data to the clients through a specified protocol of the instance network. When a movable node in the wireless network leaves the mother pool and enters the other pool, then data losing is a scenario that will occur for sure. The vertical handoff technique offers a solution for this to maintain the lossless network changing paradigm for the movable nodes in the heterogeneous network. Some vertical handoff techniques take much longer time to complete the handoff process, and this enables the system to stream the data through the buffer system present in the instance network or from another network. This not only slows down the whole process of vertical handoff, but also sometimes data loss is happening due to vertical handoff. To deal with this kind of poor quality of service during the vertical handoff, this research paper uses the data parceling technique. Streaming data from the server to the moving node in the network is divided into many parts, and then, they are loaded to ongoing threads handling the vertical handoff. The data allocation process of the system is powered with the Hungarian data assessment technique to perform the lossless vertical handoff successfully.

Keywords: Data parceling · Hungarian assessment model · Vertical handoff · Wireless network

1 Introduction

A level of quality has to be maintained for a vertical handoff, which has to be governed by a set of QoS parameters that ensure good quality without any drops in connectivity or user inconvenience. A couple of parameters that vertical handoff has to adhere to provide a respectable user experience such as network bandwidth, power consumption, power dissipation, and network conditions.

For example, a device is connected to a WLAN AP while on the move, and as it starts moving away from the limited range of the said access point, then the signal received by the device would keep receding as the device got further away, until it reaches a threshold value designated by the QoS parameter. As soon as the devices approach the threshold value, the cellular card is activated and begins to synchronize

© Springer Nature Singapore Pte Ltd. 2021
P. Deshpande et al. (eds.), *Next Generation Information Processing System*,
Advances in Intelligent Systems and Computing 1162,
https://doi.org/10.1007/978-981-15-4851-2_5

with the device to prepare for the handoff. Relative signal strength is the parameter of great importance as it judges the cellular signal for a potential of an overloaded or distant signal that is no longer advantageous.

Therefore, the QoS management system makes a decision based on the parameters to pre-empt and predict a vertical handoff situation and accordingly prepares for it. This preparation helps achieve a seamless connection that is devoid of any drops and delays experienced due to mobility of the user or inability or reduced strength of the signal. As no operator can provide high bandwidth, low latency connection with a wide-area data service, vertical handoffs are an essential part of any heterogeneous wireless network. Consequently, the optimum solution is to use a congregation of wireless networks to provide efficient all-round connections over a large geographical area. Quality too dictates the probability of a handoff to occur as the parameters help make viable decisions to handover a poorly performing network in terms of bandwidth, relative signal strength, power consumption, etc. For example, if the network is using up significant amount of power and a network with less dissipation is in range, a vertical handoff will occur to maintain a considerable amount of quality for the user. Therefore, vertical handoffs play a very important role to maintain another network QoS in a heterogeneous wireless network.

In this research paper, Sect. 2 discusses some related work, and Sect. 3 presents the design of our approach. A deep analysis of the result is performed in Sect. 4, and Sect. 5 provides the conclusion of this research paper.

2 Related Work

Chen et al. [1] introduce the concept of QoE or quality of experience in a heterogeneous wireless access network. The authors, during their preliminary investigations, infer that it is QoS, that is, usually considered in conventional vertical handoff schemes in a network. But as QoS falls short of its objective of enhanced user satisfaction, the researchers present QoE or quality of experience as a factor in the decision process of a vertical handoff. The authors also hypothesized the use of Q-learning deployed network to ascertain the mutuality between QoS and QoE that exists in a heterogeneous network.

Cai [2] explores the concept of an emergency pre-emption technique that is based on a time-threshold for upward vertical handoffs in cellular networks embedded with WLAN. The researchers modified the TUPS by revaluating the preferences of cellular cells and the types of channels to redesign a new emergency pre-emption technique. This technique is an improvement over the ones previously envisioned, and in comparison with the traditional methods, the presented method produces astonishing results in WLAN-embedded cellular networks.

Kulkarni and Singh [3] concern with the approach for network selection in a vertical handoff situation. The researchers attempt to modify the selection process by introducing a ranking system that can precisely alter the criteria and assure the anticipated QoS resides well inside the boundaries while maintaining the critical criteria to ensure optimum performance. In a heterogeneous wireless network, to be able to take benefit from the advantages of all the types of networks, the vertical handoff has to be heavily optimized for switching. Zineb et al. [4] proposes a novel amalgamation of an

ANN to the decision-making aspect of a vertical handoff, to increase the efficiency and accuracy of the handover. The authors reflect that the most important objective in a 5G cognitive radio inside a heterogeneous network is to manage the connectivity and furnish a respectable QoS. Lahby and Sekkaki [5] explore the realm of telecommunication, the 4G deployment, and their use in facilitating easier and almost constant connectivity offered by these emerging technologies. 4G has enabled a user to be on the move and experience high data transfer rates. This is only possible in a heterogeneous network when the algorithm identifies the network with the best access and switches to that network. The fundamental challenge in this process is the framework that ascertains the network to be switched.

Chopra et al. [6] propose a dispersed version of the Hungarian method for solving the well-known LSAP. The proposed algorithm allows a selection of robots to collectively compute the ideal solution to the LSAP without any administrator or shared memory. The authors prove that under asynchronous implementation, all robots convene to a collective optimal assignment within O(r3) iterations. By running simulation experiments over multiple instances of the LSAP with varying problem sizes, the researchers show that the average number of iterations for convergence is a lot more compact than the theoretic worst-case bound of O(r3).

Cao et al. [7] introduce a version of the Hungarian-based methods for querying business process model repositories. The start point of the work is to map the elements of two process models first and then calculate their structural similarity. The authors use the petri-net throughout the paper for illustration, where we simplify the mapping of the elements to the place nodes mapping since the text similarity of the transition nodes is mature in information retrieval field. The researchers design two-place context similarity from two perspectives of the commonality and the graph edit distance, and then, the problem of searching the best place mapping combination is transformed to the classical assignment problem which could be solved by the Hungarian algorithm.

Xie et al. [8] explore the global alignment of biomolecular networks which is important for the development of network medicine and the study of species evolution. In this work, the authors present a hybrid algorithm (HGA) that combines Hungarian algorithm and greedy algorithm for more accurate alignment but with decreased computation time compared to HA. Specifically, these experiments show that HGA aligns large networks well with an acceptable computation time, whereas HA fails.

3 Contribution

The proposed model of maintaining quality of service parameters using Hungarian and the data parceling technique can be depicted in the Fig. 1. The vertical handoff process is triggered with an avalanche effect on changing of the network that is mentioned in our past edition [9]. As the data is requested from any client in the heterogeneous wireless network from the server, server starts streaming the data to the client, this is what happens in the general scenario of the client-server architecture in heterogeneous wireless network. So, on vertical handoff situation, these kinds of architecture utterly fail to switch over the data streaming without any data dropping.

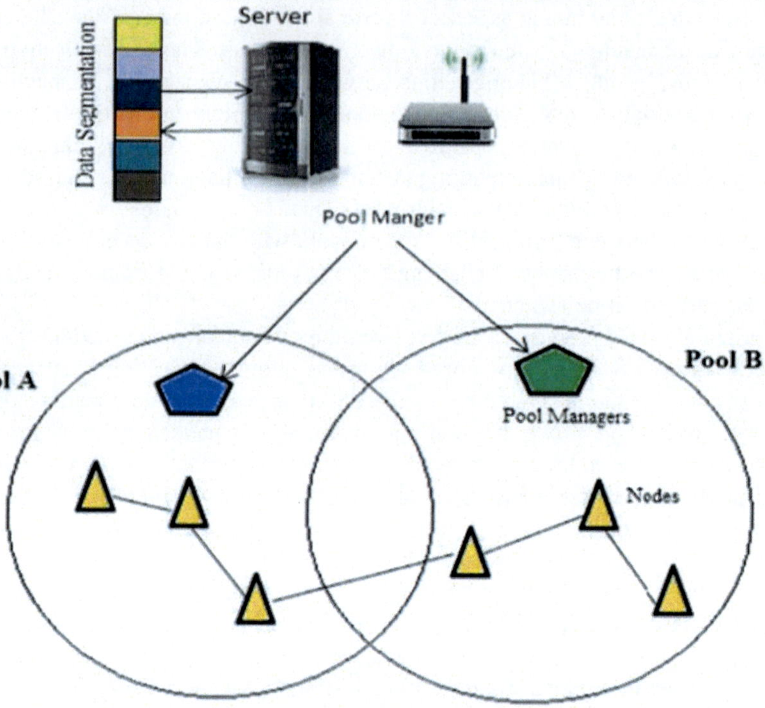

Fig. 1. Proposed model overview for data parceling technique

So, to maintain the QoS in data streaming, proposed model uses the Hungarian assessment model which works on the segmented parts of the data for the many numbers of threads. The data segmentation process is carried out as soon as data streaming starts, where data is divided into n number of parts using its byte array. This partition of the data is purely depending on its size. The process of segmentation can be shown in Eqs. 1 and 2.

$$X = \{0, 1, 2, 3, 4, \ldots(k-1)\} \tag{1}$$

$$f(Pt) = \frac{\sum_{i=0}^{i=k-1} x_i}{n} \tag{2}$$

$$f(sg) = \int_{i=0}^{n} f(pt)/n \tag{3}$$

where *X*—The byte array of the data, *k*—The length of the array, *n*—Number data segments, *f*(*Pt*)—Data partition function, and *f*(*sg*)—Data segmentation function.

The data segmentation process can be depicted in the below-shown Algorithm 1.

Algorithm 1: Data segmentation
//Input: Byte Array A
//Output: Segmented vector S_L
Function: Data_Segmentation (A)

1) Start
2) B_L = Size of A/n
   (B_L = Block Length, n = Number of Data segments)
3) T_B [ ] (Temp Block array)
4) ***for*** $i = 0$ to size of A
5) $T_B[k{+}{+}] = A[i]$
6) If $k = B_L$
7) ADD T_B to S_L
8) end For
9) ADD T_B to S_L
10) ***return*** S_L
11) stop

This segmented data is now streaming to the client node which requested the same. Data is streaming segment by segment based on the Hungarian data allocation model, which eventually takes the data streaming decision based on the pattern that takes considerably less time in streaming.

This data parceling technique includes both the data segmentation and Hungarian data allocation model, which eventually triggers on verge of vertical handoff decision. Suppose if there are n segments that are there, some $(n - x)$ segments are been allocated to the client before the handoff could take place. On handoff $(n - x + 1)$th segment starts allocating to the node with the help of the pool managers of the heterogeneous network. So, this process is kept alive till n segments are being allocated to client to deliver the complete data.

Hungarian optimal time estimation for data allocation process can be shown in the below Algorithm 2.

Algorithm 2: Hungarian Time Estimation for Data
Allocation process
//Input: No. of Threads C_r
// No. of data Segments C_l
//Time set $T_{set} = (C_{r1}C_{l1}, C_{r2}C_{l2}, C_{r3}C_{l3} \ldots C_{rn}C_{ln})$
//Output: total time
Function: Hungarian _Model (C_r, C_l, T_{set})

1) Start
2) total time = 0
3) ***for*** $i = 0$ to size of C_l
4) ***for*** $j = 0$ to size of C_r
5) ***if*** $j = 0$ then
6) ***if*** C_{rj} is engaged

7) $T_{C_{rj}C_{li}} = T_{C_{rj}C_{li}} + T_{set}C_{rj}C_{li}$
Where [T: Time]
8) *else*
9) *if* C_{rj} is engaged
10) *if* $T_{pC_{rj}C_{l}} > T_{cC_{rj}cC_{li}}$
11) $T_{C_{rj}C_{li}} = T_{pC_{rj}C_{l}} + W_t + T_{set}C_{rj}C_{li}$
Where [W_t = Waiting Time]
12) total time = total time + $T_{C_{rj}C_{li}}$
13) end *for*
14) end *for*
15) ***return*** total time
16) stop

4 Results and Discussion

Unlike most of the vertical handoff in a wireless network, scenarios are weaved in a simulation environment, whereas the proposed model is deployed in real-time physical environment. The model uses the Java-based windows machine for the very purpose which enables the model to measure most of the parameters in real-time environments of wireless heterogeneous network.

To measure the performance of the model, throughput is considered as the best parameter, where throughput is measured as the number of bytes transferred in a given time. For the purpose, proposed model measures the throughput which is tabulated below along with the other methodologies as mentioned in [10].

The graph in Fig. 2 indicates that our deployed system in real time using Hungarian and data parceling technique provides more throughput compared to another routing process of [10] (Table 1).

This is because the proposed model efficiently handles the data by segmenting and then parceling to the client node easily. This ease of process continues even on a vertical handoff situation because of the segmented data and that is powered with the Hungarian model which mainly tends to decrease the processing time drastically and transfer the huge amount of data.

Because of the data parceling technique, the end to end delay (average latency) of the vertical handoff diminishes drastically. When average latency of the proposed model is measured along with the other techniques as mentioned in [11], the obtained results are shown in Table 2. The plot in Fig. 3 clearly indicates that due to data parceling technique data moves faster and it crosses the hurdle of vertical handoff too with ease.

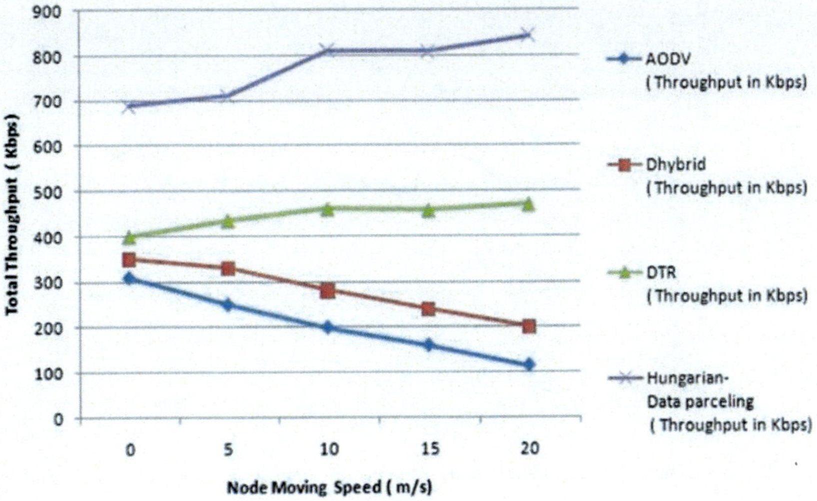

Fig. 2. Throughput versus mobility comparison with Hungarian_Data parceling technique and other methods

Table 1. Throughput versus mobility comparison with Hungarian_Data parceling technique and other methods

Node moving speed (m/s)	AODV (throughout in kbps)	Dhybrid (throughout in kbps)	DTR (throughout in kbps)	Hungarian data parceling (throughout in kbps)
0	310	350	400	690
5	250	330	435	712
10	200	280	460	812
15	160	240	457	810
20	115	200	470	844

Table 2. Average latency comparison with Hungarian_Data parceling technique and other methods

Packet size (byte)	ETT metric	iETT metric	Hungarian_Data parceling
500	1.25	0.8	0.45
1000	1.5	1.2	0.8
1500	1.7	1.4	1
2000	1.8	1.5	1.11

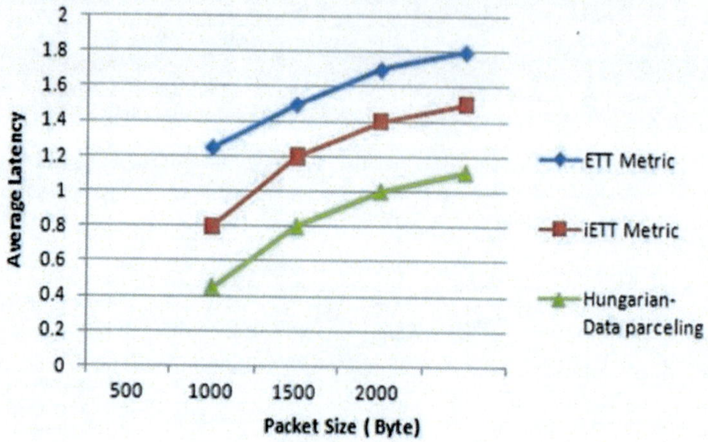

Fig. 3. Average latency comparison with Hungarian_Data parceling technique and other methods

5 Conclusion

This research paper not only deals with the vertical handoff and also maintains the lossless scenario. This is mainly because of the core architecture of the system, where data is segmented into different parts and then they are loaded to the independent threads handling by the Hungarian model. In the proposed model, Hungarian model is optimized to get the least data allocation time for the different generations. The whole scenario is deployed under the real-time environment which obtains the best average latency and throughput values when compared with that of [10, 11].

References

1. Chen, J., Wang, Y., Li, Y., Wang, E.: QoE-aware intelligent vertical handoff scheme over heterogeneous wireless access networks. IEEE Access (2017). https://doi.org/10.1109/access.2018.2853730
2. Cai, J.: A time-threshold-based emergency pre-emption scheme for vertical handoff in Cellular/WLAN interworking. In: 8th IEEE International Conference on Software Engineering and Service Science (2017)
3. Kulkarni, S., Singh, P.: Enhancement of handover decision in heterogeneous networks. In: International Conference on Signal Processing and Communication (2017)
4. Zineb, A., Ayadi, M., Tabbane, S.: QoE-based vertical handover decision management for cognitive networks using ANN. In: Sixth International Conference on Communications and Networking (ComNet) (2017)
5. Lahby, M., Sekkaki, A.: Optimal vertical handover based on TOPSIS algorithm and utility function in heterogeneous wireless networks. In: International Symposium on Networks, Computers, and Communications (2017)
6. Chopra, S., Notarstefano, G., Rice, M., Egerstedt, M.: Distributed version of the hungarian method for a multirobot assignment. IEEE Trans. Robot. **33**(4), 932–947 (2017)

7. Cao, B., Wang. J., Fan, J., Yin, J., Dong, T.: Querying similar process models based on the Hungarian algorithm. IEEE Trans. Serv Comput. **14**(8) (2015)
8. Xie, J., Xiang, C., Ma, J., Tan, J., Wen, T., Lei, J., Nie, Q.: An adaptive hybrid algorithm for global network alignment. IEEE/ACM Trans. Comput. Biol. Bioinform. **13**(3), 483–493 (2016)
9. Mali, G., Gautam, D.: Avalanche effect based vertical handoff system for wireless communication. In: ICCASP (2018)
10. Shen, H., Li, Z., Qiu, C.: A distributed three-hop routing protocol to increase the capacity of hybrid wireless networks, pp. 1536–1233 (c). IEEE (2015)
11. Biaz, S., Qi, B., Ji, Y.: Improving expected transmission time metric in multi-rate multi-hop networks. In: 2008 5th IEEE Consumer Communications and Networking Conference (IEEE CCNC), Las Vegas, NV, pp. 533–537 (2008)

Study of Three-Way Solution for Node Movement in Mobile Ad Hoc Networks (MANETS)

Severin Gbetondji Michoagan[1(✉)], S. M. Mali[1,2], and Sharad Gore[1]

[1] Department of Computer Science, Savitribai Phule Pune University, Pune, India
{sevegbeton, shankarmali007, sharaddgore}@gmail.com
[2] Dr. Vishwanath Karad, MIT World Peace University, Pune, India

Abstract. Mobile ad hoc network (MANET) is a subset of computer wireless networks and in its simplest form is the collection of mobile stations connected with each other by radio links to form a network. MANET is a self-configured, self-organized network where the movement of the mobile stations is random during the network. The positions of the nodes are unpredictable and have direct impact over network efficiency and its performance. For MANET network monitoring, we need to trace the motion of the node, and for controlling, we need to decide the optimal node movement. The focus for this paper is the study of three-way solutions of modeling node movement in MANET such as movement via trajectories, facilities for random mobility, and movement via programming interface. This study is followed by routing protocols performance analysis. Three routing protocols DSR, AODV, and OLSR are compared with segment-based trajectory as node movement model. OPNET is the simulation tool use, and the performances of the three routing protocols are analyzed based on the network metrics: throughput, end-to-end delay, and network load. The comparison analysis will be carried out, and at last, the conclusion will be presented.

Keywords: MANET · Multihop · Topology · DSR · OLSR · AODV · QOS · OPNET

1 Introduction

A mobile ad hoc network is composed of autonomous mobile wireless devices interconnected among themselves without any centralized assistance to form a network [1]. MANET is also called as peer-to-peer, multihop networks where data packets are routed from a source host to the receiving host with the help of intermediate hosts for the hosts which are not in the transmission range of each other [2]. The network can be formed anywhere and at any time because there is no need for human assistance and infrastructure to start the network [3]. MANET is easy to deploy, economically cheaper as compared with wired networks [2]. This type of network is used in military battlefield, or in cases of unpredictable events such as hurricanes, earthquakes, or sensor networks, where nodes are constantly moving.

© Springer Nature Singapore Pte Ltd. 2021
P. Deshpande et al. (eds.), *Next Generation Information Processing System*,
Advances in Intelligent Systems and Computing 1162,
https://doi.org/10.1007/978-981-15-4851-2_6

To design good mobile ad hoc networks [4], we need to address various challenges such as:

- Dynamic topologies.
- Limited power constraints of mobile nodes.
- Bandwidth-constrained and variable-capacity link.
- Limited physical security.

In MANET, the nodes are part of the network only for the duration of the communication session [5]. However, it is very challenging to have proper topology during the communication session why nodes are on move. Nodes mobility patterns play a vital role in determining an overall performance of MANET network. Therefore, an efficient and effective node movement patterns need to be studied in advance depending of the resources and requirement of the network.

The goal of the proposed work in this paper is to provide researchers with clear categories of node movement patterns which we have named it as three-way solutions node movement patterns along with their benefits and drawbacks. Additionally, we aim to provide researchers with guideline to help them to choose node movement patterns properly to avoid ambiguity in designing MANET network [6].

The remaining of the paper is structured as follows: Sect. 2 explains the impact caused by node movement on MANET. Section 3 is the study of the three-way solutions node movement patterns. Section 4 describes the simulation setup with OPNET and presents the results. Finally, Sect. 5 is a conclusion which gives some concluding remarks.

2 Impact of Node Movement on MANET

The performance of mobile ad hoc network is influenced by the movement of the communicating nodes. Node movement patterns have direct impact over neighbor selection, network performance, and protocols performance.

2.1 Impact on Routing Protocols

Routing protocols provides mechanisms to forward traffic from source node to the destination node [7].

Node movement causes wireless link failures due to the dynamic pattern in which nodes are connected. Also the movement of the node can cause route failures and route changes leading to packets losses and delayed acknowledgments [8]. Due to the frequent updates for changes in topologies, an effective route which has been established for packets to get deliver can be interrupted as mobile nodes move from one place to another [9].

2.2 Impact on Security

The openness nature of the MANET network and the dynamic move of the mobile nodes creates challenges for security. The communicating nodes are self-dependent, randomly moving in any direction and any new nodes can join the network and leave at anytime, which offer special opportunities to attackers [10]. The random nature of MANET without any fixed infrastructure, no central authority to regulate the network, makes it harder to detect malicious behaviors of the node and authenticity of a message. The key management for proactive and preventive measure is not effective leading to the increasing of the network vulnerabilities.

2.3 Impact on Quality of Service (QoS)

With the structure network like wired network, the uniformity of the QoS can be maintained. But in case of MANET network, the QoS is not uniform.

Guarantee the quality of service (QoS), is very important key factor for the correct transmission of the multimedia network traffic [11]. Ensure QoS refers to maintain a wide set of metrics such as throughput, packet loss, delay, jitter, error rate, and so on. In MANET, the dynamically changing of the network topologies as nodes are highly mobile makes it more difficult to have a guarantee of the quality of service as compared in fixed network.

3 Study of the Three-Way Solutions for Node Movement Patterns

Nodes movement plays an important trait why designing mobile ad hoc networks. Node movement follows basic attributes such as: the starting point of the node, its direction, velocity, and speed changes over time [12].

3.1 Movement via Trajectories

A trajectory is a sequence of consecutive line segments.

In MANET node, trajectory reflects to the actual movement path of mobile node during a simulation or a communication network. It is also defined as a continuous trajectory of a mobile node [13].

3.1.1 Segment-Based Trajectory

In segment-based trajectory, the movement paths of the mobile devices are defined as a series of path segments determined by predefined points. There are of two types: fixed-interval trajectory and variable-interval trajectory [9] (Fig. 1).

In **fixed-interval trajectory**, a mobile node traverses or navigates every path segment with the same time period, whereas in **variable-interval** trajectory, any point along the path trajectory describes its own altitude, with wait time, segment traversal time, and orientation. The wait time makes a mobile device to pause during each segment point before traversing the next segment point (Fig. 2).

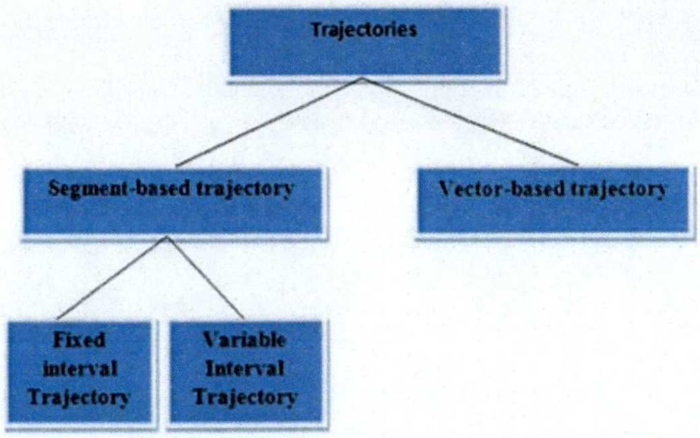

Fig. 1. Types of node trajectories

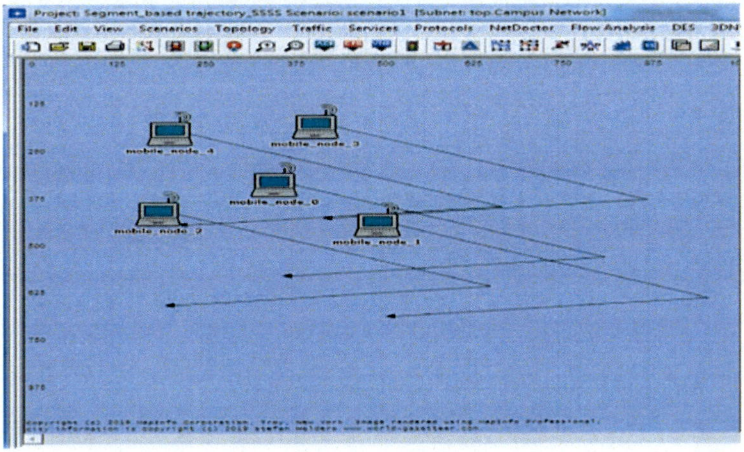

Fig. 2. Node with segment-based trajectory

3.1.2 Vector-Based Trajectory

In contrast to segment-based trajectory, there is no explicit end point for vector-based trajectory. Vector-based trajectory follows the circular path around the earth. The path is determined by the attitude, ground speed, and ascent rate attributes of the mobile devices (Fig. 3).

3.2 Movement via Random Mobility

A node has a state of a random mobility in which it chooses a target direction arbitrary and moves according to that direction with a randomly chosen speed and time duration [3]. Various random nodes nobility has been provided in the literature [14, 15], and they are: random waypoint, random walk, and random direction.

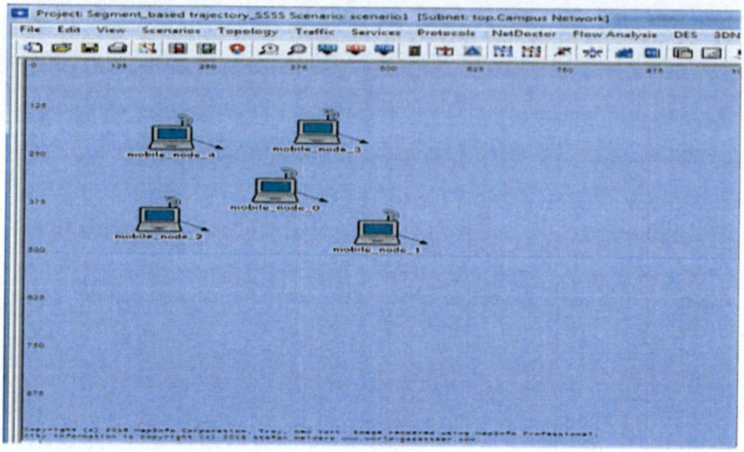

Fig. 3. Nodes with vector-based trajectory

3.2.1 Random Waypoint Model

According to the literature survey [16, 17], random waypoint (RWP) node movement patterns are of widespread use by several authors in ad hoc wireless networking community. This node movement model produces more realistic movement of mobile devices to human movement patterns.

Figure 4 explains the notion of random waypoint model. In RWP, user arbitrary chooses any target end point (waypoint) in the system area, traverses with constant movement speed in the same direction to this point, why in this point, user pauses for a while, and then again chooses a new target end point.

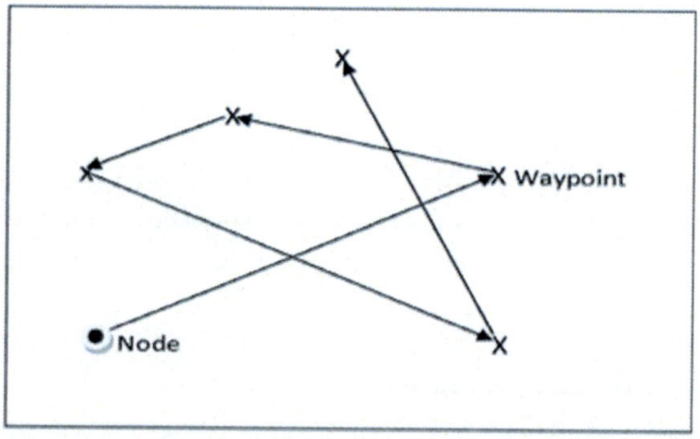

Fig. 4. Illustration of random waypoint

3.2.2 Random Walk Model

In 1926, Einstein has proposed random walk mobility to evaluate the unpredictable movement of particles in physics [18]. This movement pattern is also called as Brownian movement. In this movement pattern, the speed of the node is not important;

however, the focus is more on the direction on the move. In random walk model, a mobile node moves from its current position to a randomly direction along with a randomly selected speed. The speed and direction are selected from predefined ranges, [min_Speed, max_speed] and [0, 2π], respectively [13].

3.2.3 Random Direction Mobility Model

The random direction node mobility model is as per the literature [19, 20], the most used synthetic mobility model besides random waypoint mobility model. In this model, node movement is considered as a straight walk segments along with random direction with constant speed and optional pause [16]. Random direction mobility model has been used to erase the density wave problem in random way point model which is the clustering of nodes in the center of the simulation area [17]. To cope with this situation, each node moves why choosing its random direction between 0 and 2π, with velocity of v from its initial state pointing the border of simulation area without stop during this period. The node waits for some time period, chooses new direction uniformly in (0, π) after hitting the border of the simulation area and then starts its next movement progress.

3.3 Movement via Programming Interface

Node movement in MANET can be modeled via programming interface. This method is the most flexible. The position of communicating nodes consists of three components: 'x' position, 'y' position, and 'y' position, and can be programmed depending on the simulation model use. For example, in OPNET Modeler [13], node position can be dynamically changed via programming interfaces, but one needs to create a custom process model or modify a current process model, because some extra code for dynamically changing positions is required. The op_ima_obj_attr_set_dbl() function is used to dynamically change the attributes related to node position during simulation in OPNET Modeler.

It is vital to select an adequate node movement pattern, because node movement patterns play a salient role in designing of the network and hence affect its performance (Table 1).

4 Performance Evaluation

The performance of the three routing protocols such as: AODV, DSR, and OLSR are evaluated using segment-base trajectory with fixed-interval as node movement.

4.1 Routing Protocols

4.1.1 DSR (Dynamic Source Routing)

The dynamic source routing (DSR) [21] is a reactive protocol with a route discovery cycle for route finding.

DRS is a source routing protocol, because instead of data packets being sent hop by hop, data packets contain strictly source routes information that specify each node

Table 1. Comparative studies of the three-way solutions for the node movement

Node movement types	Movement model name		Benefits	Drawbacks
Random movement	Random walk movement		– Memory-less mobility process	– Generate unrealistic movement of mobile nodes, such as a sudden stop and sharp turns
	Random waypoint movement		– All kinds of MANET's routing protocols can be simulated, due to its ease of operations	– Cause density wave problem: Nodes get clustered in one part of the simulation area – Geographic restrictions of mobile node movement – Spatial Dependency of Velocity
	Random direction movement		– Reduce density wave problem caused by random waypoint model – Network partitions are more portable	– The probability for the mobile node to choose a new destination that is located in the simulation area or a destination which requires travel through the middle of the simulation area is high
Movement via trajectories	Segment-based trajectory	Fixed-interval trajectory	– Node movement can be predicted – Good quality of service	– Node energy consumption is high
		Variable interval	– Node energy consumption is less	– Low quality of service
	Vector-based trajectory		– Increase in packet processing	– Reduce the efficiency of the packet
Movement via programming interface	programmable movement		– High quality of service – Security can be maintained	– The computational cost is high – Consume more memory

along the path to the end node. In DSR, every node keeps record of the path of which it is aware of in its route cache. When a node receives a route request (RREQ) for path, it refers to its route cache to check if it already contains the required route information, and once the route is found, the node replies with the route reply (RREP) message with the route information.

DSR works under two principles mechanisms: discovery and maintenance of the path, which simultaneously allow mobile hosts to find out the path and preserve the source routes to the destination.

4.1.2 AODV (Ad Hoc on-Demand Distance Vector)

The ad hoc on-demand distance vector (AODV) routing protocol [22] is network-centric algorithm adapted to work with mobile ad hoc networks and to overcome bandwidth constrained and restrained battery energy of the mobile nodes. AODV works relatively as Bellman–Ford distance vector algorithm. It is an on-demand algorithm, that is, it determines the paths to the destination if only if a host is willing to transmit data packet to that destination, and after route being detected, it maintain the paths until there is no needed for it.

4.1.3 OLSR (Optimized Link State Routing Algorithm)

The optimized link state routing (OLSR) protocol [23] is a variation of tradition link state routing protocol, modified to improve operation in mobile ad hoc networks. The enhanced features of OLSR are the use of multipoint relays (MPRs) to reduce the overhead of network flow and the size of the link state updates. Due to the proactive nature of OLSR, each node consistently maintains the route to every other node in the network.

4.2 Simulation Environment

The movement of the node is modeled with segment-based trajectory with fixed-interval segment traversing time. Two fixed segments trajectories are defined, and each node in the communication channel is assigned to the trajectory defined and the node moves according to that trajectory. The mobile node takes the same time period to traverse each path segment. We have considered 10, 20, and 30 minute of the time use to traverse each segment path, and we have conducted performance evaluation of DSR, AODV, and OLSR base of three network parameters: the average throughput, average end-to-end delay, and average network load. The simulations are done using network simulator (OPNET) and MANET as network model. The move area of the mobile nodes is within 500 m × 500 m (Figs. 5, 6 and Table 2).

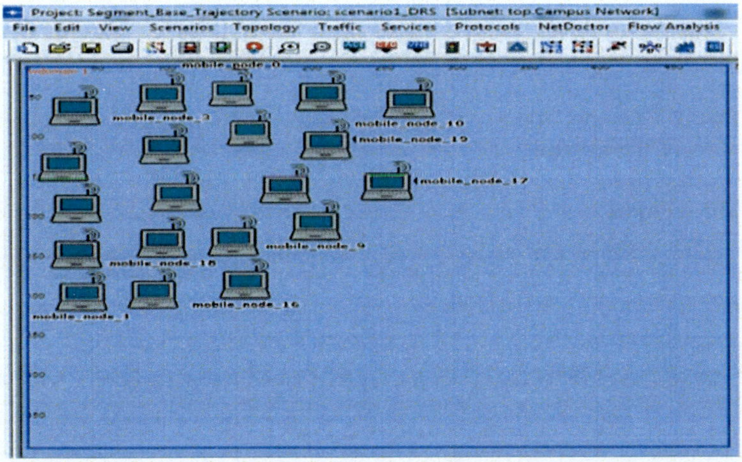

Fig. 5. Simulation window for 20 nodes without segment-based trajectory defined

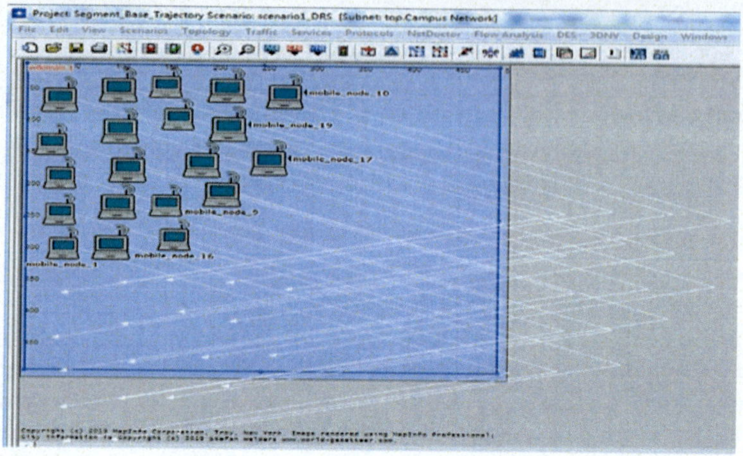

Fig. 6. Simulation results after segment-base trajectory get defined and assigned to all nodes

Table 2. Simulation parameters

Properties	Values
Simulation	OPNET
Coverage	500 m × 500 m
Number of nodes	20
Trajectory name	Trajectory 1_Trajectory Fix
Trajectory type	Fixed interval
Time for traversing each trajectory segment	10, 20 and 30 m
Simulation time	600 s
Packet interval time	5 packets/s
Packet start time	1 s
Packet size	512 bits
Routing protocols	DSR, AODV, OLSR

4.3 Simulation Result and Analysis

Table 3 below shows the statistic results for the three routing protocols based on the configuration of the network.

4.3.1 Throughput

The throughput is the average rate in (bits/s) at any given instant of time at which over a communication network, the data packets is delivered successfully from one node to another [20].

Table 3. Statistics results for the three protocols

Segment traversing time (min)	10	20	30
Average throughput (bits/s)			
DSR	6916	6702	6881
AODV	240,427	246,935	242,747
OLSR	205,635	205,635	205,635
Average end-to-end delay (s)			
DSR	0.001194	0.001218	0.001219
AODV	0.0007954	0.0008192	0.0008464
OLSR	0.00031228	0.00031228	0.00031228
Average network load (bits/s)			
DSR	6358	6158	6340
AODV	18,073	18,204	18,495
OLSR	15,205	15,205	15,205

Figure 7 shows the results for the three protocols for the three variation of the segment traversing time.

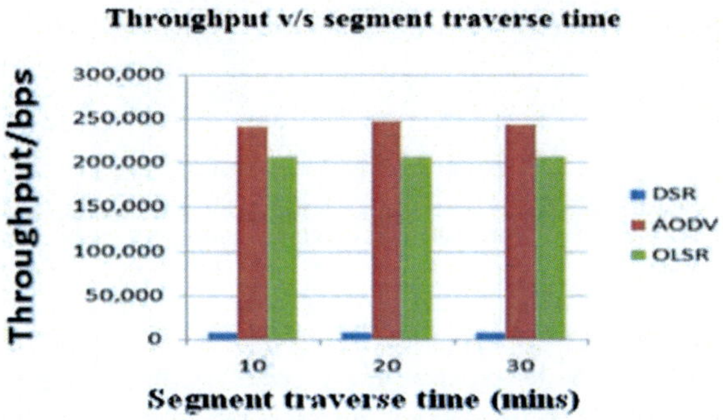

Fig. 7. Average for throughput versus segment traverse time

The average throughput of DSR remains constant, very low during 10, 20, and 30 mins of segment traversing time. But, AODV and OLSR present highest throughput compared to DSR. In the communication channel, throughput is directly proportional to the ratio of total data packet sent; lower data packets transmitted results to lower throughput.

4.3.2 End-to-End Delay

In communication channel, a data packet starts its journey in a host (the source), passes a series of router, and ends its journey in another host (destination). During its journey of traversing from one node (host or router) to the subsequent node (host or router) along the way, the packet suffers from several types of delays, at each node. It includes node processing delay, transmission delay, queering delay, and propagation delay. Together, these delays are called as total nodal delay.

$$\text{Nodal delay} = \text{node processing delay} + \text{transmission delay} \\ + \text{queering delay} + \text{propagation delay}.$$

End-to-End delay (s) is the nodal delay accumulated during the transfer of the packet from source to destination [13].

Figure 8 indicates that the average end-to-end delay for DSR, and AODV is greater than OLSR. The increase of the end-to-end delay for DSR and AODV is due to the fact that both the routing algorism are searching for the alternate paths based upon their functionality mechanism which leads to packets lost. Hence, both protocols require more additional time to transmit data packets to the destination node via alternate paths in comparison with the original path.

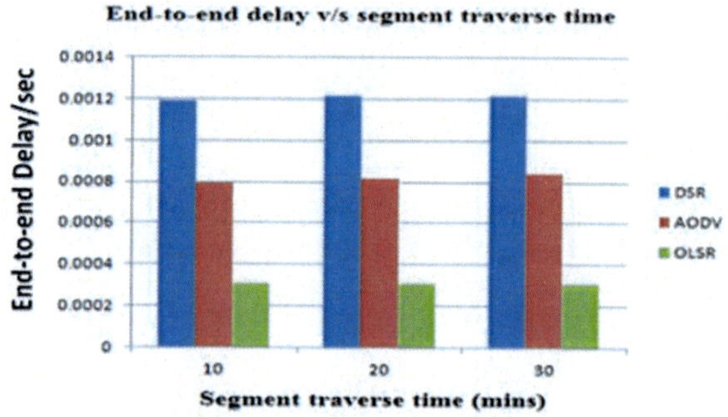

Fig. 8. Average end-to-end delay versus segment traverse time

4.3.3 Network Load

Network load is the ability to balance traffic over multiple routes without using complex routing protocols [7]. In a communication network, when the load is greater than the resource that the system can handle, congestion occurs. To reduce the congestion, either we increase the resources by increasing the bandwidth, or we decrease the load refusing service to some users, or lowering service to all users, and having users implement their service demands in a more predictable way.

Figure 9 shows the static results of the three algorithms. There are no large differences in network overload between the three segments traverse times for the three protocols. All present almost constant network load during the three segments traverse time. But, the network load is more considered for AODV and OLSR as compared to DSR. The increase of network load for both protocols reveals the presence of congestion meaning that the load is greater than the resources that the system can handle.

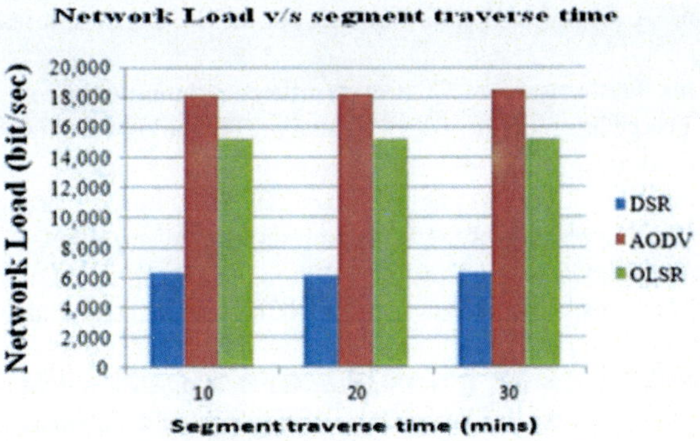

Fig. 9. Average network load versus segment traverse time

5 Conclusion

The ultimate advantage of mobile ad hoc networks over wired networks is mobility. Wired connection uses cables for transmission, whereas mobile ad hoc networks use wireless spectrum. Out of many important issues that affect the performance of MANET, node movement is extremely the most affecting issue. Here, we have described three techniques used to model node movement in mobile wireless ad hoc network and also the studies of the comparative analysis for performance evaluation of DSR, AODV, and OLSR using segment-based trajectory (fixed interval). Focus is mainly done on the network parameters such as: throughput, end-to-end delay, and network load. With the analysis results, we have found out that the throughput is very low for DSR that means the average message transmitted to the end node during the simulation time is very less. Therefore, DSR does not give great throughput via segment-base trajectory with fixed-interval in case of our studies. As the performance is a critical issue why designing a communication network, a connected network is only workable when its performance has been measured and compared. In the future work, the tracing of the node motion can be used to monitor the MANET network in order to detect malicious behavior of the moving node.

References

1. Kumar, S., Basavaraju, T.G., Puttamadappa, C.: Ad hoc Mobile Wireless Networks, Principle, Protocols, and Application. Aeubach Publications, Taylor & Francis Group (2008)
2. Johnson, D.B., Maltz, D.A., Broch, J.: DSR: The dynamic source routing protocol for multi-hop wireless adhoc networks. Computer Science Department, Carnegie Mellon University Pittsburgh, PA, 15213–3891 (March, 2001). http://www.monarch.cs.cmu.edu/
3. Zadina, A., Fevensa, T., Bdirib, T.: Impact of varying node velocity and HELLO interval duration on position-based stable routing in mobile ad hoc networks. In: The 11th International Conference on Future Networks and Communications (FNC 2016)
4. Basagni, S., Conti, M., Giordana, S., Stoymenovic, I.: Mob. Ad hoc Netw. (2004)
5. Tanenbaum, A.S.: Computer Networks, 4th Edn. Prentice-Hall, Inc (2003)

6. Camp, T., et al.: A survey of mobility models for ad hoc network research. WCMC **2**(5) (2002)
7. Ahuja, A., et al.: Performance of TCP over different routing protocols in mobile ad-hoc networks. In: Proceedings of the IEEE Vehicular Technology Conference (VTC 2000), Tokyo, Japan (2000)
8. Anastasi, G., Borgia, E., Conti, M., Gregori, E.: IEEE 802.11 ad hoc networks: performance measurements. In: Proceedings of the Workshop on Mobile and Wireless Networks (MWN 2003), Providence, Rhode Island (2003)
9. Lu, Z., Yang, H.: Unlocking the Power of OPNET Modeler. Cambridge University Press (2012)
10. Sanzgiri, K., Dahill, B., Levine, B.N., Shields, C., Belding-Royer, E.M.: A secure routing protocol for ad hoc networks. In: Proceedings of the 10th IEEE International Conference on Network Protocols (ICNP), Paris (2002)
11. Papastergiou, G., et al.: Deep-space transport protocol: a novel transport scheme for space DTNs. Comput. Commun. (COMCOM). Elsevier Science (2009)
12. Camp, T., et al.: A survey of mobility models for ad hoc network research. In: Wireless Communication and Mobile Computing (Special Issue on Mobile Ad Hoc Networking: Research, Trends and Applications) vol. 2, pp. 483–502 (2002)
13. Spyropoulos, T., et al.: An analytical study of fundamental mobility properties for encounter-based protocols. IJAACS (2008)
14. Wang, K., H., Li, B.: Group mobility and partition prediction in wireless ad-hoc networks. In: IEEE International Conference on Communications (ICC 2002) vol. 2, pp. 1017–1021. IEEE (2002)
15. Stojmenovic, I.: Position Based Routing in Ad hoc Mobile Networks. IEEE Commun. Mag. **40**(7), 128–134 (2002)
16. Roy, R.R.: Handbook of Mobile Ad Hoc Networks for Mobility Models. Springer Science + Business Media, LLC (2011)
17. Bettstetter, C., Hartenstein, H., Perez-Costa, X.: Stochastic properties of the random waypoint mobility model: epoch length, direction distribution and cell-change rate. In: Proceedings of the 5th ACM International Workshop, MSWIM2002 (2002)
18. Sanchez, M., Manzoni, P.: Anejos: a java based simulator for ad-hoc networks. Future Gener. Comput Syst **17**(5), 573–583 (2001)
19. http://doi.ieeecomputersociety.org
20. Bettstetter, C. et al. The node distribution of the random waypoint mobility model for wireless ad hoc networks. IEEE Trans. Mob. Comput. (2003)
21. Timcenko, V., Stojanovic, M., Rakas, S.B.: MANET routing protocols vs. mobility models: performance analysis and comparison. In: Proceedings of the 9th WSEAS International Conference on Applied Informatics and Communications (AIC '09)
22. Perkins, C., Royer, E.: Ad-Hoc on-demand distance vector routing. In: Proceedings of the Second IEEE Workshop on Mobile Computing Systems and Applications, WMCSA'99, pp. 90–100 (February, 1999)
23. Clausen, T., Jacquet, P., Laouiti, A., Muhlethaler, P., Qayyum, A., Viennot, L.: Optimized link state routing protocol. In: Proceedings of IEEE INMIC, Lahore, Pakistan (2001)

IoT Capable Mechanism for Crowd Analysis

Kanchan R. Mangrule(✉), H. T. Ingale, S. K. Chaudhari, and Anil J. Patil

Department of Electronics and Telecommunication, Godavari College of Engineering, Jalgaon, India
kanchanmangrule@gmail.com

Abstract. This paper describes a crowd analysis of different activities using surveillance videos is an important topic for communal security. This paper also describes the detection of dangerous crowds if the weapon is present in the crowd. In our study, we are using raspberry pi 3 board for the development of a system that consists of ARMv8 CPU that detects the human heads and provides a count of humans in the region using Open CV-Python. The direction of the movement of the person can be achieved by human tracking. Generally, there are three different stages algorithm for computer-based crowd analysis, (1) people counting, (2) people tracking, and (3) crowd behavior analysis. This project is made for security purposes where there is a possibility of a dangerous crowd, for example, mall, railway station, shopping center. In our method, we are used CNN to trained dangerous weapons and DNN used for human detection. This method not only detects the direction of the crowd but also detects if the crowd is dangerous or not. In this method, also count the total number of human and it also gives confidence score that means, in how many percents it is related to original people. In this way, we could have prevented many deaths and injuries.

Keywords: Video surveillance · Crowd density · Dangerous weapon detection · Crowd tracking

1 Introduction

Nowadays, managing the crowd in crowded areas properly is very important. Also, we are detecting weapons if anyone is carrying a weapon in a crowded area. The human eye cannot observe multiple cameras at the time. Sometimes, during an election, a celebration of the festival and in-mall situation goes out of control at that time the human eye cannot predict properly. Hence, we are using a camera to monitor the crowd. Thus, for continuously monitoring the crowd for a long period we must use an automated technique. Previous paper can make system for a small group of objects. In the present work, we made a system for unlimited crowd with high accuracy and propose algorithms for their application to a variety of problems. There are different methods used for moving object detection. The general method is background elimination under the situation of fixed cameras. Generally, crowd analysis is done by four steps, crowd behavior, crowd estimation, crowd tracking, and crowd motion detection. This is shown in Fig. 1. Videos of crowd scenes present difficult problems in computer

© Springer Nature Singapore Pte Ltd. 2021
P. Deshpande et al. (eds.), *Next Generation Information Processing System*,
Advances in Intelligent Systems and Computing 1162,
https://doi.org/10.1007/978-981-15-4851-2_7

vision. Nowadays, managing the crowd in crowded areas properly is very important. Thus, for continuously monitoring the crowd for a long period, we must use an automated technique. In our project, we made a system for unlimited crowd with high accuracy and propose algorithms for their application to a variety of problems. During election, if anyone can carry a weapon, we are detecting weapons and can take action. We immediately catch the weapon-carrying person from crowded areas and save other people. 26/11 Mumbai attack happened in 2008. At that time, Kasab entered into railway station carrying a gun. If our system was present at that time, then it would have detected the gun at the railway station and prevented loss of a number of people. Hence, I selected this topic for security purposes of a crowded area, i.e., airport, mall, railway station, etc.

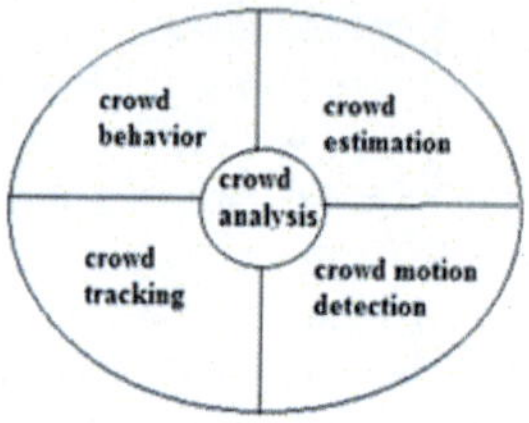

Fig. 1. Crowd analysis

2 Related Work

The previously used system consists of only edge detection of a crowd, but in our system, we present head detection, tracking, and motion detection. It also detects a number of people gathered in which direction according to the camera. Authors in [1] discussed the method for the training of human head detection using haar cascade classifier and for the direction of movement of the person, which can be archived by human tracking. Here, we detect the person's head by using haar features and Adaboost algorithm. DNN has more accuracy than the haar feature. Authors in [2] discussed method based on SIFT feature, which is used to develop a crowd abnormality detection methodology. Authors in [3] journal discussed methods for SS abnormal events of the crowd that has been proposed. According to the projected AMC (adjacency-matrix primarily based clustering) algorithm, optical flow is used as a primary consideration to cluster human crowds into teams [3]. This method consists of four modules for crowd analysis and detection of the abnormal crowd: (i) Background/Foreground Modeling, (ii) Blob Analysis, (iii) Crowd Detection, and (iv) Abnormal Crowd Tracking [4]. In this paper, the detection of abnormal crowd method is based on the change of energy-level distribution. This method can be used to reduce the camera perspective effect and detect crowd abnormal behavior in time [5]. In this paper, the results show that MDT-based anomaly detection outperforms all other approaches. A novel framework for anomaly detection in jammed scenes is given [6]. In this paper, work is made for identifying five crowd behaviors (bottlenecks, fountainheads, lanes, arches, and blocking) in visual scenes [7]. This paper detects the normal or abnormal behavior in crowd. Preprocessing for abnormal crowd behavior detection use different stages. For feature extraction preprocessing is one of the important steps. This paper described that preprocessing consists of four different analysis using no. of an algorithm for each analysis. Here it uses different levels for analysis which are (1) pixel-level analysis (2) texture level analysis (3) object-level analysis (4) frame-level analysis [8]. HoG is used for trained gun and knife but its accuracy is less as compared to CNN [9]. Recently, IoT technology emerged as a key technology for remote detection of humans. This technology will play a key role in future technology for crowd management [10–12].

2.1 Comparative Table

No.	Reported Approach	Our Approach
1.	Haar cascade classifier used to detect human head with only edges	We are using DNN to train humans and get better accuracy than the haar cascade classifier. It also gives confidence score which means, in how many percent it is related to original people. This is shown in Fig. 6
2.	Work on static images	Work on live images
3.	HoG is used for trained gun and knife but its accuracy is less as compared to CNN	We are using CNN to detect the Gun. CNN gives immediate results and gets very fast results. It gives Fps is maximum. This is shown in Fig. 7
4.	Result displays on screen only	In the present work, we detect the gun and immediately send mail to the authorized person that somebody is carrying a gun. Hence we prevent injuries and bad incidents. This is shown in Fig. 8

3 Methodology

3.1 Block Diagram Description

Block diagram of crowd analysis is shown in Fig. 2. It is based on of raspberry pi module. The system consists of web camera, power supply.

The input to the system is images captured by the web camera of the crowd at the mall, railway station. Festive area Hare, we can use the camera present at the system or external camera. We are using a web camera. It is connected to the port of raspberry pi. The controller will send a “1” signal to the active camera. The image processing is done by the open CV. Raspberry pi is a mini-computer that allows image processing. In this system, we are using DNN for human detection and CNN is used for weapon detection. This system also required an internet connection which gives no. of people present in crowd and its direction. If the weapon is present in the crowd it immediately sends mail to an authorized person and also announces that somebody is carrying a weapon.

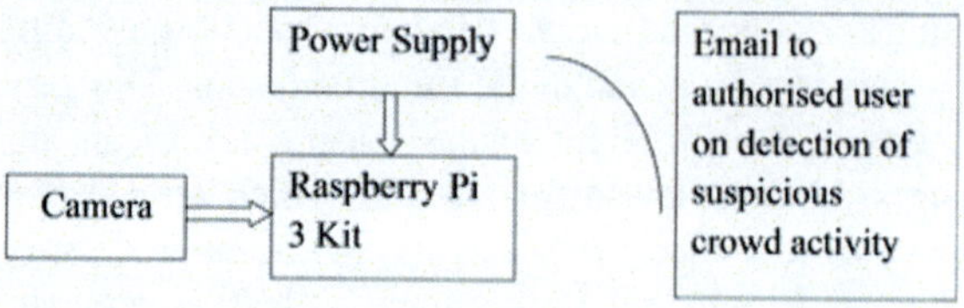

Fig. 2. Block diagram of the system

3.2 Hardware Description

Web Camera The system consists of a web camera. The input to the system is images captured by the web camera of the crowd at the mall, railway station. Festive area Hare, we can use the camera present at the system or external camera. We are using a web camera. It is connected to the port of raspberry pi. The controller will send a "1" signal to the active camera. The image processing is done by the open CV.

Raspberry Pi Raspberry pi is a mini-computer that allows image processing. This system also requires an internet connection which gives no. of people present in crowd and its direction. Our system also detects dangerous weapons, and for this dangerous weapon we are also using DNN. Accuracy of DNN is better than haar cascade classifier.

Power Supply The power requirement of the hardware in traffic light control system using raspberry pi is the power supply of 5 V, 3 A with respect to Ground.

4 Software

The system will perform in the following steps (Fig. 3).

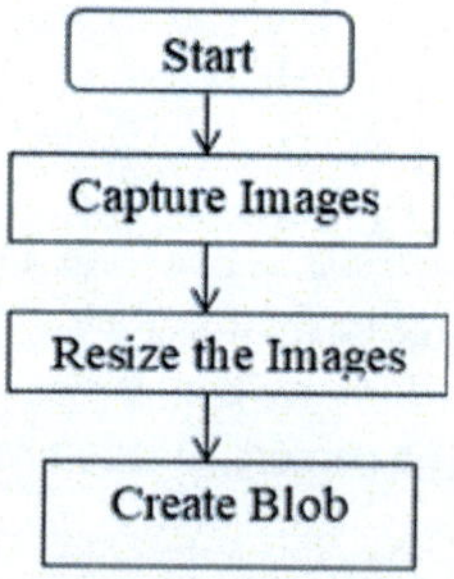

Fig. 3. Flowchart

Algorithm Steps for Flowchart 1

1. Start the computer system and the raspberry pi.
2. Capture the images with the help of web camera which is connected externally to the system.
3. The captured images are not of suitable resolution. So, resize the images by open CV software. Thus, adjust its resolution.
4. After the resolution create the blob of 3 * 3.

The flowchart also has the second stage as follows (Fig. 4).

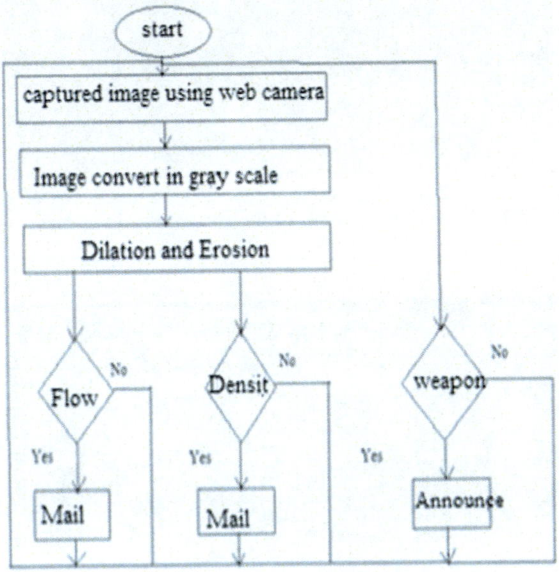

Fig. 4. Flowchart 2

Algorithm steps for Flowchart 2

1. Start
2. Captured the image using web camera.
3. Images convert into grayscale because color information does not help us identify an important edge.
4. Dilation adds pixels to the boundaries of the object in an image. Erosion removes pixel on object boundaries.
5. Crowd analysis is done in python using DNN. Detect human head and weapon using DNN and announce if a gun is detected.
6. This system runs until power is on.

5 Results and Discussion

In this figure, we show raspberry pi 3B+ model for processing and web camera for the captured image. The following image shows when power supply is connected. Insert the microSD card into the slot on the underside of the raspberry pi (Fig. 5).

Fig. 5. Hardware Setup of the Proposed Mechanism

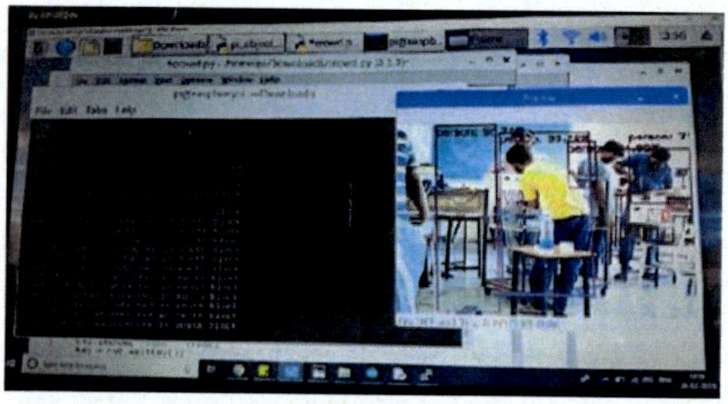

Fig. 6. Image detection 3

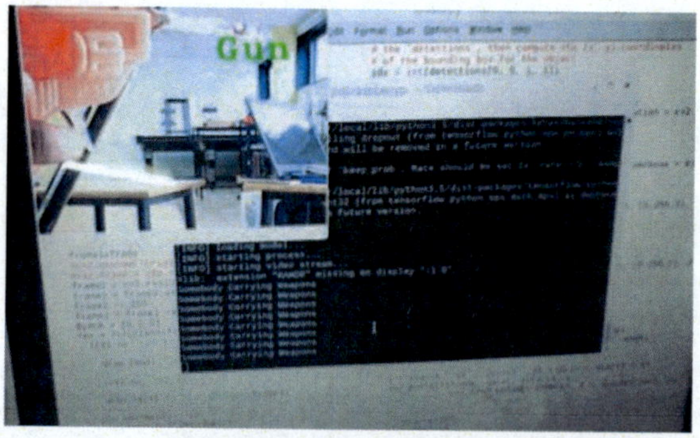

Fig. 7. Image detection 5

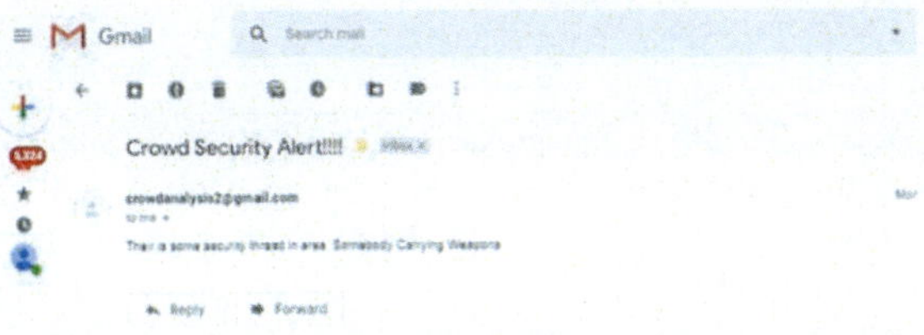

Fig. 8. Security alert message on mail

Web camera is connected to camera port between the audio port and HDMI port.

First, we scan the IP address of raspberry pi with the help of an advance IP scanner. This IP address of raspberry pi is in the putty configuration. Connect laptop with raspberry pi using VNC viewer. When the web camera captures images, they are converted into suitable form with the help of open CV.

Our system has recognized people. It also gives confidence score which means, in how many percent it is related to original people. Also, it shows the direction (north and south) of people according to the position web camera.

CNN is used for trained weapon, i.e., Gun. There are 600 images used for training. The following image shows that the weapon is present in crowd.

If weapon gun is present in the crowd, then a mail is immediately sent to an authorized person and also announces that somebody is carrying a weapon on speaker. Speaker is connected to the raspberry pi audio port. This is very useful for security purposes. Hence the proposed system is useful in mall, railway station, park, etc.

6 Conclusion

The proposed system provides the number of people present in the north and south directions. For crowd analysis, video surveillance is one of the most effective methods. The proposed work detects dangerous crowd and crowd density. For more safety, we detect not only the crowd behavior but also detect the weapons, for example, Gun. It also announced the weapon is present. In our method, we use DNN for human detection and CNN for gun detection. This method gives better accuracy than other methods.

Future Scope

Crowd analysis is required for safety purposes. The proposed work detect the human and count number of people and its motion. Also, it detects the dangerous weapon in the crowd and if the gun is present then it is announced. Hence, in this way, we are avoiding injuries and stampede. In the future, we also detect chain snatching in the crowd and immediately catch thieves. But training the machine is more time-consuming. However, for training, a large number of data is required in deep learning related methods. This is more useful for investigation purposes.

References

1. Syed Ameer Abbas, S., Oliver Jayaprakash P., Anitha, M., Vinitha Jaini, C.X.: Crowd detection and management using cascade classifier on ARMv8 and OpenCV-Python. In: International Conference on Innovations in Information, Embedded and Communication Systems (ICIIECS) (2017)
2. Güler, P.: Automated Crowd Behavior Analysis for Video Surveillance Applications. The Graduate School of Informatics of the Middle East Technical University, September (2012)
3. Chen, D.-Y., Huang, P.-C.: Motion based unusual event detection in human crowd. J. Viss. Common. Image (2011). Journal Homepage: www.elsevier.com/locate/jvci
4. Santhiya, G., Sankaragomathi, K., Selvarani, S., Niranjil Kumar, A.: Abnormal crowd tracking and motion analysis. In: IEEE International Conference on Advanced Communication Control and Computing Technologies (ICACCCT) (2014)
5. Zhang, X., Zhang, Q., Hu, S., Guo, C., Yu, H.: Energy level-based abnormal crowd behavior detection. MDPI. Published: 1 February (2018)
6. Mahadevan, V. Li Viral, W., Nuno Vasconcelos, B.: Anomaly detection in crowded scenes. In: IEEE Conference on Computer Vision and Pattern Recognition, San Francisco (2010)
7. Solmaz, B., Moore, B.E., Shah, M.: Identifying behaviors in crowd scenes using stability analysis for dynamical systems. IEEE Trans. Pattern Anal. Mach. Intell. **34**(10), 2064–2070 (2012)
8. Rohit, K., Mistree, K., Lavji, J.: A review on abnormal crowd behavior detection. In: International Conference on Innovations in Information, Embedded and Communication Systems (ICIIECS) (2017)
9. Vajhala, R., Maddineni, R., Yeruva, P.R.: Weapon detection in sureveillance camera images (2016)
10. Deshpande, P., Iyer, B.: Research directions in the internet of every things (IoET). In: International Conference on Computing, Communication and Automation (ICCCA), pp. 1353–1357 (2017)
11. Patil, N., Iyer, B.: Health monitoring and tracking system for soldiers using internet of things (IoT). In: 2017 International Conference on Computing, Communication and Automation, pp. 1347–1352 (2017)
12. Iyer, B., Patil, N.: IoT enabled tracking and monitoring sensor for military applications. Int. J. Syst. Assur. Eng. Manag. **9**, 1294–1301 (2018). https://doi.org/10.1007/s13198-018-0727-8

Probability Analysis of Vehicular Traffic at City Intersection

Jyoti Motilal Sapkale[1](✉), Vijay D. Chaudhari[2], H. V. Dhande[2], and A. J. Patil[3]

[1] M.Tech student (VLSI & Embedded systems), Godavari College of Engineering, Jalgaon, Maharashtra, India
sapkalejyoti999@gmail.com

[2] Asst. Prof. E&TC Engg Dept, GF's Godavari College of Engineering, Jalgaon, India
{vinuda.chaudhari,hemrajd99}@gmail.com

[3] Principal, Shri. G. D. College of Engineering, Jalgaon, India
anilj48@gmail.com

Abstract. Nowadays, congestion in traffic is a serious issue all over the world. The traffic congestion is caused because of large red light delays. The delay of the respective light is coded hardly in the traffic light and also it is not dependent on traffic density. The existing system varies the particular light delay time by taking the vehicle count using IR sensors which has several disadvantages. This project presents the system based on raspberry pi. It includes a high-resolution camera. It captures images of vehicles. It performs the blob detection of a vehicle. It gives a separate count of vehicles and people too. This recorded vehicle count data is used in the future to analyze traffic conditions at respective traffic lights connected to the system. For appropriate analysis, the raspberry pi will work on the information to send correct signal into the LED lights. However, to solve the problem of emergency vehicles stuck in the overcrowded roads, a portable controller device is designed. The system will give the vehicle count by the deep neural technique. After vehicle detection and its count, the system will apply conditional probability to glow the green signal for a specific time period on a particular side according to the vehicle count.

Keywords: Traffic control · Raspberry pi · Image processing · Vehicle counting · Python

1 Introduction

The second more popular country in the world is India. It is the fast growing economy. Infrastructure growth is slow as compared to the growth in number of vehicles, due to space and cost constraint. The Indian traffic is also chaotic and non-lane based. This traffic congestion affects the transportation system in cities. Rapidly increasing the number of automobiles and the constantly rising number of road users are not accompanied with promoted infrastructures with sufficient resources. Some of the solutions were offered by constructing new roads, implementing flyovers, and bypass roads and establishing road rehabilitation. The traffic lights consist of three universal

© Springer Nature Singapore Pte Ltd. 2021
P. Deshpande et al. (eds.), *Next Generation Information Processing System*, Advances in Intelligent Systems and Computing 1162,
https://doi.org/10.1007/978-981-15-4851-2_8

red, yellow, and green colored lights. They have coding of when to perform the ON and OFF operation. This coding is fixed. It does not look for whether there is a crowd of vehicle or not. It gives the same time for more traffic as well as for less.

In this system, we have controlled the traffic lights time according to the traffic density. We are using here deep neural network which was not used yet, it detects vehicles like a car, bus or other with the confidence score in percentage. It can do this based on the vehicle-related data given to it. We have developed a system which can be broadly applied over the city or country. We also are using python software. It is a very simple programming language. The errors can be easily removed. We are performing many functions, so the code is complex. Python programming language is useful for it. The programming goes easy through this language.

2 Related Work

Traffic lights green, red, and yellow are signaling devices that are developed since 1912 to control the traffic flows. The green light allows traffic to pass in the indicated direction, the yellow light gives warning to the vehicles to take for a short stop, and the red signal stops any traffic from passing forward [1]. All the free parameters of traffic signals such as split times, cycle times are adjusted online using model predictive control. By using the data obtained from microscopic simulation, these parameters are calibrated. After that, this simple model is converted into mixed logical dynamic system. Thus, the mixed-integer dynamic optimization is applied to the MPC framework by which time is adjusted [2]. The project is concerned with the development and implementation of Sensor-based Traffic Light System with Dynamic Control which in turn reduces the Average Trip Waiting Time (ATWT). It consists of IR sensors; Low Power embedded controllers, comparators, and storage devices [3]. To assign a time for green as well as red signal based on traffic density, the system was developed which consists of PIR (Proximity Infrared Sensors) sensors. After calculating the density, the glowing time for the green signal is allotted by Arduino which green signal glow time [4]. An intelligent and very smart traffic light control system was developed based on (WSN) Wireless Sensors Network. The system consists of WSN and a control box. This system provides a traffic safety [5]. The system was proposed for controlling the traffic light by using image processing. The images are captured by electronic sensors. A camera will collect this data of sequence. Further, this sequence will be analyzed by using digital image processing. The traffic light signals are controlled according to the traffic condition on the road [6]. The automatic green signal time control system is also proposed by using image processing. The system will allow the variable time to the specific side according to the traffic density using raspberry pi. For image processing, open CV software is used. The system also provides the facility for emergency vehicles to pass first as the first priority [7]. The system containing IR transmitter and IR receiver which were mounted on each side of the road was developed. It used raspberry pi as a controller which takes a decision of allowing the time to the green or red signal based on the vehicle count. Likewise, the traffic-related data is also recorded which will help in the future for the analysis of the traffic control [8]. The technique of edge detection is also used to control the traffic at the city intersection. The images were captured with the help of the camera which were converted into the greyscale after that into black and white images. These obtained images are now compared with the reference image means an empty road. Here, gradient-based and Laplacian-based edge detection technique has been used. The timing of the traffic signal is allotted by microcontroller. For the image processing, MATLAB software has been used [9]. The system was developed which can capture the images through web camera, then performing the image processing on it. It uses the candy edge detection technique. It captures the images by 40 s. It gives the time to the traffic signal according to the vehicle count. It gives time to the green as well as red traffic signal [10].

The system using canny edge detection technique takes the CCTV footage for the image processing. Further, the system was developed which converts the static camera into the video camera for the system to be more readable. The images captured in the video form are used as a input to the system by video object detector which is used as a detector. These are converted into number of frames which are used for counting and adjusting the time of the red or green signal [11]. The system developed to reduce the vehicle waiting time by using VANET. It is vehicle communication on the basis of which vehicle density is determined and allotted a particular time [12].

The time allocation to the emergency vehicles as per the priority is proposed in the system [13]. In [14], the authors proposed a system that will control the traffic using probability concept depending on the number of vehicle arrival at the traffic intersection. The existing system will vary the time of the traffic light only in a specific time with same rotation sequence. Recently, IoT-based systems emerged as the best alternative for the existing systems [15–17]. Such systems are becoming very popular due to its remote accessing possibilities, compact size, energy efficiency, and low cost. After studying the above mentioned literature, we have prepared a performance comparison of various methods used till now and their concerned results too in Table 1.

Table 1. Performance comparison table with other methods

No	Methods used in previous papers	Paper results
1.	Use of gradient-based and Laplacian-based edge detection technique using MATLB software	The images were captured by camera and converted into the greyscale after that into black and white images. By comparing with empty road signal time is allotted
2.	Use of candy edge detection technique for image processing	It captures the images by 40 s and gives traffic signal according to the vehicle count
3.	Traffic Light Control System to be designed based on the Vehicle density in VANET	The traffic light was changing the waiting time (Red signal) based upon the priority algorithm and vehicle density. Using this adaptive traffic light control system, the vehicles waiting time in traffic signals to be reduced
4.	Our developed system uses deep neural network for image processing	Our system detects the vehicles using machine learning, compare the vehicle density on each phase and allot time to the vehicle count

3 Methodology

3.1 Block Diagram Description

Figure 1 shows the block diagram of the Probability Analysis of Vehicular Traffic at City Intersecting. It is based on of raspberry pi module. The system consists of web camera, LED lights that are green and red, four two-digit 7 segment display.

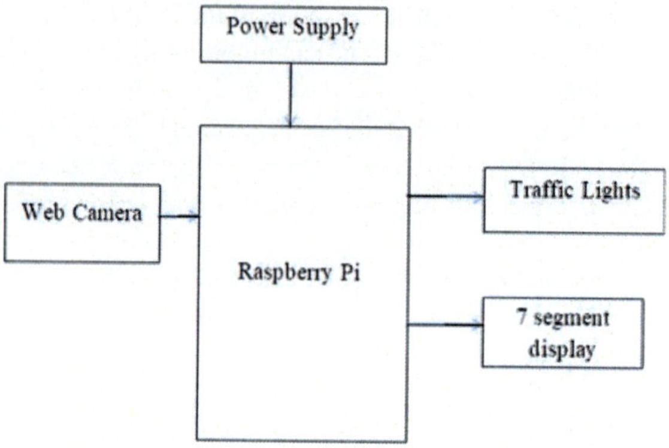

Fig. 1. Block diagram of the system

The input to the system is lights which give an indication to the vehicles whether to pass or to stop. The green LED light states the vehicles to pass. Similarly, the red LED light states to stop. LED working is based on the effect of electroluminescence. The four images captured by web camera of vehicular traffic at city intersection. Here, we can use the camera present at the system or external camera. We are using web camera. It is connected to the port of raspberry pi. The controller will send "1" signal to active camera. The image processing is done by the open CV. Raspberry pie is a mini-computer that allows image processing. For image processing, left side of each road has been considered. The system also consists of LED two-digit 7 segment display that is used which has a common anode connection. Each display is placed on one side of the road. The raspberry pi will allot the time to a specific direction. First of all, the timer has given the time of 30 s. If the traffic is more then it will be 40 s and it will also increase up to 60 s according to the vehicle count. After 10 m, again the raspberry pi will reprocess and adjust the time.

3.2 Hardware Description

Web Camera: It is used to capture the images of vehicles at a traffic intersection point. It is the external camera to the system. It is connected to the raspberry pi. Further image processing is done by a raspberry pi.

Traffic Lights: The two LEDs are green and red. Green allows vehicles to pass and red to stop them. They blink one after another for the time given by the system which is dependent on the vehicle count, calculated from the captured images.

Two-Digit 7 Segment Display: The four two-digit 7 segment display are used having a common anode connection. Each display is placed on one side of the road.
Power Supply: The power requirement of the hardware in a traffic light control system using raspberry pi is the power supply of 5 V, 3 A.
Raspberry Pi: One can use the raspberry pi as it is an open-source platform. It can perform the operations that we need. Raspberry pi has several unique features such as it provides a port for web camera connection which is the input of our system.

4 Software

Figure 2 shows flowchart 1 of the system, it tells how the system will perform.

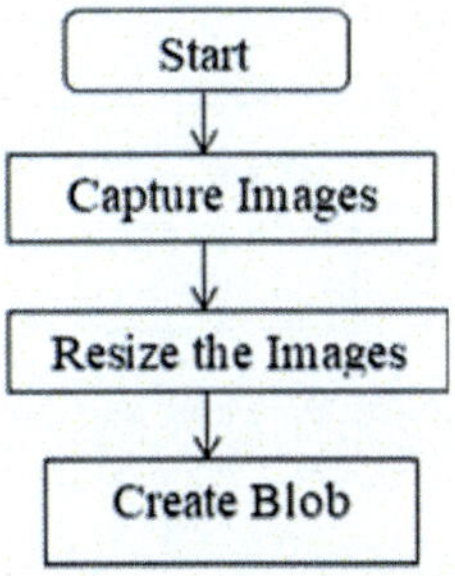

Fig. 2. Flowchart 1

Algorithm steps for flowchart 1

1. Start the computer system and the raspberry pi.
2. Capture the images with the help of web camera which is connected externally to the system.
3. The captured images are not of suitable resolution. So, resize the images by open CV software. Thus, adjust its resolution.
4. After the resolution creates the blob of 3*3.
5. The flowchart also has the second stage as following.

Figure 3 shows flowchart 1 of the system, it tells how the system will perform.

Algorithm steps for flowchart 2

1. Start the system
2. Capture images of traffic vehicles using a web camera connected to the raspberry pi.
3. Using image processing technique makes the image suitable for further process.
4. Using deep neural network count the number of vehicles.
5. Using probability decide in which direction traffic density is high.
6. Allot the time according to the traffic density in a specific phase.

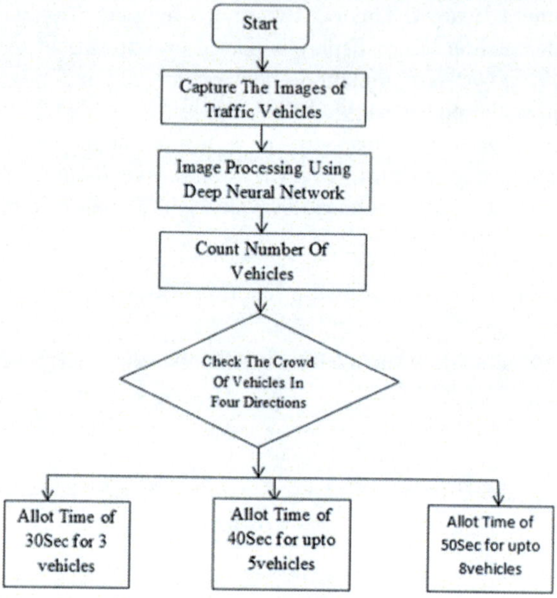

Fig. 3. Flowchart 2

5 Results and Discussion

When the images are captured by the web camera, the images are scanned and the vehicles are recognized by using deep neural networks. It gives a vehicle count.

Figure 4 shows image processing of image 1 which shows 4 counts of vehicle. It uses deep neural networks.

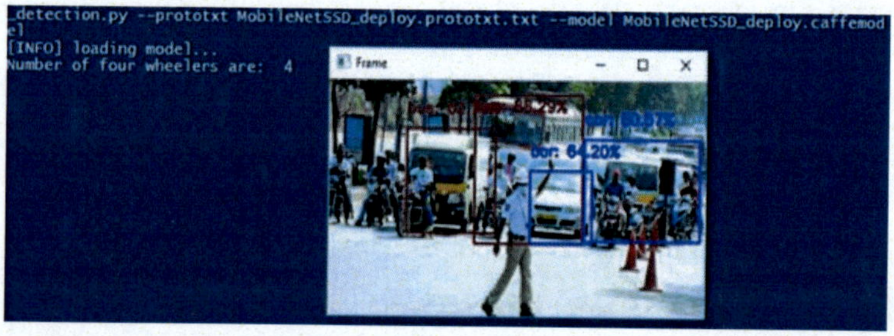

Fig. 4. Vehicle count of image 1

Figure 5 shows image processing of image 2 which shows 17 count of vehicle.

Figure 6 is the image captured on the hardware. Here we have used a toy car for practical implementation of the system. Our system is also capable of recognizing the toy car whether it is a vehicle or bus. The figure shows vehicle count 3 in which it indicates the vehicle as a car with confidence score.

Fig. 5. Vehicle count of image 2

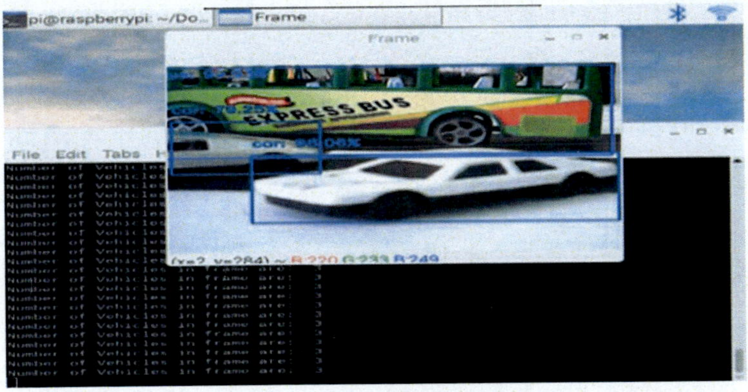

Fig. 6. Vehicle count of image 3

Table 2 shows that the time allotted for specific vehicle count.

The reference time to the system is 30 s when the vehicle count increases up to four vehicles, this time increases to the 40 s when the vehicle count increase to the eight vehicles, this time increases to the 50 s. For phase 4, since the vehicle count is 17, the time allotted to the green signal is around 60 s.

Table 2. Time allocation table (reading-1)

S. No.	No. of vehicles	Green time allotted (s)
Phase 1	3	30
Phase 2	4	40
Phase 3	8	50
Phase 4	17	60

6 Conclusion

Our system is capable of capturing and showing count of vehicles that are present at four phases near the traffic intersection in real-time. The captured images are processed with good quality image processing algorithm and edges are detected so as to get accurate vehicle count. Thus, according to the analysis of observations mentioned in the table, after getting vehicle count from four sides (phases), by using conditional probability technique, the glow timing of green and red traffic light for each side changes. In our system currently, we are detecting only cars and buses. In future, we can detect bikes which are also responsible for traffic congestion. We can develop our system in such a way that if any accident happened at the traffic intersection point, then the system will be able to inform to nearby area's police station and hospital to obtain emergency services.

References

1. Kham, N., Nwe, C.: Implementation of modem traffic light control system. Inter. J. Sci. Res. Pub. (IJSRP) **4**(6) (2014)
2. Kamal, M.A.S., Imur, J., Ohata, A., Hayakawa, T., Aihara, K.: Control of traffic signals in a model predictive control framework. In: 13th IFAC Symposium on Control in Transportation Systems, The International Federation of Automatic Control, 978-3-902823-13-7/12, pp 221–226 (2012)
3. Ghazal, B., Eikhatib, K., Chahine, K., Kherfan, M.: Smart traffic light control system, pp. 140–145 (2016). ISBN 978-1-4673-6942-8/16
4. Poyen, E.F.B., Bhakta, A.K., Durga Manohar, B., Ali, I., Rao, A.S.A.P.: Density based traffic control. Inter. J. Adv. Eng. Manag. Sci. (IJAEMS) **2**(8), 1379–1384 (2016). ISSN 2454-1311
5. Krishnaiah, G., Rajani, A., Rajesh, P.: Literature review on traffic signal control system based on wireless technology. ICDER, 63–68 (2014)
6. Choudekar, P., Banerjee, S., Muju, M.K.: Real time traffic light control using image processing. Inter. J. Comput. Sci. Eng. (IJCSE) **2**(1), 6–10 (2011). ISSN 0976-5166
7. Bhusari, S., Patil, S., Kalbhor, M.: Traffic control system using Raspberry-pi. Global J. Adv. Eng. Technol. **4**(4), 413–415 (2015). ISSN (Online) 2277-6370
8. Vidhyia, M., Elayaraja, S., Anitha, M., Divya M., Divya Barathi, S.: Traffic light control system using Raspberry-pi. Asian J. Electr. Sci. (AJES) **5**(1), 8–12 (2016). ISSN 2249-6297
9. Ramteke, M.D., Pote, H.P., Ukey, A., Ugemuge, P., Gonnade, S.: Edge detection based adaptive traffic control system. Inter. J. Recent Innov. Trends Comput. Commun. (IJRITCC) **4**(4), 323–332 (2016). ISSN 2321-8169
10. Tahmid, T., Hossain, E.: Density based smart traffic control system using canny edge detection algorithm for congregating traffic information. In: EICT. IEEE-978-1-5386-2307-7/17 (2017)
11. Vijayaraj, J., Loganathan, D.: Traffic congestion control of vehicles based on edge detection using image processing. Inter. J. Pure Appl. Math. (IJPAM). **119**(14), 1407–1418 (2018). ISSN 1314-3395
12. Balasubramani, S., John Aravindhar, D.: Design traffic light control system based on location information and vehicle density in VANET. IJRTE **7**(5S4) (2019). ISSN 2277-3878

13. Chaudhari, V.D., Patil, A.J.: Prioritized ViU departure at traffic intersection using internet of things. In: Iyer, B. et al. (eds.) Computing in Engineering and Technology, Advances in Intelligent Systems and Computing, vol. 1025, pp. 267–276. Springer Nature Singapore Pte Ltd (2020)
14. Sapkale, J.M., Chaudhari, V.D., Patil, A.J.: Vehicular traffic monitoring at city intersection using probability. Inter. J. Innov. Eng. Sci. (IJIES) **4**(10), 82–84 (2019). ISSN 2456-3463
15. Deshpande, P., Iyer, B.: Research directions in the internet of every things (IoET). In: International Conference on Computing, Communication and Automation (ICCCA), pp. 1353–1357 (2017)
16. Patil, N., Iyer, B.: Health monitoring and tracking system for soldiers using internet of things (IoT). In: 2017 International Conference on Computing, Communication and Automation, pp. 1347–1352 (2017)
17. Iyer, B., Patil, N.: IoT enabled tracking and monitoring sensor for military applications. Int. J. Syst. Assur. Eng. Manag. **9**, 1294–1301 (2018). https://doi.org/10.1007/s13198-018-0727-8

Check for updates

Automated Malware Identifier and Analyzer

Monica Malik(✉) and Bhawna Narwal

Indira Gandhi Delhi Technical University for Women (IGDTUW), Kashmere Gate, Delhi, India
monicamalik17@gmail.com, bhawnanarwal@igdtuw.ac.in

Abstract. Malware analysis is an essential part for any kind of crimeware analysis required in an organization. At present, there are so many malware variants available in the market with the target to sneak into IT exoskeleton of company. Millions of noxious applications and programs are invented every month. Most of the malware are so heavily disguised to be able to hide their original intent. While there are anti-malware softwares and firewalls available for help, but sometimes they are not enough, that is where malware analysis comes into the picture.

Keywords: Cyber-attacks · Malware · Metasploit · Security · VMWare

1 Introduction

There is an ever-growing arms race between the security defenders and the malware writers in introducing new techniques of malware detection and its evasion [1–5]. This provides increased automation and coverage through reducing user input on specific malware signatures. A malware is malicious code that is created with the intent of harming people, devices, and data. In broad sense, it is a set of instructions developed by an attacker to carry out their evil purposes on the victim's machine. In order to secure cyber world, it is mandatory to stop harmful files and malware access in the computer systems. Detection [6, 7] is first step in malware analysis, and second phase is preventive [6, 8–16]. In the era of cyber security and cyber world, defensive move is very important to achieve secure environment [7, 17–19].

This paper is divided into four phases Malware Development Phase, Malware Exploitation Phase, Efficiency Evaluation using OSINT, and the last Detection and Analysis Phase. Malware is created using Metasploit [20] in Kali. Then, the malware is downloaded on the victim's machine, i.e., on Windows. While creating a malware, it is bound with a picture, the picture is converted using "ICO Convert" [21], and then, archiving the malware and the icon bind both of them which is done in order to manipulate the victim. When the victim opens the image, the malware starts running in the background. This is how binding the malware makes it less suspicious and helps to manipulate the victim easily, and this comes under social engineering [22] attacks where the attackers use psychological manipulation to trick the victims that makes them to make security mistakes or make them give away their sensitive information to the attacker. VirusTotal [7, 23] depicts the difference between detecting the malware before and after binding (Table 1).

© Springer Nature Singapore Pte Ltd. 2021
P. Deshpande et al. (eds.), *Next Generation Information Processing System*,
Advances in Intelligent Systems and Computing 1162,
https://doi.org/10.1007/978-981-15-4851-2_9

Table 1. System configuration

Host machine	
Model	Lenovo ideapad 300
Processor	Intel(R) Core™ i7-6500U CPU @ 2.50 GHz 2.59 GHz
RAM	8.0 GB
System type	64-bit operating system
Operating system	Windows 10
Software	PyCharm professional
Language	Python 2.7.16
Tools/Packages	Capstone-Windows Metasploit OSINT VirusTotal
Hypervisor	
Version	VMware workstation 15.x
Processor	64-bit operating system
Virtual machine	
Kali Linux	v64-bit, 2 GB RAM, Hard Disk: 20 GB
Windows 7	v64-bit, 1 GB RAM, Hard Disk: 40 GB

2 Process of Automated Malware Identification and Analysis

2.1 Malware Development Phase

Step 1: Open Kali on VMWare/VirtualBox, run Metasploit [20] on Kali, and then find the IP address of local machine using command “ifconfig.”

Step 2: Create malware using “msfvenom” command [24], specifying protocol, host IP, port number, file type, and output location and name the malware as you want, let us say “monica.exe” and start apache.

Step 3: Turn off Firewall in Windows. Use “Win + Run” and run “firewall.cpl.”

Step 4: Turn off Windows Defender.

Step 5: Change proxy settings in Windows. Set Kali IP in the proxy settings.

Step 6: Write Kali’s IP address in search bar with malware name, i.e., “192.168.75.132/monica.exe.” Download it on Windows system.

Step 7: Download an icon image for binding the malware with it (image can be any, as per your choice).

Step 8: Convert the image into an icon. Go to “ICO Convert.” Upload the image you downloaded. Save the converted image.

Step 9: Add to archive both icon and the malware.

Step 10: Change the General settings while archiving the files. Got to advance and change the Advanced SFX Options to Absolute path.

Step 11: Browse the converted image with extension “ICO” and select the malicious file that is to be bound with the image.

Step 12: Change the Update Setting and go to Setup and change Setup Settings by entering files to be bound and click OK!

Step 13: Run the bounded file as Administrator and click Yes.

Step 14: Go through installation setup as shown in Fig. 1 and malware is installed.

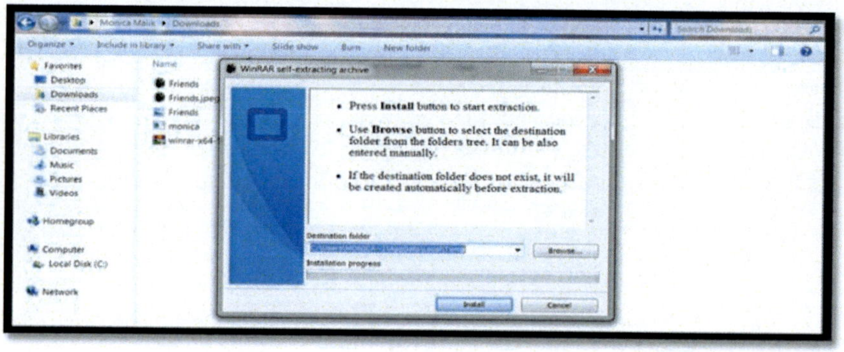

Fig. 1. Malware installation setup

2.2 Malware Exploitation Phase

Step 1: Exploit the victim's computer by with the help of Metasploit [20] using the command on Kali (use exploit/handler) as shown in Fig. 2.

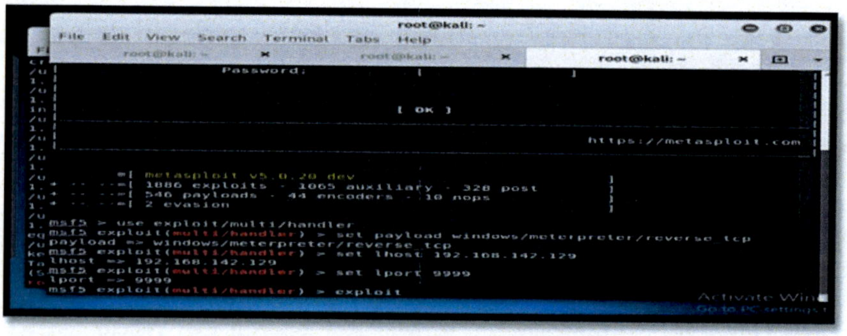

Fig. 2. Exploitation using malware

Step 2: Run "msfconsole" on attacker machine.

Step 3: Exploit using meterpreter by setting port and host.

2.3 Efficiency Evaluation Using OSINT

Step 1: Check whether malware is malicious or not on "VirusTotal" it is an open source intelligence tool (OSINT). Go to www.virustotal.com [23].

2.3.1 For Malware After Binding

(a) Select the bounded malware and upload it on VirusTotal.
(b) Results reveal that VirusTotal is working and display that created malware is malicious and observes the bounded summary as well as shown in Figs. 3 and 4.

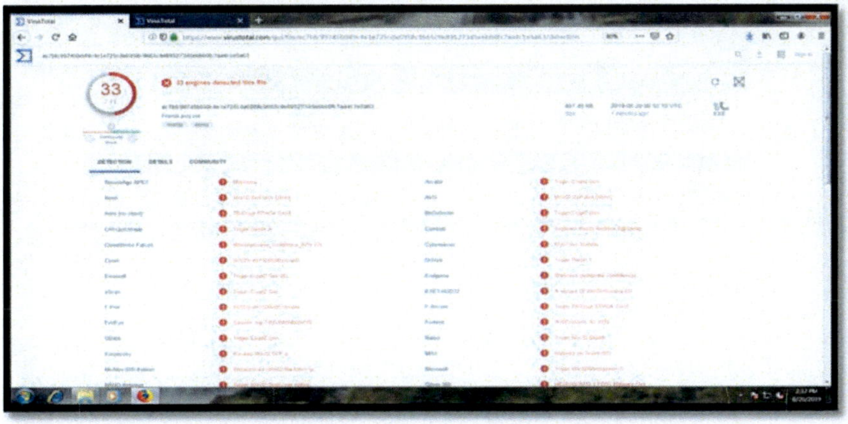

Fig. 3. VirusTotal results for bounded malware

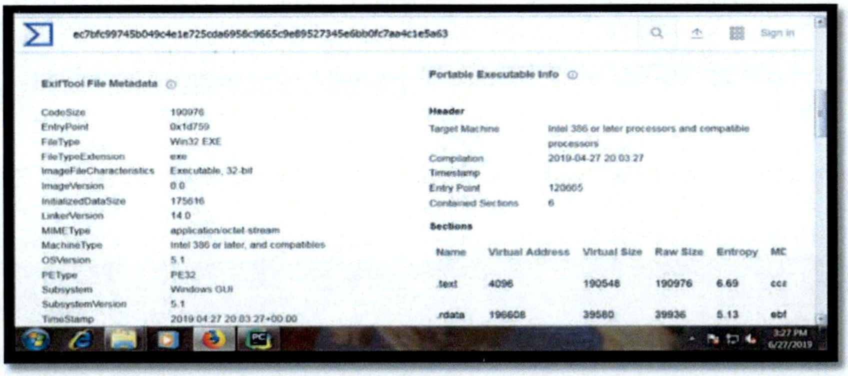

Fig. 4. Summary of bounded malware

2.3.2 For Malware Before Binding

(a) Select the unbounded malware and upload it on VirusTotal.
(b) Observe the result for unbounded malware and its bounded summary as well as shown in Figs. 5 and 6.

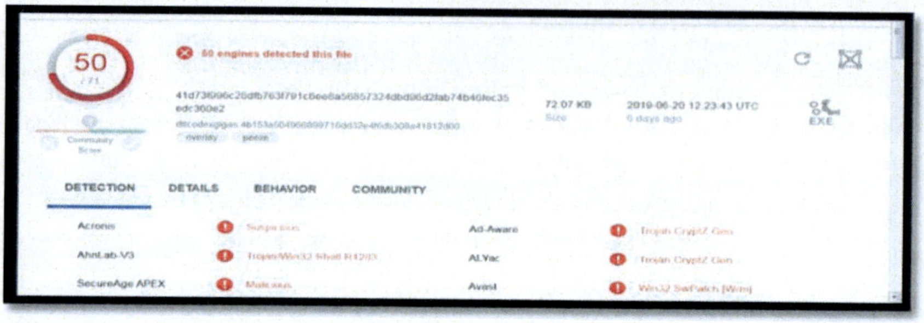

Fig. 5. VirusTotal results for unbounded malware

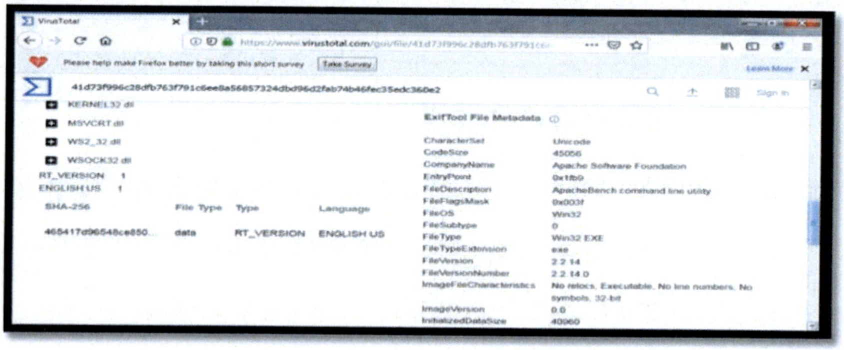

Fig. 6. Summary of unbounded malware

Step 2: Get system information of victim's system in meterpreter framework by running "sysinfo" command. Then, run the "getsystem" command. To capture the keystrokes of victim's system, run the "keyscan" command as shown in Fig. 7.

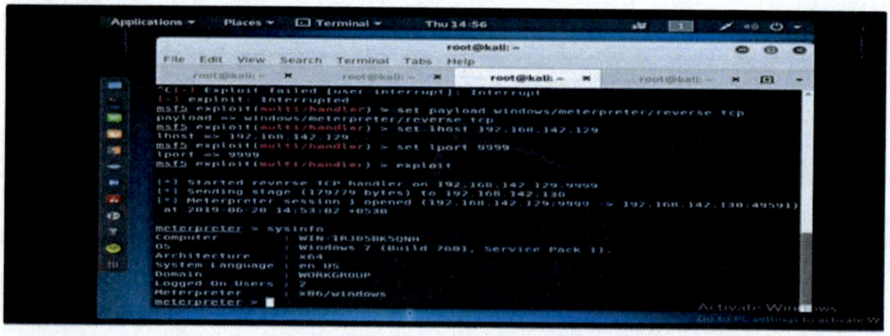

Fig. 7. Getting system info of victim's in meterpreter framework

2.4 Detection and Analysis Phase

Step 1: Download Python [25] version 2.7.16 and then install pip.
Step 2: Install "Capstone-Windows" [26] for the pyew.
Step 3: Open PyCharm. Run "pyew.py" and "pyew_core.py"
Step 4: Run the pyew.py command [27].
Step 5: A block of hexadecimal data is displayed in the pyew's prompt as shown in Fig. 8. After that, write "pyew.py a.html" and the prompt is the (hexadecimal) offset in the file.

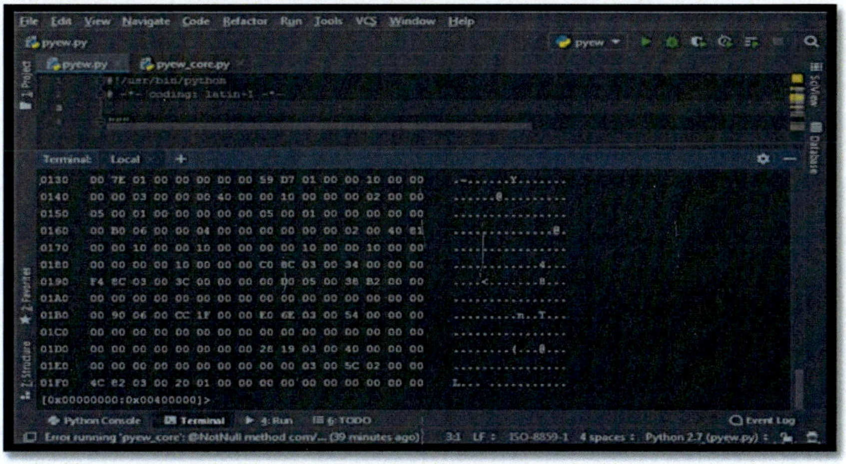

Fig. 8. Pyew.py command's hexadecimal output

Step 6: Move to the offset 100 in decimal (note that you can use hexadecimal offsets) to see the hexadecimal dump of the block at this specific position by issuing the command "s 100" ("seek to position 100") and "x" (hexadecimal dump).
Step 7: To change the block size and to see a smaller hexadecimal dump use "pyew. bsize = 64" and the "x":
Step 8: To check the URL availability in malware, issue the command "chkurl."

3 Limitations and Future Work

In this project, the tool used named "Pyew" was not able to check many things like on which systems it will not work, what are the requirements of a malware to attack on the system. Many more features can be added in the future like the tool will be able to produce the hash value of a malware and can compare the complexity of the malware using the tool like how and what features it can compromise on the victim's machine.

Other features can be added to the self-build tool like to what extent the malware before binding and after binding makes a difference. Recently, cloud computing along with IoT is becoming very popular [28–30]. The proposed systems may be implemented in cloud-based IoT systems.

4 Conclusion

Malware after binding was unrecognizable by many engines as compared to the malware after binding. Non-bounded malware was recognized by 50 engines out of 71. However, the bounded malware was recognized by only 33 engines out of 71. It shows the malware before binding was recognizable and was reported earlier also on the engines. Similarly, the anti-viruses or anti-malwares work the same, and sometimes they ignore the bounded malware, or say anti-malwares are not capable of detecting the malware sometimes after binding them with an image which makes the attacker to compromise the victim's machine easily by manipulating the victim.

References

1. Bulazel, A., Yener, B.: A survey on automated dynamic malware analysis evasion and counter-evasion: PC, mobile, and web. In: Proceedings of the 1st Reversing and Offensive-oriented Trends Symposium, p. 2. ACM (2017)
2. Blackthorne, J., Kaiser, B., Fuller, B., Yener, B.: Environmental authentication in malware. In: IACR Cryptology ePrint Archive, vol. 928 (2017)
3. Narwal, B., Mohapatra, A.K., Usmani, K.A.: Towards a taxonomy of cyber threats against target applications. J. Stat. Manag. Syst. **22**(2), 301–325 (2019)
4. Miramirkhani, N., Appini, M.P., Nikiforakis, N., Polychronakis, M.: Spotless sandboxes: evading malware analysis systems using wear-and-tear artifacts. In: 2017 IEEE Symposium on Security and Privacy (SP), pp. 1009–1024. IEEE (2017)
5. Rani S., Narwal B., Mohapatra A.K.: RREQ flood attack and its mitigation in ad hoc network. In: Panda B., Sharma S., Roy N. (eds.) Data Science and Analytics. REDSET 2017. Communications in Computer and Information Science, vol. 799. Springer, Singapore (2017)
6. Sudhodanan, A., Carbone, R., Compagna, L., Dolgin, N., Armando, A., Morelli, U.: Large-scale analysis & detection of authentication cross-site request forgeries. In: IEEE European Symposium on Security and Privacy (EuroS&P), pp. 350–365. IEEE (2017)
7. Masri, R., Aldwairi, M.: Automated malicious advertisement detection using VirusTotal, URLVoid, and TrendMicro. In: 8th International Conference on Information and Communication Systems (ICICS), pp. 336–341. IEEE (2017)
8. Polino, M., Continella, A., Mariani, S., D'Alessio, S., Fontana, L., Gritti, F., Zanero, S.: Measuring and defeating anti-instrumentation-equipped malware. In: International Conference on Detection of Intrusions and Malware, and Vulnerability Assessment, pp. 73–96. Springer, Cham (2017)
9. Narwal, B., Mohapatra, A.K.: A review on authentication protocols in Wireless Body Area Networks (WBAN). In: 2018 3rd International Conference on Contemporary Computing and Informatics (IC3I), Gurgaon, India, pp. 227–232 (2018)

10. Tanabe, R., Ueno, W., Ishii, K., Yoshioka, K., Matsumoto, T., Kasama, T., Rossow, C.: Evasive malware via identifier implanting. In: International Conference on Detection of Intrusions and Malware, and Vulnerability Assessment, pp. 162–184. Springer, Cham (2018)
11. Narwal, B.: Fake news in digital media. In: International Conference on Advances in Computing, Communication Control and Networking (ICACCCN), Greater Noida, India, pp. 977–981 (2018)
12. Narwal, B., Mohapatra, A.K.: SEEMAKA: Secured energy-efficient mutual authentication and key agreement scheme for wireless body area networks. Wireless Pers. Commun. (2020). https://doi.org/10.1007/s11277-020-07304-3
13. Narwal, B., Mohapatra, A.K.: SALMAKA: Secured, anonymity preserving and lightweight mutual authentication and key agreement scheme for wireless body area networks. Int. J. Sens. Wireless Commun. Control, 2020 (forthcoming)
14. Narwal, B., Mohapatra, A.K.: Secured secret sharing and reconstruction algorithm for organizations. In: 2018 3rd International Conference on Contemporary Computing and Informatics (IC3I), Gurgaon, India, pp. 223–226 (2018)
15. Bharadwaj, P., Pal, H., Narwal, B.: Proposing a Key Escrow Mechanism for Real-Time access to End-to-End encryption systems in the Interest of Law Enforcement. In: 2018 3rd International Conference on Contemporary Computing and Informatics (IC3I), Gurgaon, India, pp. 233–237 (2018)
16. Narwal, B.: Security analysis and verification of authenticated mobile payment protocols. In: 2019 4th International Conference on Information Systems and Computer Networks (ISCON), Mathura, India, pp. 202–207 (2019)
17. Dhawan, S., Narwal, B.: Unfolding the mystery of ransomware. In: International Conference on Innovative Computing and Communications, pp. 25–32. Springer, Singapore (2019)
18. Narwal, B., Mohapatra, A.K.: Performance analysis of QoS parameters during vertical handover process between Wi-Fi and WiMAX networks. In: International Conference on Recent Developments in Science, Engineering and Technology, pp. 330–344. Springer, Singapore (2017)
19. Narwal, B., Mohapatra, A.: Energy efficient vertical handover algorithm for heterogeneous wireless networks. Int. J. Control Theor. Appl. **9**(19), 9221–9225 (2016)
20. Metasploit Homepage. https://www.metasploit.com/. Last accessed 16 Oct 2019
21. ICO Convert Homepage. https://icoconvert.com/. Last accessed 14 Sept 2019
22. Krombholz, K., Hobel, H., Huber, M., Weippl, E.: Advanced social engineering attacks. J. Inf. Secur. Appl. **22**, 113–122 (2015)
23. VirtusTotal Homepage. https://www.virustotal.com/gui/home/upload. Last accessed 11 Oct 2019
24. Msfvenom Homepage. https://www.offensive-security.com/metasploit-unleashed/msfvenom/. Last accessed 07 Sept 2019
25. Python Homepage. https://www.python.org/. Last accessed 15 Oct 2019
26. Capstone course Homepage. https://en.wikipedia.org/wiki/Capstone_course. Last accessed 24 Oct 2019
27. Pyew Homepage. https://github.com/joxeankoret/pyew. Last accessed 20 Oct 2019
28. Deshpande P., Sharma S.C., Peddoju S.K.: Implementation of a private cloud: a case study. In: Proceedings of the Third International Conference on Soft Computing for Problem Solving. Advances in Intelligent Systems and Computing, vol. 259. Springer, New Delhi (2014)

29. Deshpande, P., Sharma, S.C., Peddoju, S.K., et al.: Security and service assurance issues in cloud environment. Int. J. Syst. Assur. Eng. Manag. **9**, 194–207 (2018). https://doi.org/10.1007/s13198-016-0525-0
30. Deshpande, P.: Cloud of everything (CLeT): the next-generation computing paradigm. In: Computing in Engineering and Technology. Advances in Intelligent Systems and Computing, vol. 1025. Springer, Singapore (2020)

Performance Study of Spin Field-Effect Transistor Based on Cobalt-Modified Iron Oxide Ferromagnetic Electrode

Neetu Gyanchandani[1(✉)], Santosh Pawar[2], Prashant Maheshwary[1], and Kailash Nemade[3]

[1] Department of Electronics, JD College of Engineering and Management, Nagpur 441501, India
gyanineetu@gmail.com

[2] School of Engineering, Dr. A.P.J. Abdul Kalam University, Indore 452016, India

[3] Department of Physics, Indira Mahavidyalaya, Kalamb 445401, India
krnemade@gmail.com

Abstract. Spintronics-based field-effect transistors (s-FET) are a new category of devices, which is an improvement over ordinary transistor by adding the properties of magnetoresistance. The conductivity of s-FET can be controlled by the spin degree of freedom of an electron, which results in extremely low power consumption and low heat dissipation. In the present work, a primary attempt is made to analyze the performance of s-FET designed on two-dimensional electron gas substrate. Superconducting quantum interference device (SQUID) is employed to analyze the magnetic properties of ferromagnetic contacts that cobalt-modified iron oxide. The role of spin polarization in the spin transport phenomenon of s-FET is also analyzed. It is proved that for the higher possible value of spin polarization, spin current also increases. For the value of spin polarization ($p = 0.8$), strong enhancement was observed in the spin current. The switching action in s-FET is checked as a function of gate voltage, and it shows a strong dependence on the gate voltage.

Keywords: Spintronics · Field-effect transistor · Two-dimensional electron gas substrate

1 Introduction

Spintronics is the next version of electronics, which utilizes the spin degree of freedom in the device fabrication process for memory, logic and switching applications [1]. In this process, s-FET is the most studied fundamental device by researchers due to its outstanding features like ultralow power consumption and novel logic design. The s-FET first discussed by Datta and Das by experimenting with the external electric field to control the spin orientation of the spin-polarized current in two-dimensional electron gas [2]. Researchers across the globe tried to realize s-FET and recently succeeded in achieving it with feasibility through experimental as well as theoretical approaches [3–6]. The working mechanism of s-FET comprises the modulation of the source to drain

© Springer Nature Singapore Pte Ltd. 2021
P. Deshpande et al. (eds.), *Next Generation Information Processing System*, Advances in Intelligent Systems and Computing 1162,
https://doi.org/10.1007/978-981-15-4851-2_10

conductance by controlling the gate voltage. In s-FET, ferromagnetic source drains, injects and detects the spin-polarized current [7]. The key motivation behind to explore s-FET is its outstanding features such as low power dissipation and less gate charging time.

The attractiveness of spin-valve transistor is that only selective electrons with precise spin polarization allow to pass through the device. These transistors found potential application in various field such as signal processing, automation and memory devices [8]. Kumar et al. investigated the magnetic field sensor-based spin-valve transistor. In this study, ferromagnet–semiconductor hybrid structure was used to fabricate spin-valve transistor. In this transistor, hot electron transport through the spin valve and collected at Schottky diode with energy and momentum selection [9]. The working of s-FET is based on the principle of modulation of resistance by controlling the spins of the carriers in ferromagnetic contacts. Mostly, in the s-FET spin transport is controlled by using the gate voltage. Koo et al. studied the s-FET based on InAs heterostructure. In this work, it is demonstrated the electrical injection and detection of spin-polarized electrons in s-FET. It is also observed that the conductance of s-FET is controlled by gate voltage [10]. Xiao et al. investigated s-FET and concludes that conductance oscillation of the s-FET mainly depends on interfacial barriers and mainly on spin–orbit coupling precession [11].

Presently, spin-valve and spin field-effect transistors devices facing an issue like low spin injection efficiency, resistance mismatch, spin relaxation and the spread of spin precession angles [12].

In this article, the performance of s-FET designed with cobalt-modified iron oxide ferromagnetic electrode on two-dimensional electron gas substrate is analyzed. During this study, focus was on switching action in s-FET as a function of gate voltage. The spin transport phenomenon is the main aspect of s-FET designing and its study, so it is checked by sweeping an in-plane magnetic field by varying gate voltage. To understand the role of spin polarization in the spin transport phenomenon of s-FET, different values of spin polarization are taken into consideration.

2 Experimental

In the present work, the prototype of spin field-effect transistor (s-FET) is developed. The necessary components require for the fabrication of s-FET are two-dimensional electron gas (2-DEG) substrate, insulating layer, gate and ferromagnetic materials. The materials required for the purpose were readily procured from IndiaMART-Manufacture and Supplier. The important ferromagnetic contacts necessary for s-FET are prepared in the laboratory and checked for structural purity (using X-ray diffraction analysis) and magnetic properties using SQUID.

For the macroscopic magnetic properties of the cobalt-modified iron oxide ferromagnetic contacts, SQUID technique is employed to study magnetic behavior. SQUID measurements were taken at 300 and 315 K. SQUID technique is the standard measurement technique, which generally employs for highly sensitive magnetization studies. In the present study, SQUID measurements were performed on Quantum Design SQUID magnetometer MPMS-5.

In the present work, the two contacts of cobalt-modified iron oxide ferromagnets (FM) are deposited on 2-DEG substrate as shown in Fig. 1a by using the chemical vapor deposition technique. These two contacts work as source (S) and drain (D) for s-FET. After the deposition of FM contacts on 2-DEG substrate, the insulator layer of polyvinyl acetate (indicated in figure by I) is fixed between the FM contacts as shown in Fig. 1a. After this step, gate (G) contact is deposited on top of the insulator. With these simple steps, the prototype of s-FET is developed. Figure 1b shows the scanning electron microscopy image of s-FET-top view (scale bar, 5 μm) of as-fabricated s-FET.

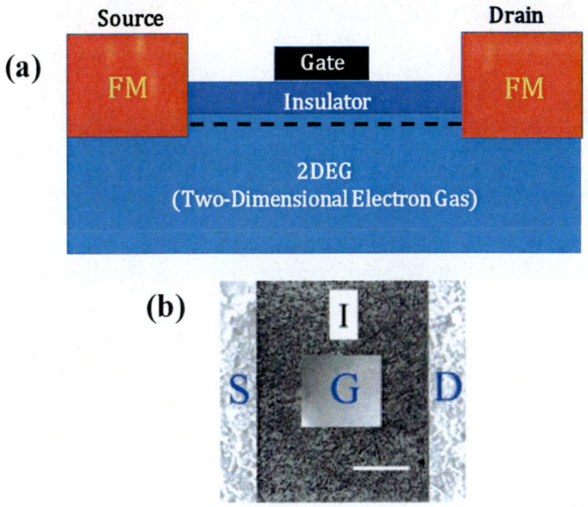

Fig. 1. **a** Architecture of s-FET and **b** scanning electron microscopy image of s-FET-top view (scale bar, 5 μm)

In the present work, all measurements of s-FET were performed in a cryostat having provision to control temperature and magnetic field. Keithley 6221 was used as the current source. For the detection of voltage, Keithley 2182A nanovoltmeter was utilized whereas to control the gate voltage, Keithley 2612 voltage source meter was used.

3 Results and Discussion

In the present work, cobalt-modified iron oxide ferromagnetic contacts are used for the fabrication of s-FET. The magnetic properties of FM contacts are determined by a superconducting quantum interference device (SQUID). The magnetic properties of FM contacts in device are studied at two temperatures 300 and 315 K. The FM contacts show typical magnetic response as given in Fig. 2. The coercive force for FM contact materials was found to the order of ~129 Oe with a small deviation for both temperatures. This characterization confirms the proper working of FM contacts in the device.

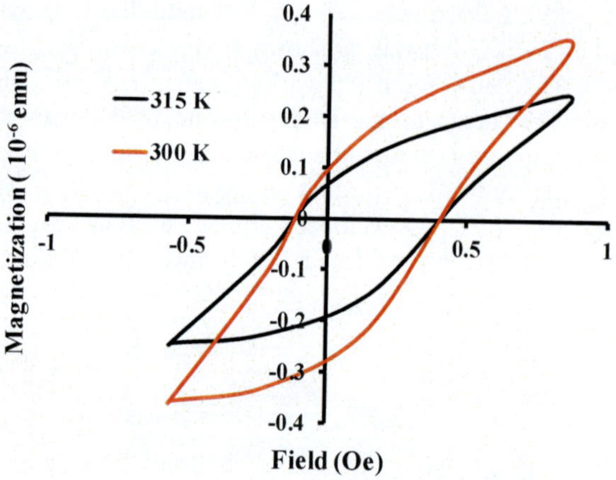

Fig. 2. Magnetic properties of FM contacts in s-FET at different temperatures

In the present work to study spin current, the concept of diffusion dynamics of electrons in a ring is used, which firstly employed by Jung et al. The result of this study gives equation, which co-relates the electronic current and spin current as follows in Eq. (1),

$$I_s = I_e \frac{p}{2} \exp\left(-\frac{L_{\text{Source}}}{\lambda_N}\right) \tag{1}$$

where I_s is spin current, I_e is electronic current, p is spin polarization and its value ranges between 0.1 and 0.9 for ferromagnetic materials, L_{source} is length of tinny magnets in ferromagnetic material (generally, its value is 10 nm) and λ_N is spin-flip length and its value is $\lambda_N = 1$ µm [13].

Spin current is a key phenomenon in spintronics which can apply useful torque in spintronic devices [14]. In traditional electronics, spin property of the electron is ignored. But, in spintronics, the spin of electron plays a very crucial role in device fabrication and its performance [15]. This kind of device has a faster switching time and lower power consumption than conventional electronic devices. This is because spin operates faster with lower energy than charge [16]. Generally, the spin current is produced by sending an electron current through a polarized layer of ferromagnetic (FM) materials with an architecture of FM/two-terminal magnetic tunnel junction (MTJ)/FM [17, 18].

Herein, this research variation of spin current and electronic current with voltage and spin polarization is verified. To check the dependence of spin current on spin polarization and electronic current, all remaining parameters are considered as constant such as $\lambda_N = 1$ µm and $L_{\text{source}} = 10$ nm.

Spin polarization (p) is the degree to which the spin of elementary particles (that is an electron in this case) is aligned with a given direction. Basically, the spin polarization is a dimensionless quantity, which depends on the density of states of the spin-up and spin-down bands at the Fermi level [19]. In ferromagnetic materials (cobalt-modified iron oxide), the possible values are $p < 1$. In the present work, the spin current is estimated for four different values of spin polarization p (0.2, 0.4, 0.6 and 0.8). In Fig. 3a–d, it is observed that the spin polarization factor has a significant influence on spin current. The values of spin current are increased with an increase in spin polarization. The highest value of spin current is associated with $p = 0.8$.

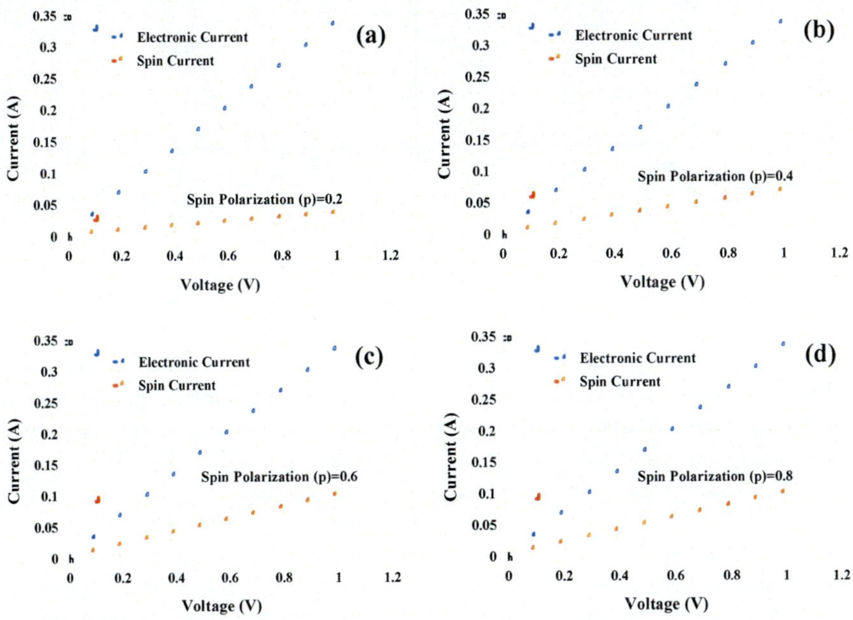

Fig. 3. Variation of electronic current and spin current with applied voltage for different values of spin polarization **a** $p = 0.2$, **b** $p = 0.4$, **c** $p = 0.6$ and **d** $p = 0.8$

In Eq. (1) by simply putting the values of spin polarization, it is concluded that during device fabrication, proper optimization of spin polarization for ferromagnetic materials is necessary. Spin polarization is the key parameter to control the performance of the device.

Figure 4 shows the modulation of spin valve by controlling the gate voltage at 300 K. From the plot, it is clear that below zero gate voltage, we observed almost constant value of resistance. In other words, below 0 V, the spin-valve signal has no change in amplitude. Between 0 and 0.425 V, a significant decrease is observed in resistance value and it becomes almost zero at 0.5 V. This characteristic of the curve, plotted between gate voltage and change in resistance, indicates that as-fabricated s-FET has resistance modulation behavior, which can be controlled by using gate voltage. This accomplishment of work indicates that s-FET has switching application as it shows strong modulation and vanishing spin current from on to off state with the application of gate voltage. The switching potential of s-FET is attributed to the effective spin transport between FM contacts through 2-DEG channel and low interfacial scattering.

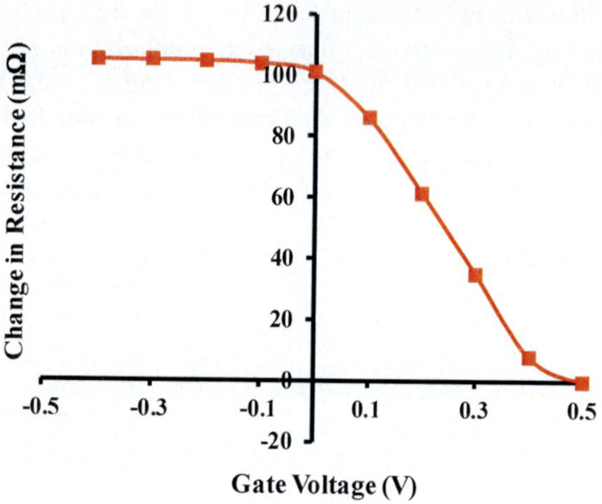

Fig. 4. Modulation of spin valve by controlling gate voltage at 300 K

The working mechanism of s-FET is mainly divided into three processes, first is the injection of spin-polarized current of the electron into the two-dimensional electron gas channel from FM contacts. The second process is the transport of electrons through two-dimensional electron gas channel without losing the spin direction. The third and important process is the detection of spin-polarized current into FM drain contact. Formation of two-dimensional electron gas channel is very important aspect of s-FET. The main challenging thing in this process is the transport of electrons through two-dimensional electron gas channel without losing the spin direction. Many factors affect the spin direction in two-dimensional electron gas/semiconductor substrate such as temperature, defects in semiconductors, material surface, volume impurities, phonons scattering and diffusive transport (electrons scatter continuously due to lattice).

To study the spin transport phenomenon in s-FET, spin-valve behavior was checked at 300 K, in addition to sweeping an in-plane magnetic field (aligned parallel). The use of the magnetic field produces magnetoresistance, which plays a key role in controlling of spin transport phenomenon. During measurement, the current source was fixed at 45 μA. From Fig. 4a–d, it is clearly observed that spin-valve signal can be precisely controlled using the gate voltage around room temperature. This indicates considerable spin transport which can be achieved in as-fabricated s-FET. In other words, spin current can be altered precisely by altering gate voltage. In Fig. 5a–c, for $V_g = -0.3$ V, 0 V and 0.3 V, respectively, the two resistance states can be clearly observed in plots. One is a high resistance state associated with the parallel magnetization, and second is a low resistance state associated with the antiparallel configuration [20]. In Fig. 5a, b, the baseline resistance has almost the same values (−18.5 Ω), whereas in Fig. 5c, it is observed that for $V_g = 0.3$ V positive value of gate voltage, baseline resistance value becomes −13.77 Ω for the same experimental conditions [21]. This is because the gate positive voltage attracts electrons to the interface at insulator/gate. Further, if the value of gate voltage increases to $V_g = 0.6$ V (Fig. 5d), the two resistance states which are associated with high and low value of resistance become invisible.

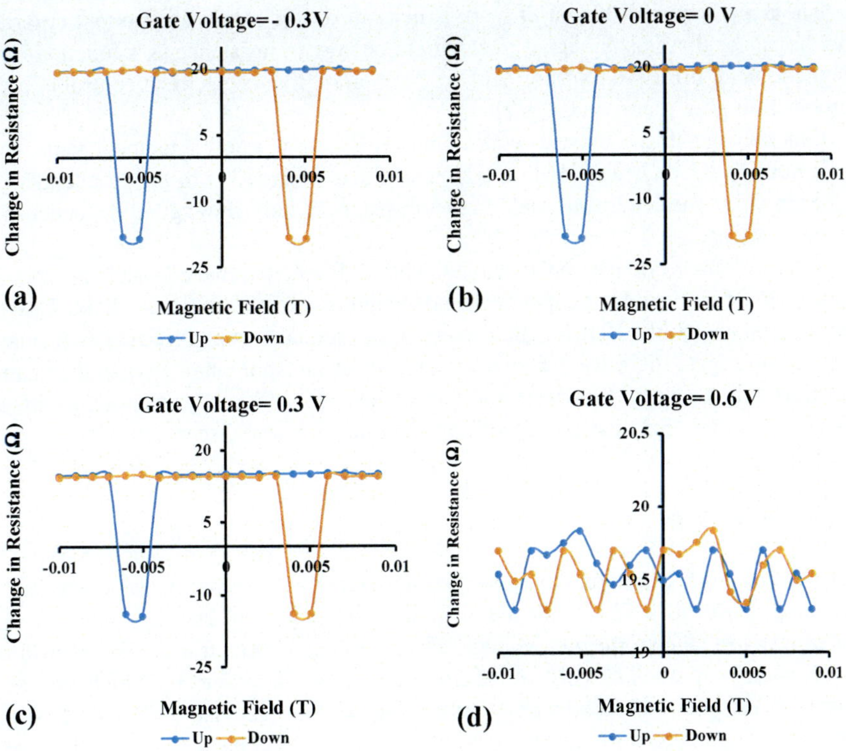

Fig. 5. Spin-valve magnetoresistance variation at 300 K for different gate voltages **a** −0.3 V, **b** 0 V, **c** 0.3 V and **d** 0.6 V

In s-FET, the injected electrons into 2-DEG channel are spin-selected. Most probably the injected electrons have single spin because the source of injected electrons is ferromagnetic contact, in the present work cobalt-modified iron oxide used as ferromagnetic contact. As the source of injected electrons is ferromagnetic contact, it does not show equilibrium between two spin states. Hence, the majority of electron injected in channel has spin up or down.

The spin direction of injected electrons can be altered by the gate voltage. According to the Rashba effect, the spin precession angle of electron in semiconductor channel depends on the applied voltage. Therefore, by using Rashba effect it is possible to change direction of electrons in semiconductor channel, and in other words, it is possible to achieve on and off state in s-FET by changing direction of electron.

Thus, when direction of electron in ferromagnetic drain is parallel with the majority spin electrons in channel, the current can pass through the drain, and hence, on-state of s-FET is created. If the spin direction between the electrons in the channel is antiparallel to the direction of the drain, the current cannot pass through the device, which results in sharp decrease in drain current. This state of s-FET is considered as off state. Therefore, by altering the gate voltage, it is possible to achieve on–off states because of different precession angles of electrons as a function of applied gate voltages.

The main accomplishment of work is that our device satisfies important criteria of spin transistor that is spin signal modulation, change in resistance as a function of the magnetic field, diffusive transport and very important that is electrical control of a spin current through a gate voltage [22].

Our findings in the present work about s-FET show good agreement with recent work published by Danker et al. in which s-FET is designed with the graphene/MoS2 [6]. Spin-valve measurements and magnetoresistance study show good dependence on the gate voltage.

The present work can be extended with different parameters such as spin-flip length, spin polarization, temperature and magnitude of the magnetic field. Use different components like Heusler alloys contact, hexagonal boron nitride (as an insulator) and ultra-thin gate dielectric for effective control of the spin parameters in the channel will also help to study furthering the present research [23]. Nanotechnology of graphene and its modification is created opportunities for new discoveries [24].

4 Conclusions

The present work concludes that the researchers have successfully demonstrated the switching action in s-FET. This action to achieve on and off states is precisely controlled using the electronic gate voltage. The switching potential in s-FET is attributed to magnetic properties of cobalt-modified iron oxide FM contacts, which are determined by using a superconducting quantum interference device (SQUID). It is also attributed to the efficient transport of spin through 2-DEG channel and low interfacial scattering. The spin transport phenomenon in s-FET was checked at 300 K, and it exhibited strong dependence on sweeping an in-plane magnetic field and gate voltage.

Further, it is confirmed that the spin polarization value of FM contacts plays a very crucial in the performance of s-FET. In present work, it is proved that with higher possible value of spin polarization (that is $p < 1$) can improve device performance. In this case, at $p = 0.8$, which shows significant enhancement in spin current is achieved.

Further scope to improve the performance of present work, there are many possibilities such as to improve spin-flip length, optimization of temperature, use of efficient FM contacts, insulating layer and gate contact.

Acknowledgements. Prof. (Mrs.) Neetu Gyanchandani is very much thankful to Dr. S. R. Choudhary, Principal, JD College of Engineering and Management, Nagpur, for providing necessary academic help.

References

1. Awschalom, D.D., Flatte, M.E.: Challenges for semiconductor spintronics. Nat. Phys. **3**, 153–159 (2007)
2. Datta, S., Das, B.: Electronic analog of the electro-optic modulator. Appl. Phys. Lett. **56**, 665–667 (1990)

3. Choi, W.Y., Kim, H.J., Chang, J., Han, S.H., Koo, H.C.: Ballistic spin Hall transistor using a heterostructure channel and its application to logic devices. J. Electron. Mater. **46**, 3894–3898 (2016)
4. Kim, J.H., Bae, J., Min, B.C., Kim, H., Chang, J., Koo, H.C.: All-electric spin transistor using perpendicular spins. J. Magn. Magn. Mater. **403**, 77–80 (2016)
5. Lin, X., Su, L., Si, Z., Zhang, Y., Bournel, A., Zhang, Y., Klein, J., Fert, A., Zhao, W.: Gate-driven pure spin current in graphene. Phys. Rev. Appl. **8**, 34006–34011 (2017)
6. Dankert, A., Dash, S.P.: Electrical gate control of spin current in van der Waals heterostructures at room temperature. Nature Commun. **8**, 16093–16099 (2017)
7. Koo, H.C., Han, S.H., Chang, J.Y., Kim, H.J., Choi, J.W.: Complementary Spin Transistor Logic Circuit. U.S. Patent (2012)
8. Krishnan, R.: Spin valve transistors. Int. J. Innov. Res. Adv. Eng. 1, 118–122 (2014)
9. Kumar, P.S.A., Lodder, J.C.: The spin-valve transistor. J. Phys. D Appl. Phys. **33**, 2911–2920 (2000)
10. Koo, H.C., Kwon, J.H., Eom, J., Chang, J., Han, S.H., Johnson, M.: Control of spin precession in a spin-injected field effect transistor. Science **325**, 1515–1518 (2009)
11. Xiao, Y., Zhu, R., Deng, W.: Ballistic transport in extended Datta-Das spin field effect transistors. Solid State Commun. **151**, 1214–1219 (2011)
12. Kum, H., Heo, J., Jahangir, S., Banerjee, A., Guo, W., Bhattacharya, P.: Room temperature single GaN nanowire spin valves with FeCo/MgO tunnel contacts. Appl. Phys. Lett. **100**, 182402–182407 (2012)
13. Jung, S., Lee, H.: Spin-current-induced charge current. Phys. Rev. B **71**, 25341–25348 (2005)
14. Liu, L., Pai, C., Li, Y., Tseng, H.W., Ralph, D.C., Buhrman, R.A.: Spin-torque switching with the giant spin Hall effect of tantalum. Science **336**, 555–558 (2012)
15. Ando, Y.: Spintronics technology and device development. Jpn. J. Appl. Phys. **54**, 070101–070111 (2015)
16. Sharma, P.: How to create a spin current. Science **307**, 531–533 (2005)
17. Hirsch, J.E.: Spin hall Effect. Phys. Rev. Lett. **83**, 1834–1839 (1999)
18. Valenzuela, S.O., Tinkham, M.: Direct electronic measurement of the spin Hall effect. Nature **442**, 176–179 (2006)
19. Sakuraba, Y., Hattori, M., Oogane, M., Ando, Y.: Giant tunneling magnetoresistance in $Co_2MnSi/Al–O/Co_2MnSi$ magnetic tunnel junctions. Appl. Phys. Lett. **88**, 192508–192514 (2006)
20. Kasai, S., Hirayama, S., Takahashi, Y.K., Mitani, S., Hono, K., Adachi, H.: Thermal engineering of non-local resistance in lateral spin valves. Appl. Phys. Lett. **104**, 162410–162415 (2014)
21. Garzon, S., Zutic, I., Webb, R.A.: Temperature-dependent asymmetry of the nonlocal spin-injection resistance: evidence for spin nonconserving interface scattering. Phys. Rev. Lett. **194**, 176601–176608 (2005)
22. Sugahara, S., Nitta, J.: Spin-transistor electronics: an overview and outlook. Proc. IEEE **98**, 2124–2154 (2010)
23. Yamaguchi, T., Moriya, R., Oki, S., Yamada, S.: Spin injection into multilayer graphene from highly spin-polarized Co2FeSi Heusler alloy. Appl. Phys. Exp. **9**, 63006–63012 (2016)
24. Lazic, P., Belashchenko, K.D., Zutic, I.: Effective gating and tunable magnetic proximity effects in two-dimensional heterostructures. Phys. Rev. B **93**, 241401–241407 (2016)

Check for updates

Study of Mechanized Recognition of Driver's Smartphone Exploiting Common Vehicle-Riding Actions

Kadiyala Yaswanth[1(✉)], Rajasekhar Manda[1], and Durgesh Nandan[2]

[1] Department of ECE, Aditya Engineering College, Surampalem, India
yashyaswanth996@gmail.com, rajasekharm@aec.edu.in
[2] Accendere Knowledge Management Services Pvt. Ltd., CL Educate Ltd., New Delhi, India
durgeshnandano51@gmail.com

Abstract. Distracted driving due to using smartphone like texting, browsing Web, etc. increases the risk of accidents. To prevent this distracted driving, many suggestions have been proposed, but out of them, none addressed completely and efficiently to prevent this distracted driving. This work presents a concept called as mechanized recognition of driver's smart phone exploiting common vehicle-riding actions to overcome above said deficiency concept. The fusion of the driver's smartphone with phone's sensory provides the information related to rider's actions. This information can be obtained by sequence of steps such as walking toward the vehicle, opening the door from driver's side, closing the door, standing near the vehicle, entering into it, sitting, and kicking of the engine. The recognition of the smartphone depends on the position of the smartphone placed in the vehicle. This concept identifies the driver's smartphone just before it leaves out of the parked location. It differentiates between the seated rows by detecting the electromagnetic (EM) spikes occurring when the vehicle starts. By conducting all these sequences of steps, this concept will effectively identify the driver's smartphone and which efficiently prevent distracted driving.

Keywords: Detector · Automatic identification of driver's smartphone · EMF fluctuations · Electronic devices · Motorizes · Sensors · Entrance detector · Vehicle door closing sound

1 Introduction

For intelligent transportation system, we have applied a number of sensors which are widely used in automatic vehicle identification system (AVIS). The AVIS can be employed in extending the existing technology. The system may include tool fares, speed detectors, etc. For the present system, we have applied a flexible system called radio frequency identification (RFID) traceability system. The RFID signals is transformed into microwaves which are used to scan the vehicle's number plate at the checking point. The information on the barcode is decoded at this stage. The retrieved information is then sent to the main server. In the main server, the information is

© Springer Nature Singapore Pte Ltd. 2021
P. Deshpande et al. (eds.), *Next Generation Information Processing System*,
Advances in Intelligent Systems and Computing 1162,
https://doi.org/10.1007/978-981-15-4851-2_11

verified by the analyzer to store it in a secondary storage device and sends the receipt to the owner of the vehicle. The speed detector sensor records the speed and stores that data with the help of microcontroller [1]. Most of the vehicle manufacturing industries adopted logistic management. It is important to monitor the situation to control any material consumption and to insure the supply of the material without any mistakes or delay. As many defects exist due to workers, transportation, and unnecessary inventory, they must be reduced. With the advances in RFID, it is integrated with an antenna with electronic circuitry to develop a transponder that will pooled by remote interrogator such that it gives an echo back for the identification number. The activities like identification at a distance and tracking real-time control can be done through a RFID system [2]. Automobile transportation has become one of the common activity in the field of speed and technology. This eventually is also a reason for high probability of traffic rule violation by scofflaw individuals as the drivers. A thorough observation is needed for this purpose. In order to make this possible, modern and up-to-date technologies are needed to be employed. It should be noted that specific licenses are required to identify the real owner of a vehicle [3]. In recent studies, researchers have developed systems that can focus on analyzing individual differences in biological signals and performance data for proper detection of head movement. However, with the disadvantage of interfere driving, the reliability of the data will be decreased. Most of the methods adopted in the field of image processing and pattern-recognition have built their system based on measuring method known as PERCLOS [4].

This paper explains the recognition of driver's smartphone identification in-vehicle-riding actions automatically. The required literature review has been explored further in Sect. 2. Section 3 details the existing methodology to identify the driver's smartphone. The comparison of results which consists of the technique we used and parameter achieved is clearly explained in Sect. 4. What are the possible applications have been discussed in Sect. 5. Finally, the technique to identify the smartphone is concluded in Sect. 6.

2 Literature Review

Two new mathematical models are development for driver's behavior in single-lane car situations. By using some standard parameter identification algorithms, the optimum parameter values in each of the model can be obtained. The most basic approach is by using optimal control theory. With the help of this, the quadratic cost function can be reduced. A second model is postulated to obtain a better fit during transitions between acceleration and deceleration phases. In the method, the hypothesis is such that the driver not only tracks the car directly in front of him, but it also tracks the cars ahead of the lead car directly [5]. It also represents an active safety application for driver vehicle environment system based on diagnosis methods. For this, it needs to collect all the details about the concerned vehicle. Having a multiple correspondence analysis (MCA) provides a diagnosis which does not requires the model to be studied. It provides well-diagnosed driving situations [6]. As most of the traffic accidents are caused due to negligence of drivers, automatic driver evaluation system (ADES) aims to present a framework that integrates the various applications for driver estimation process. The system has two main phases, namely data acquisition phase and processing phase. The phases acquire sensor information from the outside world and the data is evaluated to provide valuable information. The information is used in the decision system, followed by inference engine that further evaluates the information provided by the first module. The modules finally take a decision about the actions of the driver [7]. The most safety and convenient applications are advanced driver assistance system (ADAS). This system is suitable for the different varieties of cars. The method is applied to detect empty legal parking areas in order to guide the vehicle to be parked into the selected spot. Detection can be performed both in indoor and road side parking. This system is mainly composed of computer connected Webcams that are attached to vehicles. Depending upon the movement of the vehicle, the system automatically searches for the vacant position in the parking. Once it is identified, the driver gets notified and the parking spot is monitored [8]. In several security applications, automatic face recognition algorithms are playing a key role. The enhancement of security of vehicle parking spaces can be developed here. The framework that is proposed can be divided into three separate steps which includes, recognition of vehicle, locating driver face, and finally, a robust face recognition algorithm. Face recognition algorithm aims to identify drivers based on comparison between test face image and stored database. To detect vehicles and faces, we use adaptive boosting algorithm [9]. In order to increase security and ease levels of the drivers and driving systems, the advanced integration of advanced driver assistant systems (ADAS) are inserted into vehicles. The key factors in these systems are that automatic identification of drivers [10]. Different drivers have different driving styles to follow a vehicle safely and comfortably. The different cars have to follow pedal operation patterns. The relationship between distance and velocity is modelled and approximated by a Gaussian mixture model (GMM) which is a non-linear function or statistical method [11].

3 Methodology

Driver's smartphone (DS) identification is the prime objective under neither by restricting the phone's possession nor by requiring the additional dedicated devices. The aim is also to minimize energy consumption due to limited battery capacity. During the start of the vehicle, the smartphones will not detect any significant movements.

In this concept, following main detectors have been used: Entering direction classifier (EDC), walking and standing detector (WSD), entrance detector (ETD), seated row classifier (SRC), and smartphone position classifier (SPC). These sensors check whether the driver is in the seat or not. All sensors are deactivated when the smartphone is not belongs to the driver to save the energy [12].

The identification of smartphone while driving should be accurate when the following sequence occurs. The sequence will be *walking–standing–entering–seated–engine start.* While designing the prototype, mobile energy efficiency is one of the important parameters to save energy by using accelerometer readings for the system initiators while keeping others in sleep. The inactive sensors will get activated once the driver enters the vehicle. Such sensors are useful in collecting observations from magnetometer and accelerometer. Walking and standing sensor (WSD) is mainly useful during the detection of the variance of EMF fluctuations, the magnitude of positive acceleration that is caused by sitting motion, and the sound of the closing vehicle door. The accuracy of the above concept depends on the smartphone position in the vehicle. Cases when the phone is on bag/pocket, features such as EMF and sitting motion features are used while VDCS is not applied. Similarly, when the phone is in hand, EMF and VDCS are applied but the sitting motion is not used [13, 14]. Entering direction classifier (EDC) will mention whether the driver has entered into the vehicle or not and if not, it will be inactive [15]. Seating row classifier (SRC) detector is used to categorize the driver's seated row. It is distinguished by sensing the EMF fluctuations caused due to start of vehicles. The vehicle starting process not only motorizes the engine but also provides powers to electronic devices. The front row is the major source of the EMF fluctuations because all the electronic devices are assembled at front row of the driver's seat [16].

Smart phone classifier (SPC) distinguishes between three frequent positions used by the user to hold the smartphones, which includes trouser pockets, bags, and hands. These sensors work after the ETD activates, then it acquires the clean data of the position of driver's smartphone [17–20].

The precise identification of the driver's smartphone (DS) without the need of the additional dedicated devices, under no restriction on the phone's poses, and with minimum energy consumption is the primary goal. To develop such an efficient AIDS, three assumptions are made such as no significant movement of smartphone during the start of the vehicle; the petrol running engines; and door openings of it are not controlled remotely [21].

Handling the phone's orientation physically is another way to address this issue. In order to fix the orientation, the local and global sensors are aligned such that its y-axis must point the magnetic pole that is nothing but the north pole. This approach will obtain the efficient sitting trajectories of the user. However, this method is not practically feasible. Therefore, the handheld devices must move with no physical restriction. As physical constraints are not exerting any force, quaternion algorithm can easily be applied [15, 22]. Local sensors are virtually rotated by gyroscope and magnetometers. The utilization of Euler and orthogonal matrix-based approaches under quaternion concept are advantageous over all other schemes [23, 24]. To identify the position of the smartphone, the orientation is more important and for effective identification. If the smartphone does not in the position, the sensors will be in sleep. The orientation of the smartphone can be explained by using quaternion three-dimensional space [25, 26]. The concept that used for this detection is possible only when the vehicle leaves the parking spot. It does not give the accurate result when the smartphone is in pockets or bags, this concept is applicable only when drivers should wear seat belts [27]. Figure 1 explains that the sequence of steps for identifying drivers smartphone from the initial step from user walking and standing toward the vehicle followed by user entering into the vehicle and it checks the entering direction of the user, if the direction is left, the user is driver otherwise user may be passenger. If the user is driver, then the sequence further moved to the seated rows classifier, and it checks whether the user sits in a front row or back row; if the user sits in front row, user may be driver otherwise user may be passenger. By following these sequences of steps, the smartphone of the driver identify effectively.

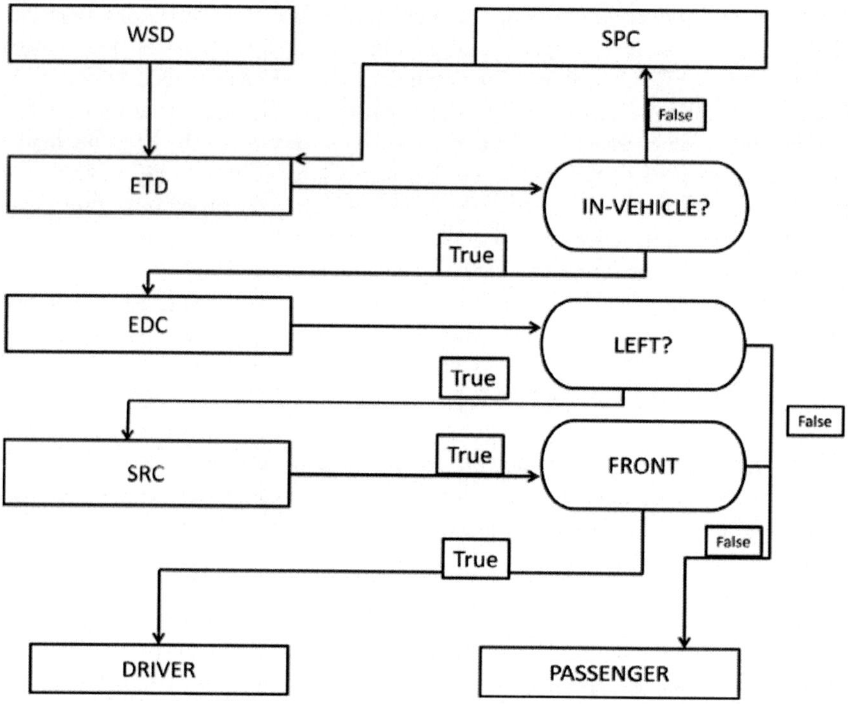

Fig. 1. Sequence of steps to identify the driver's smartphone

4 Results and Discussion

From this concept of automatic recognition of smartphones exploiting in-vehicle-riding actions, different types of sensors along with their techniques are used. WSD sensor identifies the walking and standing actions of the user before entering the vehicle. SPC sensor identifies the position of the smartphone in the vehicle. ETD sensor identifies the whether the user entered into the vehicle or not. EDC sensor identifies that in which direction the user will enter into the vehicle. SRC sensor differentiates the rows of the vehicle and checks whether the driver sitting in the driving seat or not. The SRC sensors will depend on the EMF fluctuations of the electronic devices. By all this technique, we can identify the driver's smartphone while exploiting vehicle-riding actions. The below tabular form explained about each technology using for to recognize the smartphone in-vehicle-riding actions (Table 1).

Table 1. Outcome of each sensor worked to detect smartphone of driver

Experimental setup	Parameter achieved
Walking and standing (WSD) sensor	WSD sensor detects the concept of walking and standing position of the driver before entering into the vehicle; it can be initiated by activating of inactive sensors depend upon the position of the drive [13, 14]
Smartphone position classifier (SPC) sensor	SPC sensor detects the concept of the position of the smartphone whether it may in drivers' hand or pocket or any other bags; based upon these, the sensor will be activated which is useful for further identification of driver's smartphone [17]
Entrance detector (ETD) sensor	ETD sensor detects the concept of whether the driver entered into the vehicle or not; based upon the variance of EMF fluctuations, accelerations caused by sitting motions and the sound that occurred when the vehicle door is closed [18, 19]
Entering direction classifier (EDC) sensor	EDC sensor detects the concept that the driver's direction while entering into the vehicle, it may be either left or right. It will be active when the driver enters the direction of the left side of the vehicle otherwise it will be inactive [15]
Seat row classifier (SRC) sensor	SRC sensor detects the concept of differentiating the rows in a vehicle based on EMF fluctuations caused due to the starting of the vehicle and motorizes the electronic devices which are densely populated in the front row, and based upon this, it classifies whether it is a front or back row of a vehicle [16]

When the driver came near to the car, WSD sensor observes that driver coming near to the car and then ETD sensor detects the driver entering into the car. The EDC sensor gives the direction in which the driver enters into the car and SPC gives seat position seated by driver. If all sensors are activated, then it confirms that the driver seated in the driving seat. Now, based upon the position of the smartphone, we can easily detect whether driver using smartphone or not. The flowchart (Fig. 2) represents that the driver entering into car to identification of smartphone.

5 Applications

To check whether the driver is following the traffic rules or not, especially using the phone while driving. This application mainly identifies texting or browsing of a smartphone (using of smartphone) which may distract the driver's attention while driving and may cause accidents. To avoid this, the above concept is used and this easily identifies the user who is using the phone while riding the vehicle. It reduces the accidents because major accidents occurred when driver gets distracted in driving while using phone. Recently, RF-based systems are gaining its popularity in remote detection of objects. Such systems are gaining popularity due to its detection accuracy [28, 29].

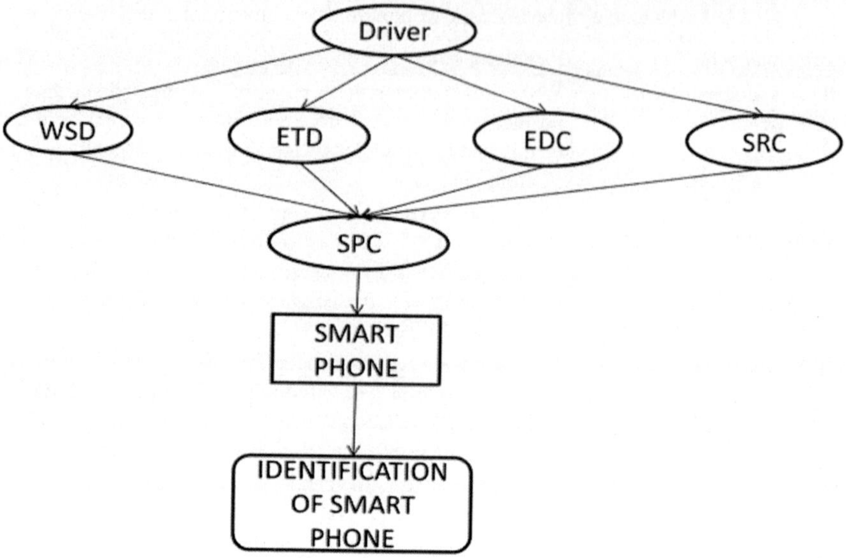

Fig. 2. Diagrammatic representation of identification of smartphone based of sensor activation

6 Conclusion

This paper discussed about a system which is used to ensure the safety of drivers and prevents road accidents. This paper proposed a number of sensors for this purpose. The solution for proposed system is automatic identification of driver's smartphone. This system detects if a person is using a smartphone or not, by observing the motions of the person with the help of sensors. This solution proves to be helpful to avoid many road accidents. Different tests were conducted by adopting this technique using different cars. It proves to be an efficient solution which safeguards the life of people.

The energy required for this process is a bit more. To be a completely efficient solution, this problem should also be addressed in the nearby future. Hence, an energy efficient system for automatic identification of driver's smartphone is expected to be developed in the future.

References

1. Ajay Kumar Reddy, G.S., Jagadesh Chandra, S.V., Naresh Kumar Reddy, B.: Developing the fabricated system of automatic vehicle identification using RFID based poultry traceability system. In: International Conference on Information Communication and Embedded Systems, pp. 1–6 (2014)
2. Su, W., Hu, K., Zhang, L., Ma, L.: A RFID based material supply management system in automatic vehicle assembly streamline. In: Information Technology and Computer Science (ITCS 2009), pp. 259–262. IEEE Press (2009)

3. Tashk, A., Helfroush, M., Karimi, V.: An automatic traffic control system based on simultaneous Persian license plate recognition and driver fingerprint identification. In: Telecommunications Forum (TELFOR), pp. 1729–1732 (2012)
4. Hong, T., Qin, H., Sun, Q.: An improved real time eye state identification system in driver drowsiness detection. IEEE In: International Conference on Control and Automation (ICCA2007), Guangzhou, CHINA, May 30–June 1 (2007)
5. Burnham, G.O., Seo, J., Bekey, G.A.: Identification of human driver models in car following. IEEE Trans. Autom. Cont. **19**, 911–915 (1974)
6. Sonnerat, D., Tricot, N., Popieul, J.: Driver's environment identification using automatic classification methods. In: Active Safety Application Intelligent Vehicle Symposium. IEEE (2002)
7. Kaplan, K., Kurtul, C., Akin, H.L.: ADES: automatic driver evaluation system. In: 2012 IEEE International Conference on Vehicular Electronics and Safety (ICVES), pp. 442–447 (2012)
8. Krasner, G., Katz, E.: Automatic parking identification and vehicle guidance with road awareness. In: IEEE International Conference on the Science of Electrical Engineering (ICSEE)
9. Mahmood, Z., Ali, T., Khattak, S., Khan, S.U., Yang, L.T.: Automatic vehicle detection and driver identification framework for secure vehicle parking. In: 13th International Conference on Frontiers of Information Technology (FIT), pp. 6–11 (2015)
10. Martínez, M., Echanobe, J., del Campo, I.: Driver identification and impostor detection based on driving behavior signals. In: 2016 IEEE 19th International Conference on Intelligent Transportation Systems (ITSC), pp. 372–378 (2016)
11. Miyajima, C., Nishiwaki, Y., Ozawa, K., et al.: Driver modeling based on driving behavior and its evaluation in driver identification. Proc. IEEE **95**(2), 427–437 (2007)
12. Park, H., Ahn, D.H., Park, T., Shin, K.G.: Automatic identification of driver's smartphone exploiting common vehicle-riding actions. IEEE Trans. Mob. Comput. 17(2), 265–278 (2017)
13. Park, J.-g., Patel, A., Curtis, D., Teller, S., Ledlie, J.: Online pose classification and walking speed estimation using handheld devices. In: Proceedings of the 2012 ACM Conference on Ubiquitous Computing, pp. 113–122. ACM (2012)
14. Electromagnetic capability: www.iso.org/obp/ui/#iso:std:iso:16750:-2:ed-4:v1:en
15. Madgwick, S.O., Harrison, A.J., Vaidyanathan, R.: Estimation of imu and marg orientation using a gradient descent algorithm. In: 2011 IEEE International Conference on Rehabilitation Robotics (ICORR), pp. 1–7. IEEE (2011)
16. Park, T., Shin, K.G.: Attack-tolerant localization via iterative verification of locations in sensor networks. ACM Trans. Embed. Comput. Syst. (TECS) **8**(1), 2 (2008)
17. Cho, D.-K., Mun, M., Lee, U., Kaiser, W.J., Gerla, M.: Autogait: a mobile platform that accurately estimates the distance walked. In: 2010 IEEE International Conference on Pervasive Computing and Communications (PerCom), pp. 116–124. IEEE (2010)
18. Lester, J., Choudhury, T., Borriello, G.: A practical approach to recognizing physical activities. In: Pervasive Computing, pp. 1–16. Springer, Berlin (2006)
19. Ahn, S., Kim, J.: Magnetic field design for high efficient and low emf wireless power transfer in on-line electric vehicle. In: Proceedings of the 5th European Conference on Antennas and Propagation (EUCAP), pp. 3979–3982. IEEE (2011)
20. U.S. HEV sales by Model (1999–2013): http://www.afdc.energy.gov/data/10301
21. Tarkoma, S., Siekkinen, M., Lagerspetz, E., Xiao, Y.: Smartphone Energy Consumption: Modeling and Optimization. Cambridge University Press, Cambridge (2014)

22. Mostofi, N., Elhabiby, M., El-Sheimy, N.: Indoor localization and mapping using camera and inertial measurement unit (imu). In: Position, Location and Navigation Symposium-PLANS 2014, 2014 IEEE/ION, pp. 1329–1335. IEEE (2014)
23. Shoemake, K.: Animating rotation with quaternion curves. ACM SIGGRAPH Comput. Graph. **19**(3), 245–254 (1985)
24. Kabsch, W.: A solution for the best rotation to relate two sets of vectors. Acta Crystallogr. Sect. A Cryst. Phys. Diff. Theor. Gen. Crystallogr. **32**(5), 922–923 (1976)
25. Bobick, N.: Rotating objects using quaternions. Game Dev. **2**(26), 21–31 (1998)
26. Heise, R., MacDonald, B.A.: Quaternions and motion interpolation: a tutorial. In: New Advances in Computer Graphics, pp. 229–243. Springer, Berlin (1989)
27. Yazdi, A., Lin, D., Heydari, P.: A 1.8 V three-stage 25 GHz 3 dB-BW differential non-uniform downsized distributed amplifier. IEEE ISSCC Tech. Dig. 156–158 (2005)
28. Iyer, B., Pathak, N.P., Ghosh, D.: Dual-input dual-output RF sensor for indoor human occupancy and position monitoring. IEEE Sens. J. **15**(7), 3959–3966 (2015)
29. Iyer, B., Kumar, A., Pathak, N.P., Ghosh, D.: Concurrent multi-band RF system for search and rescue of human life during natural calamities. In: IEEE MTT-S International Microwave and RF Conference, pp. 1–4 (2013)

Animal Repellents from Agricultural Fields

P. Sreevardhan[1], B. Vidheya Raju[1], and Durgesh Nandan[2](✉)

[1] Department of Electronics and Communication Engineering, Aditya Engineering College, Surampalem, Andhra Pradesh, India
sv4484k@gmail.com, vidheyaraju.b@aec.edu.in
[2] Accendere Knowledge Management Services Pvt. Ltd., CL Educate Ltd., New Delhi, India
durgeshnandano51@gmail.com

Abstract. Crop damages inflicted by animals are one of the biggest challenges throughout the world. Animals such as pigs, monkeys, and many others may cause Spartan damage to crops. They can damage the plants by feeding on plant parts or only by organization more than the field and squashing in excess of the crops. Therefore, animals may easily cause significant yield losses and incite additional financial problems. In order to decrease the problems or damages caused by animals to the farmer which destroys the farm, there are many ways. The ways include haunting the animals, producing the sounds manually, and using chemical compounds for repelling birds and animals; some are regulated by state and federal laws while others are untested. So, the actuation of ultrasonic devices can repel hazardous animals by using the principles of bioacoustics. Humans have a hearing range of 20 Hz–20 kHz. Similarly, each and every animal as its individual hearing range like rats can hear from 200 Hz to 90 kHz; snakes can hear from 80 Hz to 1 kHz, now over the whole hearing range only in a certain range, known as a most sensitive hearing section which helps to repel animals. An up-gradation to the preexisting methods, utilizing ultrasonic frequencies has been processed to deter animals. Ultrasonic animal repelled produces different sound frequencies depending on the animal species to repel at a particular time. Hence, the high-frequency and high amount sound waves are used to drive away from the animals, which cause harm to farms. Those ultrasonic sounds can repel animals or rodents (including rats, pigs, and monkeys). For repelling different animals, separate electronic circuits will be designed to produce sound waves of different frequencies. These frequencies are time-multiplexed using Arduino UNO and radiate into the air to repel animals. These signals are to be amplified using an amplification circuit. These amplified signals are given to the speaker to produce different sounds at different time intervals. However, these frequencies do not disturb the hearing ability of persons.

Keywords: Animal repulsive · Agriculture · Webcam · Detector · Sensors · IoT · Ultrasonic sensor · Arduino

© Springer Nature Singapore Pte Ltd. 2021
P. Deshpande et al. (eds.), *Next Generation Information Processing System*,
Advances in Intelligent Systems and Computing 1162,
https://doi.org/10.1007/978-981-15-4851-2_12

1 Introduction

Andhra Pradesh is called as a rice bowl of India. In the early days, the farmers are more in number and they would produce food grains. But, in the present scenario, we should see that farmers are decreasing in number day by day. So, in order to increase that number, we identify the absence in that scenario. In this include mainly animals are done their job. Animals destroy greenfield crops. Hence, in order to catch the animals, we develop a new method. In the zoo, some of the animals are missing; in order to catch that with the help of RFID, GPS, and sensors, we track the animals [1]. The development of an IoT-based circuit would identify the presence of animals in fields and send a notification to the user [2]. The mosquito can bite humans, in order to protect from mosquitoes and develop an ultrasonic device. This device will produce ultrasonic frequencies when mosquitoes are present. The frequency range is 20 kHz within this range, and the mosquitoes are fallen [3]. For that, the appliances are a camera and a sound frequency emitting device. We place the camera at which the animals present, and we observe that. Whenever the animals come into the picture, the camera which you placed earlier can take a picture and noted it. After that has happened, the sound frequency emitting device emits the sound in the form of frequencies. And with that, the sound heard by the animals can repel with that sound. For that, the animals go away from that specific field. Different types of animals exist, the camera takes a picture of different animals, and immediately, the related sound frequency emitting device will emit the frequencies. In this, the greenfield will completely free from animals [4].

2 Literature Review

The new elevated spatial firmness PET identifier for miniature animal as well as breast imaging has been intended [5]. Sulfur mustard poison produces a characteristic toxicological effect showing a rapid decrease in white blood cells' count [6]. PC models have been hurried to assess and improve a high-goals square locator structure for a minimal effort little creature PET framework. The principle in addition to utilize GSO is rather than BGO [7]. The motivation behind the work portrayed here was to complete the structure and advancement of a high-goal volumetric PET indicator module dependent on rearrangements of our single-plane structure [8]. Our gadget comprises of a little program capable torque-engine framework that is associated with a manipulandum. Rodents are prepared to get a handle on this manipulandum and move it to at least one focuses against modified power field cautions [9]. Tracing of animals by using GPS, RFID, and Sensors: It is based on animal movements and the situation they present, and temperature is based on the sensors which are placed in zoological gardens, and they observe animal movements [10]. Tracking of animals in a digital zoo by using the smart camera: We placed a camera around the zoo, and we observe the movements. The camera captured some pictures of the animals. Based on that picture, the animals are tracked by the higher administrators who are present in the zoo [11]. Animal voice acknowledgment identification system: The animal which developed their speech identification method uses the zero-cross rate (ZCR), Mel-frequency cepstral coefficients (MFCC), and dynamic time deformation (DTW) combined methodology because of the tools for recognizing the voice of the individual animal. ZCR is employed for the tip purpose discovery of input voice specified the suppressed voice which is removed [12]. Behavioral experiments of times need continuous nursing of the responses that animals build in several things and in response to many stimuli. Though techniques are developed for specific displays to program the analysis of behaviors, together with some industrial solutions, several do not seem to be forever adequate for the actual state of affairs or sufficiently elastic to live the specified movement. We got developed and enforced a comparatively affordable system for video and perpetually extracting knowledge characterizing animal's kinetic responses [13], night vision creature discovery. The exhibited framework identifies the creatures up to 200 m from the vehicle while making not many false admonitions. For creatures that are viewed as a potential peril, progressed HMIs, for example, stamping lights which effectively lights the creatures are connected, giving the driver the fast and right data the person requires. The Autoliv night vision creature discovery framework is corresponding to presently utilized techniques for avoiding mishaps with creatures [14]. Distinguishing proof of monkey in horticultural fields is by utilizing remote sensor arranges. The arranged identification framework includes remote sensor organize which is tied by an ultrasonic echo producer to zone inedible monkeys.

The sensor hub detects the development of the monkey into the land and advises the sink hub by communicating caution tones. The new high spatial goals PET sensors for the little creature and bosom imaging have been planned [5]. Sulfur mustard toxic substance delivers a trademark toxicological impact demonstrating a quick decline in white platelets check [6]. PC reproductions have been raced to gauge and elevate a high-goals square identifier structure for a minimal effort little creature PET framework. The principle is in favorable position of utilizing GSO rather than BGO [7].

The reason for the effort expressed at this time was to complete the structure and improvement of a high-goals volumetric PET identifier component dependent on the disentanglement of our single-plane structure [8]. Our gadget comprises of a little program capable of torque-engine framework that is connected to a manipulandum. Rodents are prepared to get a handle on this manipulandum and shift it to at least solitary focuses alongside modified power ground alerts [9].

Tracking of animals by using GPS, RFID, and Sensors. This is based on animal movements and the situation they present. Temperature is based on the sensors which are placed in zoological gardens, and they observe animal movements [1]. Tracking of animals in a digital zoo by using the smart camera: We placed a camera around the zoo, and we observe the movements. The camera captured some pictures of the animals. Based on that images the animals are tracked by the higher officials which are present in the zoo [11]. Animal voice recognition identification system.

The settled creature voice acknowledgment framework utilizes the ZCR, MFCC, and DTW joint calculations as the instruments for knowing the voice of the specific creature. ZCR is utilized for the endpoint location of information influence to such an extent that the hushed voice can be disconnected [12]. Social tests regularly require relentless observing of the reactions that creatures make in various circumstances and in light of a few improvements. In spite of the fact that systems have been progressed for explicit solicitations to robotize the investigation of practices, including some business arrangements, many are not constantly satisfactory for the specific circumstance or adequately adaptable to quantify the needed development [15].

With the advancement of IoT-based smart security and monitoring, devices for agriculture gadgets are introduced. This gadget could be observed and controlled from a secluded area, and it very well may be connected in horticultural grain stores, field and cold stores for security reason [2]. The brilliant ultrasonic creepy crawly repulsing framework is driven by reaping vitality from sunlight-based boards. The ultrasonic waves have frequency in surplus of 20 kHz that is tranquil to people yet when creepy crawlies' approach in the association with ultrasonic waves, they intelligence rays from exceptional sensilla or hair present on the tests of mosquitoes which will prepare weight on the sensory method and stuff their very own ultrasonic reappearance and power them to leave that territory. The recurrence scopes of 38–44 kHz can be utilized viably to force away the mosquitoes and flies [3].

Time High (seconds) C_1 Time Low (seconds) $T_1 = 0.693 * R_2 * C_1$ 35 $T_2 = 0.693 * (R_1 + R_2)$ Time Period $T =$ Time High + Time Low $= 0.693 * (R_1 + 2 * R_2) * C_1$.

3 Existing Methodology for Animal Repellent

Jason et al. [16] natural life is regularly in charge of making boundless harm individual property, human well-being and security concerns, and other bothering issues in light of their encouraging, perching, and rearing propensities. Startling gadgets are apparatuses utilized in joined untamed life harm association to lessen the effects of creatures, yet the viability of such gadgets is frequently factor. A creature's visual and sound abilities result in how the creature will react to a boost.

Frightening devices contain lights, lasers, bioacoustics, and ultrasonic devices. By using frightening devices, total elimination of damage may not be possible, but it can reduce the damage done by the animals. Kumar et al. [17], the most common method used for control of pest is the use of pesticides. A chemical substance used for killing, avoiding, repelling, or mitigating pests is called pesticides. But it is important to know the effectiveness and toxicity of these chemicals. Moreover, these chemicals are highly toxic and act as a slow poison and harsh both to humans and wildlife. Due to the harmful belongings of pesticides to wildlife as well by way of humans, there is a request for pesticide replacements. An alternative method of pesticides includes electronic sound-producing devices. The audible frequency range for humans is 20 Hz–20 kHz, and sound above or below, this frequency is not audible to humans. Pests can detect and hear above 20 kHz frequency sound as well as feel irritated and unpleased which force them to leave and stay away from the certain boundary. The device aims to repel small birds, mosquitoes, etc.

Yahot et al. [18] plants are identified with bugs, where there are plants and there must be bugs. Nuisances are creatures that harmed the plants. One example of bugs that eat plants and harm is flying creatures. An elective that can be utilized to repulse winged creatures is by utilizing waves. Winged creatures are the creatures that are significant to ultrasonic waves. Flying creature hearing is around 29 kHz. At the point when an ultrasonic wave is radiated, then the creature will remember it as a risk.

To produce these, frequencies project an LC oscillator circuit. The amount of oscillations depends on the values of the L & C components we use. Based on the body movement, PIR sensor will detect the birds, and automatically electrical circuit will switch ON with the help of relay and starts producing ultrasonic sounds to repel birds.

Saha et al. [19] control of mosquitoes is something of most noteworthy significance in present-day with a developing number of mosquito bore infection. Yash Soni [20], malaria influences in excess of 250 million individuals and causes in excess of a million deaths every year. One significant control plan next to this and other mosquito permitted maladies are mosquito control, which expects to diminish human mosquito contact so as to defeat this issue; the electronic gadget must be intended to deliver ultrasonic sounds with the assistance of 555 clock in a stable method of activity. [21]. On the globe, the financial system of the various nations is vulnerable upon agriculture business. Disregarding monetary advancement, farming is the quality of the economy. It gives the whole nationwide production. Agribusiness meets the sustenance necessities of the people and delivers numerous simple equipments for ventures. Yields are solely being crushed by creatures. There will be a lot of misfortune to ranchers.

The methods behind to control the creatures utilizing IoT, in the widget, fundamental sensors, and electronic gadget are utilizing for this purpose. This sensor enacts the electronic gadgets, and Raspberry Pi is utilized as a server to inspect information and transmit data to the client. Parts utilized are (1) PIR sensor, (2) Ultrasonic sound repeller platform, (3) web camera, (4) ultrasonic ranging device, (5) Raspberry Pi-2 Model B+ and Language Used: (1) Linux-based Raspbian OS (2) Python (3) PTC's fixation Worx's IoT phase for M2M services The sensors and camera are associated with the GPIO header. The ultrasonic sound-based rat repeller is enacted by the server and Raspberry Pi B + GPIO header interfaces with I/O gadgets. Notwithstanding the circuit, web camera is associated with general sequential transport port of Raspberry Pi that will be open by means of IP address of server over the system. In this, python language is utilized to identify the movement of rodents utilizing a warmth sensor which gives unmistakable qualities. Considering these discrete qualities as a banner sign, URD sensor was initiated to compute the separation of rat and all the while webcam soul is actuated to catch a snap of region. Ultrasonic extending gadget and web camera are needy upon the qualities produced by the PIR sensor [2].

By utilizing sun-powered vitality, we repulse the mosquitoes; it very well may be comprehensively characterized into two classes, for example to gather the sun-powered vitality and store the collected vitality in the battery-powered battery during daytime and to drive the mosquito repeller from reaped sun-oriented vitality during the evening and morning time. The sunlight-based board gives the greatest yield voltage, and the current rating of the sun-oriented board is +18 V and 0.5 A, individually. The sun-powered board utilized in this framework charges 12 V battery-powered lead-corrosive battery of 4.5 mAh current rating. The gathered sun-powered vitality is expertly dealt with and put away in +12 V battery-powered battery by methods for a sun-powered charging circuit comprising of an LM7812 voltage controller. The yield voltage from the battery is provided to the LM7805 voltage controller which creates the directed +5 V supply to control and work the Arduino UNO. At the point when repeller is turned on by hand-off and Arduino, it produces ultrasonic sound waves which can be proficiently used to repulse the mosquitoes from the indicated zone [3].

The sticky power field opposes the development directly regarding speed. Subjects pulled the manipulandum handle past an objective position so as to get a reward. Haphazardly a ‘get preliminary’ (CT) was directed in which the subject was intentionally amazed by the unexpected expulsion of the power field. Past work has demonstrated that CT can be exceptionally helpful in deciding the control methodology utilized by human subjects. The utilization of such a power field will be helpful in models matched with neural accounts letting one to decide neural relates of engine blunder and resulting in modifying on a preliminary to preliminary premise [9].

Four modules are the joint to one locator exhibit. Every module in one locator exhibit can have eccentricity with any module in another finder cluster. The incident occasions in various module sets can be prepared simultaneously. It represents the arrangement of one finder exhibit. The strategy for joining four PS-PMTs has been arranged. With the trade-off capability of more accident impacts, the circuits rearrange the electronic handling signals. One idea in our circuit is to change the charge delicate preamplifiers. The release resistor RF for the most part is extremely enormous so the falling edge of A0 yield is delayed with timing consistent around 1 μs. We will attempt other release plans, for example, applying a simple switch and utilizing an extra non-rearranging low-gain speaker in future investigations [5].

We track the creatures by utilizing RFID, GPS, and sensors. Creature, the executives’ module, spares information of GPS, RFID, sensor hubs to the database and gives creatures’ pictures/video information. The module demonstrates the creatures’ data on a guide with their area by utilizing RFID data and demonstrates enclosure’s area data on the guide by utilizing the equation of sorting out development. The creature, the executives’ module, can give creatures’ current site to following creatures when the creatures’ leakage their confines. It demonstrates the creature following on the guide. It can way the creature area [1].

This working project starts with the mapping of the agricultural land by using the ultrasonic sensor. It produces sound waves at activated conditions. It is microcontroller-based device which works at a high frequency for humans to hear. The microcontroller-based device has two main parts—a transducer that creates an ultrasonic sound wave while the other part listens to its resonance (Fig. 1).

The ultrasonic sensor scans the farm with the help of a servo motor. The servo motor provides 180° rotation to the ultrasonic sensor in clockwise and anticlockwise directions. While scanning the farm in clockwise and anticlockwise directions four times, it takes the average value and stores that value in the microcontroller (calibration process). After the calibration process, ultrasonic sensor starts scanning for animal detection. During the scanning process, microcontroller compares the present scanned values and stored values. If the values are different, then the controller confirms the presence of the animal on the farm. The animal is detected only if the present scanned value is less than the stored value. When the controller confirms the animal presence, then it will display the message in LCD display, i.e., the animal is detected, and it enables the enable circuits. Based on the ON and OFF conditions of the enable circuits, deterrent circuits will be enabled on the time bases using time-division multiplexing which is provided by the microcontroller. The controller enables the first circuit for two seconds, the second deterrent circuit for four seconds, and a third deterrent circuit for six seconds to repel three different animals simultaneously. Rat repellent circuit

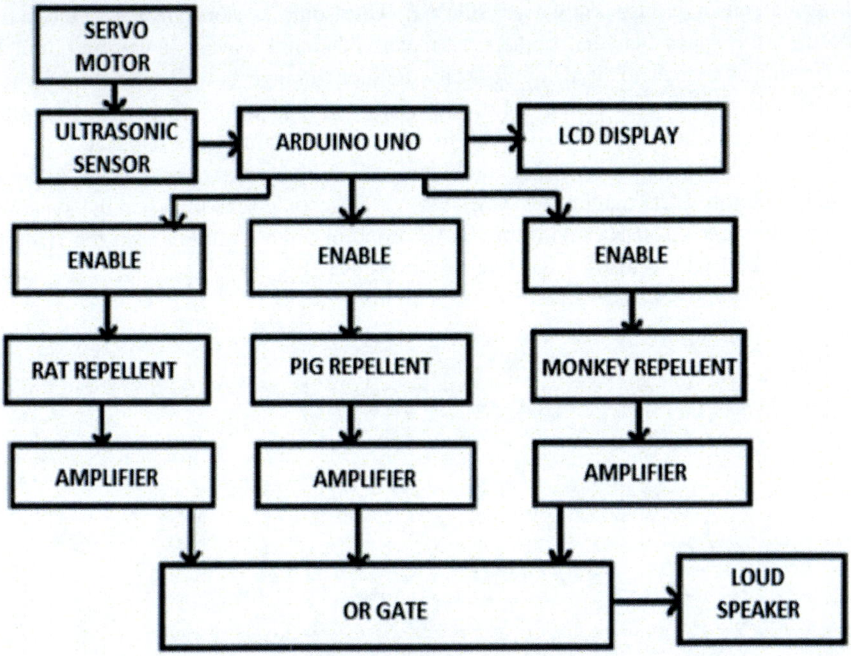

Fig. 1. Block diagram of Arduino-based animal's repellent

produces 50 kHz ultrasonic sound for repelling rats, monkey repellent circuit produces 32 kHz ultrasonic sounds for repelling monkeys, and pig repellent circuit produces 18 kHz ultrasonic sounds to repel pigs. These sounds are produced using 555 timer outputs connected to a buzzer. An LCD display is used to display the state of the project during the operation. It displays the content 'animal is detected' when the sensor detects the animal and content 'scanning the farm' when the sensor is scanning the farm. The microcontroller enables the deterrent circuits until it senses the absence of the animal on the farm. The absence of the animal is confirmed by the controller based on the mechanism followed in the sensing process. The power supply is given to all components with the help of the power supply circuit by converting AC-to-DC containing center-tapped transformer, diodes, and voltage regulator to provide 9 V to Arduino Uno, and the regulator in the Arduino Uno reduces to 5 V which is sufficient to enable the circuits and the ATmega328P microcontroller.

4 Result

Whenever the target (animal) is detected, then Arduino enables each repellent circuit for allotted time using time-division multiplexing. Each repellent circuit produces a unique frequency to repel animals. The rat repellent produces 50 kHz, monkey repellent produces 32 kHz, and pig repellent produces 16 kHz, respectively. These oscillations are converted into sound waves using the buzzer to repel animals. The ultrasonic sensor is also known as SONAR. Ultrasonic sensors are the devices that measured the space between objects by means of sound waves. It workings out by transferring away a echo signal at ultrasonic frequency range as well as waiting for it to spring back since the objective. After that some moment delays in between communication of echo and in receipt of echo that is used to evaluate the distance.

Figure 2 shown above is the whole project design which consists of a power supply, Arduino, LCD display, servo motor, and ultrasonic sensor. When the system is powered by 230 V, 50 Hz and dumping the code into the Arduino, then the Arduino initializes the LCD to display the title of the project (Fig. 3).

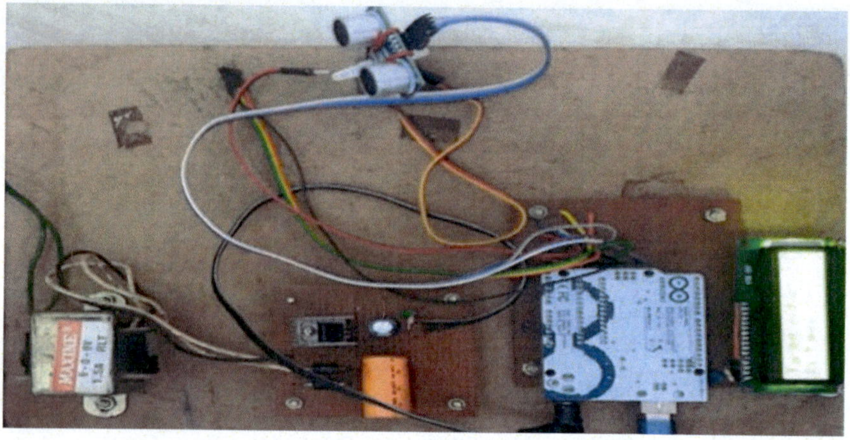

Fig. 2. Snapshot of the project design

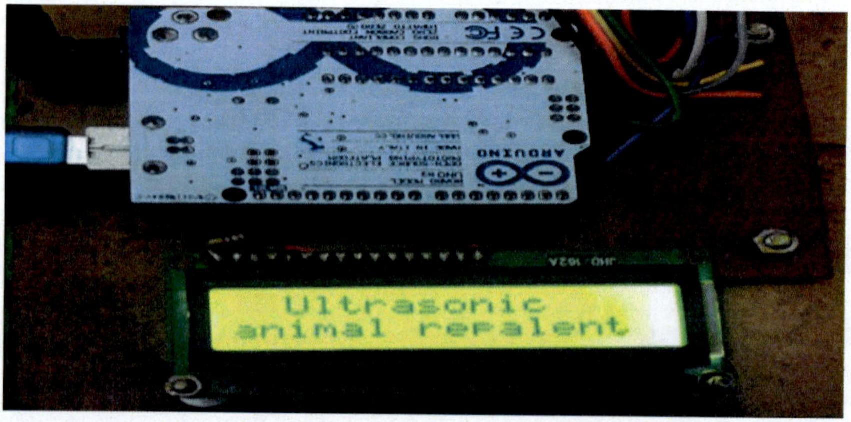

Fig. 3. Snapshot of the LCD display when it is initialized

Figure 4 gives the status of the ultrasonic sensor when it is scanning the agricultural land. This initial scanning is done to store the echo signals which are further used for the detection of animal presence through comparison.

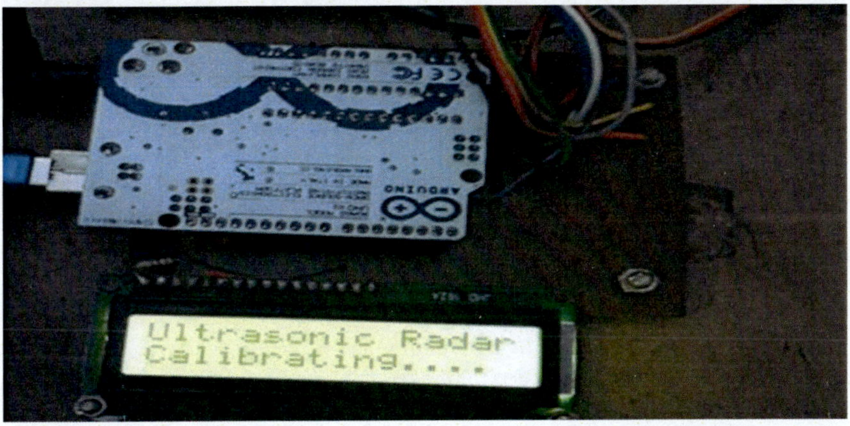

Fig. 4. Snapshot of LCD when the ultrasonic sensor scanning the farm

Figure 5 gives the status of the ultrasonic sensor when the calibration is completed. If the ultrasonic sensor completes its farm scanning, then it intimate to the Arduino which makes us knows about the calibration through LCD display (Fig. 6).

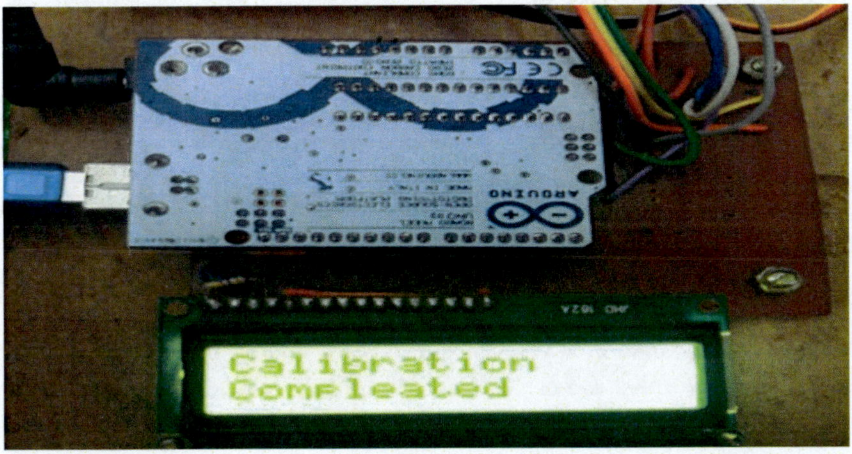

Fig. 5. Snapshot of the LCD display after completion of farm scanning

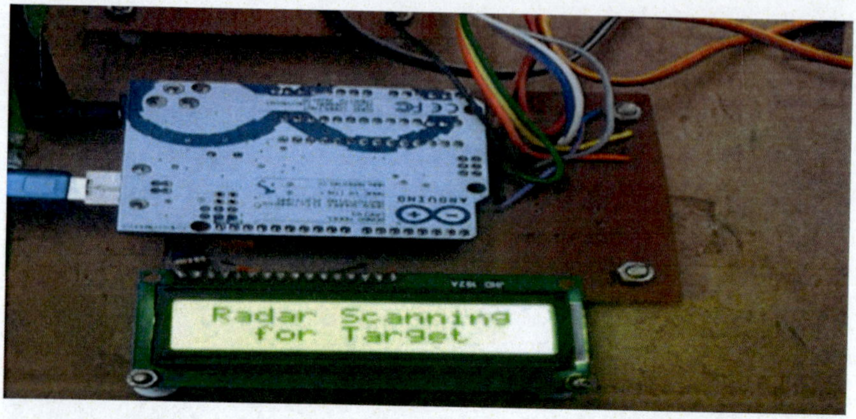

Fig. 6. Snapshot of the LCD when ultrasonic sensor is scanning for target

Whenever the calibration of the agriculture land is completed the Arduino initiates the ultrasonic sensor to scan the farm for the target (animal) detection by comparing the values obtained by the calibration.

Figure 7 gives the status of the agriculture land whether the animal is present or not. If the animal is detected, then Arduino makes us aware by displaying TARGET DETECTED, and it also displays the distance between the ultrasonic sensor and the target (Fig. 8).

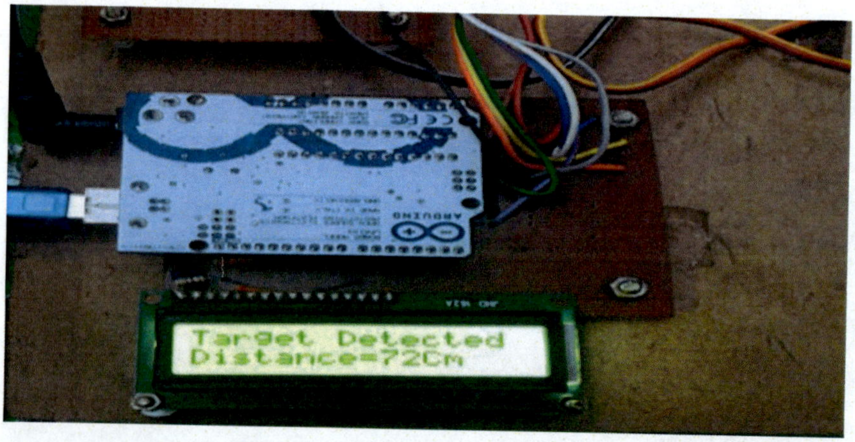

Fig. 7. Snapshot of the LCD display when the target is detected

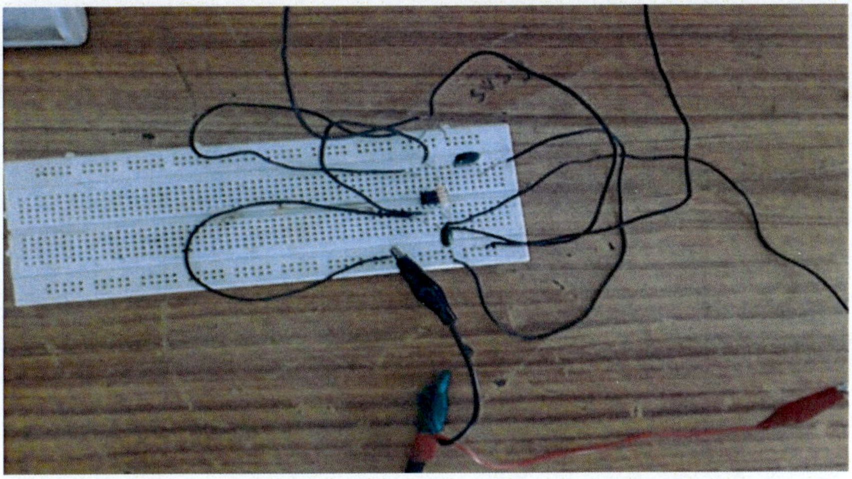

Fig. 8. Snapshot of deterrent circuits when an animal is detected

5 Applications

In case of electronics system, electronic control switches are the electronics components or devices that are capable of switch of electrical circuit, disturbing the existing, or distracting it beginning one composer to another. Classically, electronic switches used solid-state device for transistors, although space tube can be used as well in elevated electrical energy application. In this project, a transistor is used as a switch. It is used in repellent circuits; for this, we can easy to repellent the respective animals. By using Arduino, we resistor the circuits based upon our requirements.

6 Conclusion

Generally, the crops are damaged by animals which create huge loss to farmers. So, we designed a system which helps the farmers to protect the crops from the animals. By this proposed system, farmer can detect animals (rats, monkeys, pigs) with the help of servo motor which is controlled by the ultrasonic sensor. We have designed this system to repel three animals (rats, monkeys, pigs). Whenever the animal is identified by the ultrasonic device, it sends the information toward Arduino. Arduino enables the deterrent circuits (oscillator circuits made by 555 timers) which produce 50 kHz for rats, 32 kHz for monkeys, and 16 kHz for pigs through a basic enable circuit on the time-division mechanism. The frequencies generated by these repellent circuits are directed to a buzzer for sound waves. These sound waves are radiated into the air up to 100 cm to repel the animals.

References

1. Kim, S.-H., Kim, D.-H., Park H.-D.: Animal situation tracking service using RFID, GPS and sensors. In: Second International Conference on Computer and Network Technology, pp. 153–156 (2010)
2. Baranwal, T., Pateriya, P.K.: Development of IoT based smart security and monitoring devices for agriculture. In Proceedings of the 6th International Conference Cloud System and Big Data Engineering (Confluence), Noida, India, 14–15 Jan 2016, pp. 597–602 (2016)
3. Saini, S.S., Bansal, D., Brar, G.S., Sidhu, E.: Solar energy driven Arduino based smart mosquito repeller system. In: This Full-Text Paper was Peer-Reviewed and Accepted to be Presented at the IEEE Wisp NET 2016 Conference, pp. 1239–1243 (2016)
4. Shetgaonkar, A.A., Shet, V.N.: Smart animal repeller. In: International Conference on Energy, Communication, Data Analytics and Soft Computing (ICECDS-2017), pp. 3303–3308 (2017)
5. Zhang, N., Thompson, C.J., Togane, D., Cayouette, F., Nguyen, K.Q., Camborde, M.-L.: Design of front-end electronic circuits for dedicated PET detectors. In: IEEE Nuclear and Science Symposium Conference Records (2002)
6. Milosavljevic, Z., Cvetkovic, M., Dekleva, N., Radojkovic, M., Majic, V.: Interaction of magnetic field and sulphur-mustard in animals. In: Annual International conference of the IEEE Engineering in Medicine and Biology Society, vol. 12, no. 4 (1990)
7. Miyaoka, R.S., Lewellen, T.K.: Design analysis of a high-resolution detector block for a low-cost small animal positron emission imaging system, pp. 1370–1374 (1995)
8. Correia, J.A., Burnham, C.A., Kaufman, D., Brownell, A., Fischman, A.J.: Quasi-continuous detector module design for high-resolution small animal PET. IEEE Trans. Nucl. Sci. (2005)
9. Francis, J.T., Chapin, J.K.: Force field apparatus for investigating movement control in small animals. IEEE Trans. BME **51**(6), 963–964 (2004)
10. Karlsson, J., Ren, K., Li, H.: Tracking and identification of animals for a digital zoo. In: 2010 IEEE/ACM International Conference on Green Computing and Communications & 2010 IEEE/ACM International Conference on Cyber, Physical and Social Computing, pp. 510–515 (2010)
11. Yeu, C.Y.: Animal voice recognition for identification (ID) detection system. In: 2011 IEEE 7th International Colloquium on Signal Processing and its Applications
12. Kutrowski, T.M., Meydan, T., Barnes, J., Aldoumani, N., Erichsen, J.T.: Instrumentation for monitoring animal movements. In: Proceedings IEEE SENSORS, pp. 1295–1299 (2014)
13. Forslund, D., Bjarkefur, J.: Night vision animal detection. In: Proceedings of IEEE Intelligent Vehicles Symposium, pp. 737–742 (2014)
14. Radha, R., Kathiravan, K., Vineeth, V., Sanjay, v., Venkatesh, S.: Prevention of Monkey trespassing in agricultural field using application agricultural specific flooding approach in wireless sensor network. In: 2015 IEEE International Conference on Technological Innovations in ICT for Agriculture and Rural Development (TIAR 2015), pp. 106–111 (2015)
15. Stewart J.L.: Experiments with sounds in repelling mammals. In: Proceedings of the 6th Vertebrate Pest Conference, pp. 47 (1974)
16. Tiwari, DK., Alam, MA.: Electronic pest repellent: a review. https://doi.org/10.13140/rg.2.2.13557.78569 (2016)
17. Siahaan, Y., Wardijono, B.A., Mukhlis, Y.: Design of birds detector and repellent using frequency based Arduino UNO with android system. In: 2017 2nd International Conferences on Information Technology, Information Systems and Electrical Engineering (ICITISEE) (2017)

18. Saha, S.R., Ghosh, D., Mandal, K.K.: Mosquito repellent circuit. Int. J. Eng. Tec. Res. (IJETR)
19. Yash, S.: Electronic mosquito repellent. (2015)
20. https://en.m.wikipedia.org/wiki/Bioacoustics
21. https://wikivisually.com/wiki/Bioacoustics

Analysis on High-Performance Full Adders

K. V. S. S. S. S. Kavya[1], Bujjibabu Penumuchi[1](✉), and Durgesh Nandan[2]

[1] Department of ECE, Aditya Engineering College, Surampalem, East Godavari, India

[2] Accendere Knowledge Management Services Pvt. Ltd., CL Educate Ltd., New Delhi, India

{kavyakattamuri98, durgeshnandano51}@gmail.com

bujjibabu_penumuchi@aec.edu.in.com

Abstract. This paper contains the performance analysis of various available designs of full adders. It is observed that the full adder is designed for 1 bit, and later it is extended for 32 bits also. The circuit is designed by using 180 nm technology at 1.8 V supply and technology using 90 nm at 1.2 V supply using Cadence Virtuoso tools. High speed, low consumption of power, better power–delay product (PDP), layout area, better propagation delay, these are the performance parameters that are compared for various full adders. The circuit performs better in case of improvement of the full adder circuit in terms of parameters like speed and power.

Keywords: Full adder · Low power · High speed · Power–delay product · Propagation delay

1 Introduction

There is a rapid increase in the use of mobiles, PC's, integrated designs and many more. These devices are most efficient in performance parameters of their low-power consumption, high speed, delay and layout design. The full snake is one of the most significant circuits which have been a watchword in the circuits. The full snake is one of the legitimate circuits that play out the expansion activity on three one-piece paired numbers. The full snake produces a sum of the three data sources and furthermore conveys esteem. The full adder can be joined with other full adders or it can chip away at its own. The diverse rationale styles that perform better are CMOS, reciprocal pass-transistor logic (CPL) and furthermore transmission entryway full snake. These are the significant circuits. This assists in the improvement of highlights of full adders. The favourable circumstances are strength against voltage scaling and transistor measuring while high info capacitance and cradle necessities are the detriments. At the point

© Springer Nature Singapore Pte Ltd. 2021
P. Deshpande et al. (eds.), *Next Generation Information Processing System*, Advances in Intelligent Systems and Computing 1162,
https://doi.org/10.1007/978-981-15-4851-2_13

when analysed regarding speed, region, control dissemination, control postpone items, the CMOS is better than CPL [3]. The general execution examination of the framework is dictated by a snake in thinking about the basic way. Microelectronics innovation is progressed than battery innovation. Creators face a great deal of issues in structuring circuits that have rapid, low-power utilization [14]. In VLSI frameworks, the rationale circuits like microchips and computerized sign preparing (DSP) frameworks, the crucial math tasks are expansion, augmentation and duplicate gathering. These depend on adder cells. In a large portion of the cases, low voltage and low utilization of intensity are corresponding to the square of stock voltage [16].

Full adder is important for all the building blocks. To increase battery life, it is important to design a low-power consumption circuit. But some of the adder cells cannot operate efficiently at the low-power supply [18]. The inverter is compared as a nucleus of all digital designs. An inverter circuit produces an output voltage representing the opposite logic level to its input logic level. Its main function is to invert the input signal applied. We know that, by using two complementary transistors in a CMOS configuration, we can construct inverters. When its operation is well known, then it is easy to design multiplexers, microprocessors and NAND gates. And by the clear analysis of the inverter, we can design complex circuits of logic gates like NAND, NOR, XNOR. For the extension of battery life for electronic devices, it is always required the reduction of energy used for arithmetic operation [17].

By utilizing CMOS rationale, we execute a full snake rationale circuit and more than one rationale style is utilized in the crossover CMOS rationale structure. Regularly utilized CMOS rationale style utilizes the PMOS draw up transistor where the circuit is experienced commotion and low voltage swings. By utilizing an increasingly number of PMOS and NMOS circuits, the circuit progresses towards becoming bulky [4]. A 1-piece full adder circuit situated in a bootstrapped pass-transistor rationale performs lower control utilization and improved speed [5].

In this paper, a clear view of full adders is discussed. The review on full adders is described in Sect. 2. Various designs of full adders are explored in Sect. 3. In Sect. 4, the result is concluded. Section 5 explains the applications of full adders. Finally, conclusions have been done in Sect. 6.

2 Literature Work

The 1-bit full adders are designed to develop circuits having different designs. Every circuit exhibits different functions of parameters like driving capability, speed, power consumption, area.

These full adder cells are created to present circuit to configuration full snake cell that fulfils their particular application [14]. Wey et al. The full snake can give voltage swing at a low supply voltage and offers execution in power and speed than an ordinary full adder.

A new voltage design consumes low power [16]. Zhang et al. [18], the reduction in layout complexity is done by increasing transistor number in complementary CMOS. This result in very high power-efficient circuits.
Khatibzadeh et al. The adder circuit present in this procedure has lower control utilization and higher speed at low supply voltage. The phases in the circuit won't endure the debasement of its driving capabilities [9]. Goel et al. These plans depend on novel XOR/XNOR circuits where XOR and XNOR full swing yield and beat 39% improvement in PDP.
The circuits present here are most energy-efficient and perform several standard full adder designs [4]. Hernandez et al. [5], the full adder design with bootstrapped pass-transistor logic performs better in power consumption delay. Aranda et al. [2], these circuits are compared in terms of their speed, power consumption and power–delay products. It also compares for different supply voltages. Zandkarimi et al. [17], compared to other conventional cells by using Hspice simulations show the improvement in power consumption and PDP. Bhattacharyya et al. The designs so developed are with good performance of speed, area, power dissipation, power–delay and many other.
The productive coupling of solid transmission doors which are driven by the powerless CMOS inverters prompts quick exchanging of speed [3]. Katragadda et al. [8], 14 transistors are utilized to get the low-power solid full snake circuits. Thus, ascertaining the power defer result of the circuit (PDP), we measure control dissemination and execution of the system. Niranjan et al. [13], employing velocity and power utilization, control investigation of seven full adder cells announced are having low power–delay product (PDP). The half breed full adders are utilized for a structure to show better consequences of comparison. Jie et al. For structure square of 4 bits, a 1-piece half breed full snake utilizing 13 transistors with new SUM circuit and second is the new SUM circuit planned to utilize just four transistors to create the HFA sum [11]. Mewada et al. [12], the full snake is planned so that it accomplishes fast, however lower control scattering with an objective of low PDP. It has low spread deferral and PDP.

3 Existing Methodologies

The full adder generally has 3 squares. Total sign (SUM) is created by using2 XNOR modules, and module 3 is utilized to produce yield convey signal. Each module is planned exclusively so that the full snake circuit is utilized to be advanced as far as power, deferral and zone. Here, the XNOR module is utilized in the structure for power utilization. This module is designed so that the power utilization is limited in a most extreme degree. The XOR/XNOR module with 4 transistors will have an expense of low rationale. The XOR/XNOR with 6 transistors gives preferred rationale swing over 4T XOR/XNOR. The changed XNOR gives low power and fast when contrasted and 6T XOR/XNOR. By utilizing transistors yield convey engendering way is diminished by passing information help signal through a solitary transmission door. Decrease in the proliferation postponement of the transporter sign is finished by the use of

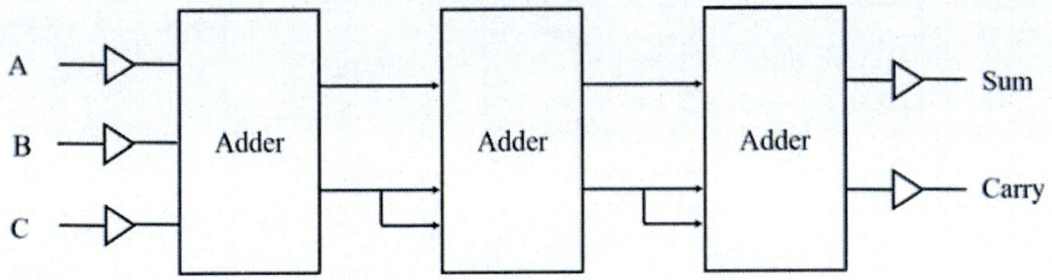

Fig. 1. Simulation test of full adder

solid transmission entryways [3] (Fig. 1). The 1 and 2 modules are connected in cascaded form, and module 3 is a multiplexer which is designed by using pass transistors. The modules 1 and 2 are XNOR gates which are designed for power usage reduction by using weak inverters. A weak inverter has small channel width which comprises of 2 transistors. The yield of the feeble transistor is utilized to create controlled inverter. The level resonator structures full swing of the yield signal. The deferral of a full adder is finished by a bearer signal. The pass transistors are known for low-power circuits. The delay of the carrier signal is reduced by reducing the carry propagation path. The controlled inverter output is XNOR of 1 and 2 modules which are degraded. To overcome this, we use pass transistors. By observing the working of the circuit, it is noticed that when the low supply voltage is given, then a single bit adder may not work properly. This may happen because of the stacking capacitance of full adder cell or inappropriate contribution to the determined cells. In the info organize and a yield arrange, three and two cradles are incorporated individually to dodge signal debasement. Legitimate stacking impact is guaranteed by yield cushions and info supports join input capacitances. After the third stage, the deferral of conveying sign expands [10]. The full adder circuit includes bits of two numbers which are in their double structure and furthermore a convey which is created in the past stage. Module 1 is XOR module, and modules 2 and 3 are utilized to deliver the entirety and convey signals. The XOR door checks the equality between given sources of info. The modules 2 and 3 have transmission doors go about as multiplexers. By comparing the full adder design with previous circuit designs, it exhibits improved PDP [1].

The full adder comprises of 2 modules. The module 1 executes three-input XOR work which is acknowledged by falling two-input XOR doors. The yield inverter is presently set at the info terminal. By the substitution of inverter, control utilization and speed are improved. The circuit is basic in structure, and the power utilization isn't little enough. For the disentanglement of the chip format, the PMOS is equivalent to NMOS. The circuit utilizes 24 transistors that were 12 transistors for some module which is in pass-transistor rationale style and 12 transistors for doing module which is in mirror type static CMOS rationale [15] (Fig. 2). The full adder has three modules. It has 3T-XNOR module with total and conveys age module. Module 1 is XNOR module which is trailed by two different modules. The whole sign is produced by module 1 and module 2.

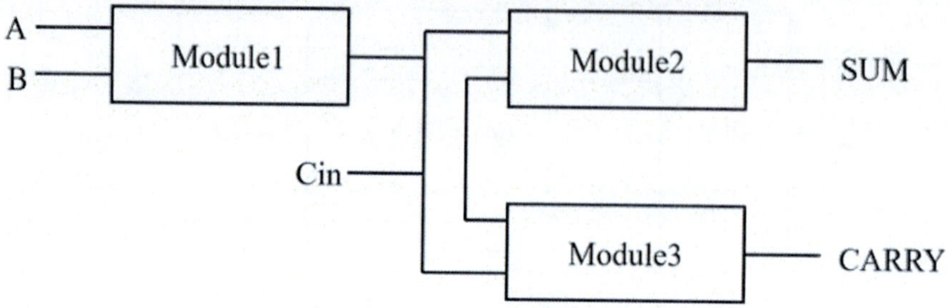

Fig. 2. Schematic structure of full adder

Module 3 creates a convey signal. The circuit is so improved as far as the format region, delay, control utilization. At the point when contrasted with other XNOR circuits, the transistors utilized which are decreased, and furthermore, the general postponement of the circuit is diminished. To lessen the general, convey proliferation way and convey spread deferral, the XOR info is given to convey age which is actualized by utilizing inverters. The subjective decrease in the spread postponement of the transporter sign is finished by utilizing solid transmission entryways. Entirety age module contains a couple of transistors NOMS and PMOS [7] (Fig. 3). The full adder comprises of three multiplexers with inverters. It has cross-breed topology, which is a mix of static CMOS and swing restoring pass-transistor logic. The decrease in the quantity of transistors, control utilization and postponement is the fundamental targets in planning full snake. The contributions for the full snake are given in a request to lessen the extent of transistors. Switch restoring pass-transistor logic (SRPL) understands the multiplexers. In this way, 20 transistors are required to understand the proposed half and half 1-piece full adder. Quality of the rationale isn't considered by and large in view of weak low and weak high flag. At each phase of the plan, it is taken into consideration in this circuit. The whole and convey are delivered by utilizing two multiplexers [6].

4 Result

In this paper, we have discussed designs of some full adders. The full adders are chosen based on performance parameters of circuits with lower power consumption, delay, layout area, power–delay product (PDP), high speed. The full adders are extended from 1 bit to 32 bits based on their circuit design. The full adders with low-power consumption and working at high speed are the most considerable adders (Table 1). Based on the graph, we can understand that the variation in power supply is less when compared to 180, 90, 45 nm and CNTFET technologies. And the variation in delay parameter is more in all the technologies. The technology which helps in developing full adders with less delay and consumes less power is used (Fig. 4; Table 2). From Fig. 5, we can understand that the power supply is the same for almost all types of full adders. The delay

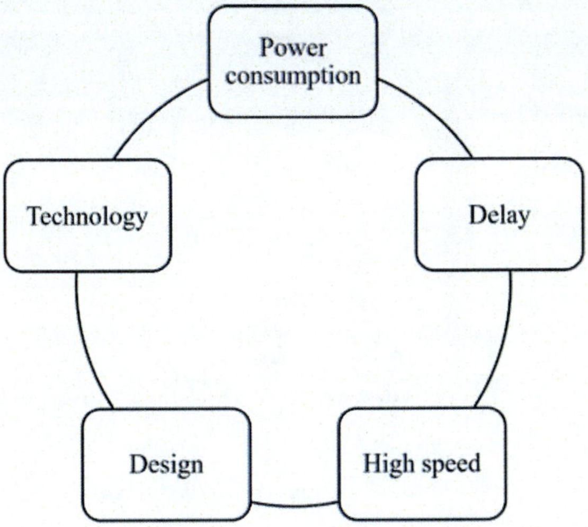

Fig. 3. Different performance parameters of full adders

is varying for every full adder, and it is different in each case. The average power is the same for some full adders, and the one which consumes less power is considered as the best full adder. The full adder with low delay is used in most of the applications.

Table 1. Various technology node based 1-bit full adder power and delay analysis

Technology	Power (uW)	Delay (ps)
180 nm	4.156	224
90 nm	1.177	91.3
45 nm	40.1	40.1
CNTFET	0.073	16.83

5 Applications

There is an improvement of 20.56% with its reported design in case of PDP using 180 nm technology at 1.8 V supply. When the supply is 1.2 V with 90 nm technology, the improvement was 27.36% This design is further extended to 3-bit using buffers [3].
XNOR modules are utilized in ultra-low power high-speed full adder. High exchanging pace can be accomplished by utilizing solid transmission doors of 43.07 ns at 1.8 V stock. A similar structure can likewise be actualized to 4-piece wave convey snake [10]. The full snake which is proposed has great execution

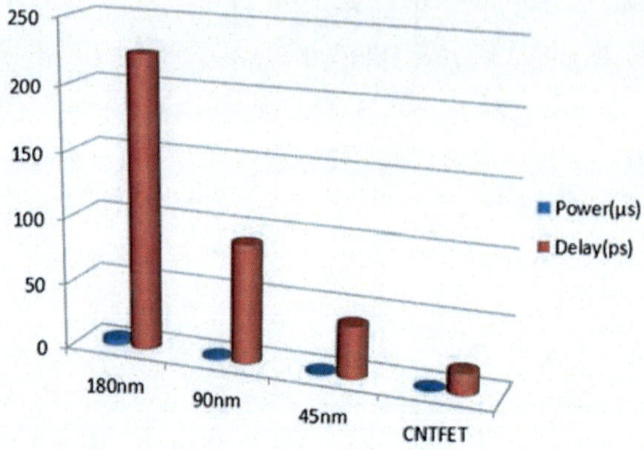

Fig. 4. Graphical representations of various technology node-based 1-bit full adder power and delay analysis

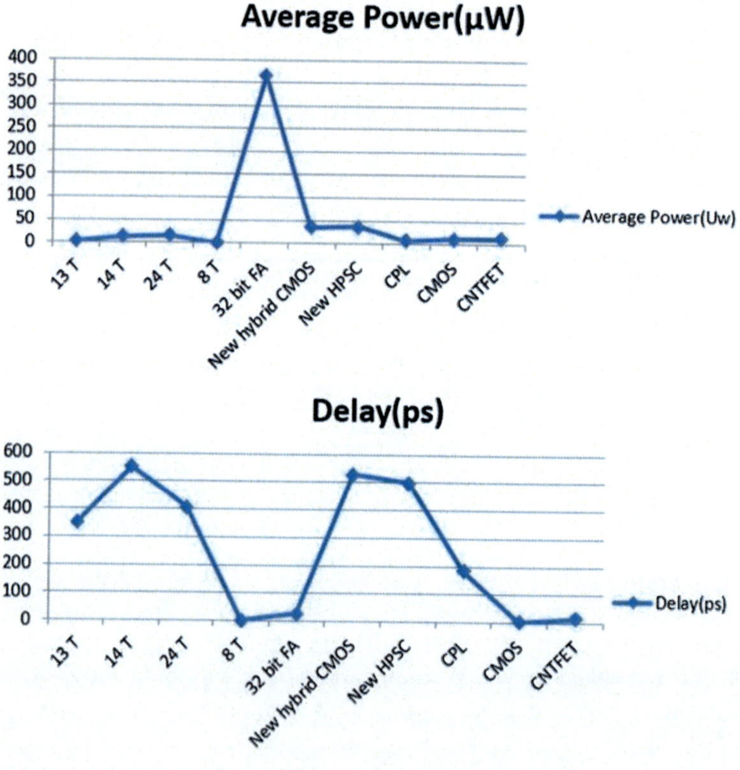

Fig. 5. Graphical representations of average power, delay comparisons of various full adders

Table 2. Comparison of full adder designs and their performance parameters

Technology	Design	Average power	Delay (ps)	Transistor count	Power supply (v)
Tanner EDA 280 nm CMOS	13T	2.09 uW	350	13	1.8
180 nm	14T	13.22 uW	551.9	14	1.8
180 nm	24T	15 uW	405	24	1.8
45 nm	8T	0.382 pW	0.793	8	1
90 nm	32-bit FA	365.26 uW	26.96	612	1.2
0.18 um	New hybrid CMOS	35.86 uW	527	24	1.8
0.18 um	New HPSC	36.45 uW	498	26	1.8
180 nm	CPL	7.719 uW	183.9	32	1.8
0.18uTSMC	CMOS [5]	11.31 nW	3.267	28	1.8
CNTFET	CNTFET [12]	12.08 uW	16.83	28	1.8

and —PDP than different structures. This structure performs better at low voltages [1]. There is an expansion in low-power utilization, time postponement and power defer item. The quick reaction of conveying yield and least power utilization is helpful for the enormous size of acknowledegment of adder [15]. In this structure, the design region of the circuit is decreased, and furthermore, we can observe better defer execution. The circuit was planned with 90 nm innovation at 1.2 V stockpile. The transistor check is likewise diminished by the utilization of XNOR entryways [7]. There is an improvement in power–delay product (PDP) and the presentation of a proposed full adder. It shows better execution at low voltages with great yields [6].

6 Conclusion

In this paper, various full adder circuits are studied and explained the extension of the 1-bit full adder to 32-bit. The standard designs are compared in terms of their performance parameters like speed, power, power–delay product, layout area, propagation delay. The transmission gates are strongly coupled to achieve speed. There is an improvement in design in terms of power–delay product (PDP) with the use of 180-nm technology at 1.8 V supply and better improvement in case of the use of 90-nm technology at 1.2 V supply. The design is further improved with 32-bit design.

References

1. Agrawal, P., Raghuvanshi, D., Gupta, M.: A low-power high-speed 16t 1-bit hybrid full adder. In: 2017 International Conference on Recent Innovations in Signal processing and Embedded Systems (RISE), pp. 348–352. IEEE (2017)
2. Aranda, M.L., Báez, R., Diaz, O.G.: Hybrid adders for high-speed arithmetic circuits: a comparison. In: 2010 7th International Conference on Electrical Engineering Computing Science and Automatic Control, pp. 546–549. IEEE (2010)
3. Bhattacharyya, P., Kundu, B., Ghosh, S., Kumar, V., Dandapat, A.: Performance analysis of a low-power high-speed hybrid 1-bit full adder circuit. IEEE Trans. Very Large Scale Integr. (VLSI) Syst. **23**(10), 2001–2008 (2014)
4. Goel, S., Gollamudi, S., Kumar, A., Bayoumi, M.: On the design of low-energy hybrid cmos 1-bit full adder cells. In: The 2004 47th Midwest Symposium on Circuits and Systems, 2004. MWSCAS'04., vol. 2, pp. II–II. IEEE (2004)
5. Hernandez, M.A., Aranda, M.L.: A low-power bootstrapped cmos full adder. In: 2005 2nd International Conference on Electrical and Electronics Engineering, pp. 243–246. IEEE (2005)
6. Janwadkar, S., Das, S.: Design and performance evaluation of hybrid full adder for extensive PDP reduction. In: 2018 3rd International Conference for Convergence in Technology (I2CT), pp. 1–6. IEEE (2018)
7. Kadu, C.P., Sharma, M.: Area-efficient high-speed hybrid 1-bit full adder circuit using modified XNOR gate. In: 2017 International Conference on Information, Communication, Instrumentation and Control (ICICIC), pp. 1–5. IEEE (2017)

8. Katragadda, R.: Analysis of low power methods in 14t full adder. In: 2015 2nd International Conference on Electronics and Communication Systems (ICECS), pp. 1210–1215. IEEE (2015)
9. Khatibzadeh, A.A., Raahemifar, K.: A 14-transistor low power high-speed full adder cell. In: CCECE 2003-Canadian Conference on Electrical and Computer Engineering. Toward a Caring and Humane Technology (Cat. No. 03CH37436), vol. 1, pp. 163–166. IEEE (2003)
10. Kumar, M., Baghel, R.: Ultra low-power high-speed single-bit hybrid full adder circuit. In: 2017 8th International Conference on Computing, Communication and Networking Technologies (ICCCNT), pp. 1–6. IEEE (2017)
11. Lee, S.J., Ruslan, S.H.: A 4-bit cmos full adder of 1-bit hybrid 13t adder with a new sum circuit. In: 2016 IEEE Student Conference on Research and Development (SCOReD), pp. 1–5. IEEE (2016)
12. Mewada, M., Zaveri, M.: A low-power high-speed hybrid full adder. In: 2016 20th International Symposium on VLSI Design and Test (VDAT), pp. 1–2. IEEE (2016)
13. Niranjan, N.K., Singh, R.B., Rizvi, N.Z.: Parametric analysis of a hybrid 1-bit full adder in UDSM and CNTFET technology. In: 2016 International Conference on Electrical, Electronics, and Optimization Techniques (ICEEOT), pp. 4267–4272. IEEE (2016)
14. Shams, A.M., Darwish, T.K., Bayoumi, M.A.: Performance analysis of low-power 1-bit cmos full adder cells. IEEE Trans. Very Large Scale Integr. (VLSI) Syst. **10**(1), 20–29 (2002)
15. Tung, C.K., Hung, Y.C., Shieh, S.H., Huang, G.S.: A low-power high-speed hybrid cmos full adder for embedded system. In: 2007 IEEE Design and Diagnostics of Electronic Circuits and Systems, pp. 1–4. IEEE (2007)
16. Wey, I.C., Huang, C.H., Chow, H.C.: A new low-voltage cmos 1-bit full adder for high performance applications. In: Proceedings. IEEE Asia-Pacific Conference on ASIC, pp. 21–24. IEEE (2002)
17. Zandkarimi, G., Navi, K., Mokari, H.: A new low power full adder cell in CMOS inverter. In: 2010 International Conference on Intelligent and Advanced Systems, pp. 1–5. IEEE (2010)
18. Zhang, M., Gu, J., Chang, C.H.: A novel hybrid pass logic with static CMOS output drive full-adder cell. In: Proceedings of the 2003 International Symposium on Circuits and Systems, 2003. ISCAS'03., vol. 5, pp. V–V. IEEE (2003)

A Clear View on Design of Low-Noise Amplifiers Using CMOS Technology

Lalitha Sowmya[1(✉)], S. Khadar Bhasha[1], and Durgesh Nandan[2]

[1] Department of ECE, Aditya Engineering College, Surampalem, East Godavari, India
simmaramya1289@gmail.com, khadarbasha.s@aec.edu.in
[2] Accendere Knowledge Management Services Pvt. Ltd., CL Educate Ltd., New Delhi, India
durgeshnandano51@gmail.com

Abstract. A detailed explanation on the design of low-noise amplifier is given in this paper. The wideband low-noise amplifiers are implemented in 0.18 μm CMOS technology. The various designs of low-noise amplifiers, such as the LNAs which reduce power dissipation, occupy less area, and consume less power, are presented in view of this paper. A low-noise amplifier design employs different methods, such as using center-tapped inductors, by interconnecting the stages, which are explained in this paper.

Keywords: Mutual coupling · Low-noise amplifier · Noise canceling · Inter-stage inductors · Frequency widening network

1 Introduction

Many numbers of low-noise amplifier designs were evolved in the CMOS technology [1–17]. The aim of the LNA is to provide wideband impedance matching to the antenna, sufficient gain, low-power consumption, and good linearity in available bandwidth. Among all the available wideband designs distributed and common gate amplifiers suffer high-noise figure. The linearity of LNA is degrading due to the additional cascading of several stages which has wide applications in the distributed LNA's. High gain and having gain flatness maximum are the benefits for distributed amplifiers when compared to the other techniques [1]. The wideband low-noise CMOS amplifier is obtained through mutual coupling which is done by using symmetrical center-tapped inductor. At the information side of low-commotion CMOS speaker, a recurrence extending system is structured through an inside tapped inductor. What is more, at the yield side of low clamor CMOS intensifier data transmission expansion is gotten through a solitary stage enhancer. The wideband low clamor enhancer depends on the innovation of 0.18 m CMOS innovation [4].

The wideband low-noise CMOS amplifier obtains 3–8 GHz bandwidth by consuming a power of 3.77 mW from the 1.8 V supply. Here we are using a smaller number of inductors in order to minimize the area requirement.

This low-commotion control intensifier is utilized in the wideband, low-power applications. A wideband clamor is dropping CMOS low-commotion intensifier for

© Springer Nature Singapore Pte Ltd. 2021
P. Deshpande et al. (eds.), *Next Generation Information Processing System*, Advances in Intelligent Systems and Computing 1162,
https://doi.org/10.1007/978-981-15-4851-2_14

3.1–10.6 GHz ultra-wideband beneficiary. An ultra-wideband commotion dropping low clamor speaker is clarified which is finished by utilizing arrangement and shunt cresting strategies. The effective data transmission for clamor dropping is improved or expanded [3].

This LNA is made using 0.18 μm CMOS advancement. The conscious uproar figure is 4.5–5.1 dB over 3.1–10.6 GHz, and power increment is 9.7 Db with −3 dB information transmission of 1.2–11.9 GHz; this noise dropping low disturbance intensifier uses a force of 20 mW from a given stock of 1.8 V supply [4]. There is publicizing potential for remote short-go correspondence with a data pace of 1 GHz/sec. In order to have these high data rates, a wide bandwidth with millimeter wave circuits is required. In order to allow low costs in mass production, CMOS millimeter wave circuits are required [5].

As the wideband frameworks are establishing to fast remote correspondence, these accomplished a lot of intrigue. The remote innovation that transmits an exceptionally low-power signal over a wide swath of radio range is ultra-wideband. The wide scope of recurrence over which the UWB framework works is 3.1–10.6 GHz. Some of the uses of ultra-wideband frameworks are imaging frameworks vehicular and ground infiltrating radars and likewise correspondence systems. We have numerous utilizations of ultra-wideband frameworks; however, every one of these applications is permitted in a recurrence scope of 3.1–10.6 GHz. In broadband correspondence framework, we need high increase data transfer capacity item which can be given by conveyed circuits utilizing transmission lines. The most favorable element is that it permits the incorporation of simple and computerized parts. Apart from the above applications, CMOS turned into a significant choice for RF applications and RF frameworks on-chip. The correct properties are offered by the inductor-degenerated LNA utilized in regular remote receivers. A run of the mill info coordinating procedure is being implemented through the inductor deteriorated system.

2 Literature Review

Sunderarajan S. Mohan et al. (2000) a method for upgrading the data transfer capacity of gigahertz wideband hardware is built up utilizing on-chip winding inductors as shunt top components. So as to encourage circuit structure precise inductance articulations were utilized in a lumped circuit inductor model [6]. Ismail and Abidi (2004) utilizing the impedance properties of LC stepping stool channels, receptive coordinating is improved to wide transmission capacities. A precise strategy is acquainted with plan wideband low clamor amplifier [7].

Wei et al. (2005) an ultra-wideband CMOS voltage-controlled oscillator with 3–5 GHz territory is planned utilizing TSMC 0.18 μm CMOS process. In this, the voltage-controlled oscillator is tunable inductor in the spot of inactive winding inductor. The dynamic inductor which is being utilized by the ultra-wideband CMOS VCO can differ the inductance between 1.5 and 7 nH with a quality factor more noteworthy than 30, with a tunable input resistor. When we contrast this topology and customary VCO, this topology is offering more favorable circumstances, for example, giving exceptionally wide tuning extent and limited chip size [8].

Lee and Mohammadi (2007) a sub-threshold low-power, low-phase LC VCOIS established using 0.18 μm CMOS process. A standard low-Q on-chip inductor is being utilized by VCO while 0.43 mW DC power under 0.45 V supply voltage is being dissipated by its core [9]. Uhrmann and Zimmermann (2007) a low-noise current pre-amplifier was established based on 0.12 μm CMOS technology. There is a chance of using these low-noise current preamplifiers in measurement instrument systems, charge amplifiers, and optical sensors [10].

Deepak Balemarthy et al. (2008) using CMOS 0.18 μm TSMC process a dual-band LNA is designed using miller capacitances the low-noise amplifier is tuned to required frequencies. It exhibits good performance at 2.1 GHz. It can be fabricated without off-chip components [11]. Ali Meaamar et al. (2009) the wideband low-noise CMOS amplifier is obtained through mutual coupling which is done by using symmetrical center-tapped inductor. At the input of low-noise CMOS amplifier, a frequency widening network is designed through center-tapped inductor. And at the output side of low-noise CMOS amplifier, bandwidth extension is obtained through single-stage amplifier. The wideband low-noise amplifier is based on the technology of 0.18 m CMOS technology. The wideband low-noise CMOS amplifier obtains 3–8 GHz bandwidth with a power consumption of 3.77 mW from 1.8 V supply. Here we are using less number of inductors in order to minimize the area requirement. This low-noise power amplifier is used in the wideband and low-power applications [12].

S. M. S Rashid and S. N. Ali (2009) using a simple passive output matching technique, a 36.1 GHz single-stage LNA is established. With the help of 0.13 μm CMOS process parameters, the circuit is stimulated in cadence specter. This low-noise amplifier after simulation provides a forward gain of 11.4 dB at 36.1 and 4.9 GHz bandwidth. It consumes 3.38 mW for 1.2 V of given supply [13]. A. R. Ximenes (2011) to the wideband low-noise amplifier, a new amplifier topology with noise cancelation technique is applied. It provides a good performance over a broad band-width [14]. Wei et al. (2012) in order to achieve good geometric scalability, different models have been presented. The geometric scalability feature has an ability to reduce elongated time in electromagnetic field simulation [15].

Tan Yang (2015) the procedure for the plan of a neural chronicle speaker cluster with ultra-low-power, low-commotion tasks is reasonable for huge scale incorporation. The intensifier is manufactured in a 90 nm CMOS process, and it involves a zone of 0.137 mm^2 on the chip area [16]. Wei-Rern Liao et al. [17] for LTE applications, a 0.5–3.5 GHz wideband CMOS low-commotion speaker was presented. The LNA configuration comprises of a typical source course intensifier with resistive input. The LNA accomplishes the increase of 17–22 dB and a commotion figure of 2.33–2.68 dB at a recurrence extends from 0.5 to 3.5 GHz. This low clamor intensifier devours an intensity of 32.8 mW for a given stockpile of 1.8 V.

3 Design Approach

In [3], a wideband LNA is planned with the door terminal one-sided with an off-chip inclination t-source. In request to diminish invert sign stream, a course framework is utilized, and the extents of transistor are painstakingly made to decrease entryway parasitic obstruction. In this, we have three reverberating frequencies because of which the circuit broadens the transmission capacity successfully. In this plan, the circuit has inbuilt arrangement opposition so no physical resistor is utilized. The inbuilt arrangement obstruction is utilized to lessen the q-factor of the yield organize, in order to improve the data transfer capacity and to improve the addition evenness. This wideband LNA is executed in 0.18 μm CMOS innovation. For this structure, a middle-tapped inductor with a coupling co-effective $0.72 < k < 0.86$ is utilized. The greatest power addition is 15.2 dB between the recurrence 3.4–4 GHz, and from the recurrence of 4.2–8 GHz, the increase levelness is superior to anything 1.5 db. This wideband low clamor speaker expends just 3.77 mW power from 1.8 V stock. Thus, it very well may be utilized in low-power applications shown in Table 1.

Table 1. Design values of the LNA

(W/L) 1	(W/L) 2	L * 1.2	L * 3.4	LL	Ls	RF (Ω)
120/0.18	40/0.18	6 nH	2.92 nH	1.31 nH	0.18 nH	1.14 k

In [18], a profoundly direct ultra-wideband LNA with the CMCC linearization system comprises of a differential course of structure in which four FETs are utilized as primary speaker and two additional FETs shapes an assistant way to produce a similar regular mode current with inverse sign to improve the linearity. A resistor in parallel to inductor and a parasitic capacitor shapes a RLC load. The capacitive piece of the information impedance is repaid by utilizing one more inductor to forestall the voltage gain decrease at high frequencies. Under this condition, the nonlinearity and commotion can be neglected [19].and the clamor commitment of the resistor winds up immaterial [20]. In order to have a reasonable correlation of linearity improvement, a shared view low clamor enhancer is structured and reproduced in a 90 nm RF CMOS process as shown in Table 2.

In [21], the choice of transistor is the underlying advance during the time spent planning the low-commotion speaker. Here we have picked PHEMT transistor which represents pseudomorphic high electron versatility transistor. The PHEMT transistor shows low ON obstruction and high OFF resistance by making the PHEMT ideal for switches and resistive FET blends. The PEMT is generally presented as a FET with a door, channel, and source. The second part in LNA configuration is DC biasing circuit. It gives a steady predisposition point to the gadget, and the biasing strategy utilized is straightforward resistive biasing which gives the least resilience. The soundness of the low clamor speaker relies upon how it opposes to oscillate, because being an intensifier it should be steady from a motion.

The third step in planning the low-commotion intensifier is steadiness examination. The coordinating of info and yield terminals is the fourth step in structuring of LNA. There are two techniques to accomplish coordinating of info and yield terminals. They are one is by utilizing lumped segments, and the other one is by utilizing small-scale strip line. The coordinating among information and yield terminals is required to accomplish the greatest power move. In this plan, we have utilized strip line structure to get a great increase and least clamor figure.

Table 2. Performance comparison

CMOS process	BW (GHz)	S11 (dB)	S21* (dB)	NF** (dB)	IIP3 (dBm)	IIP2 (dBm)	VDD (V)	Power (mW)	FoM
130 nm	4.7–11.7	<-11.9	12.4	2.9	−3	–	1.2	13.5	−4.8
180 nm	2–6.5	<-10	11	2.7	+4.4	–	1.8	7.6	−21
90 nm	2.6–10.2	<-9	12.5	3	−2.4	–	1.2	7.2	−10.9
180 nm	3.1–10.6	<-10	12.6	2.9	−4.6	–	1.8	15.2	−2.3
65 nm	0.1–10	<-11.7	24	2.6	−13.5	+5	1.2	8.6	−15.8
90 nm	2.6–13.7	<-10	11	3.8	+4.6	+69.3	1.2	12	−24

The structure comprises the negative input so as to give an exhibition more noteworthy than others by giving up a little clamor figure. At the point when the negative criticism is appropriately utilized, it safeguards the addition evenness and will give great impedance coordinating. It likewise lessens the resistance impact and improves DC and RF strength with less twisting (Table 3).

Table 3. Comparison of various technologies

CMOS technology	Bandwidth (GHz)	Gain (db)	Noise figure (db)	Power (mw)	Area (mm^2)
0.18 μm CMOS [3]	3–8	15.2	3.14–6.8	3.77	0.96
0.18 μm CMOS [2]	3.1–10.6	9.7	4.5–5.1	20	0.59
0.18μm CMOS [7]	3–10	21	2.5–4.2	30	1.8
0.18 μm CMOS [27]	2.3–9.2	9.3	4.2–8	9	1.1
0.18 μm CMOS [25]	0.5–14	10.6	3.4–5.4	52	1.6
0.18 μm CMOS [28]	2.3–9.2	9.3	4–8	9	1.1
0.18 μm CMOS [29]	2.0–4.6	9.8	2.3–5.2	12.6	0.9
0.8 μm CMOS [26]	0.6–22	8.1	4.3	52	–

In [22], two distinct LNA designs were explained. Each of the designs is meant for distinct set of parameters. One of the designs is having extremely high gain and low-noise figure. And another design is with extremely low-power dissipation. And these low-noise amplifiers are designed in 130 nm CMOS process. In both the designs, the LNAs have bandwidth of 1.9–8.1 GHz.

In one of LNA designing methods, all the stages are interconnected and the interconnection is through interstate inductors. Due to the interconnection of these stages, we could be able to achieve wide bandwidths. In this design, the first and third stages are provided with external biasing voltage, whereas the second stage is self-biased state. At the input side instead of using conventional Butterworth filter, LC high-pass filter is used. Series peaking and shunt peaking are implemented to intensify the behavior of low-noise amplifier. And the second stage is of complementary symmetry push–pull configuration; hence, the second stage provides the high gain. And the third stage is with a common source configuration where the inductive peaking technique is used [23]. Finally, the last stage flattens the gain and increases the bandwidth of the amplifier (Fig. 1).

In the second plan, the point is to lessen control dissemination. In this model, two unmistakable voltage sources are utilized for power supply a power wellspring of 0.5 V

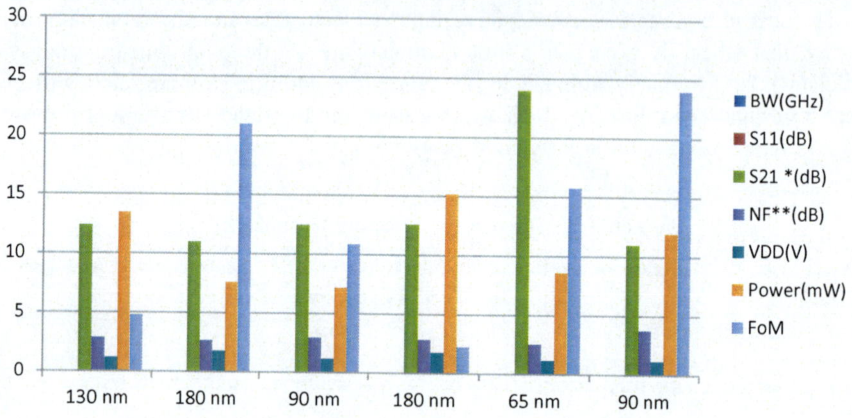

Fig. 1. Comparison of various CMOS processes

is given to first and third stages. The power supply to the subsequent stage is given as 0.8 V by utilizing forward body predisposition from the start and third stages and decrease in stock voltage is obtained [22].

The double-band low clamor intensifier comprises of two normal source stages and indent channels. The primary stage gives high increase at low frequencies. The principal organizes and furthermore utilizes capacitance and inductance in order to give wide info coordinating. The capacitor is associated in parallel, and the inductor is associated with an arrangement to the FET. A current reuse topology is utilized constantly arrange so that to limit the utilization of intensity and to improve gain. The resistive input and arrangement inductors are utilized to have simultaneous impedance coordinating and wide transfer speed. The score channels are utilized to accomplish the simultaneous double-band LNA recurrence reaction [24] (Fig. 2).

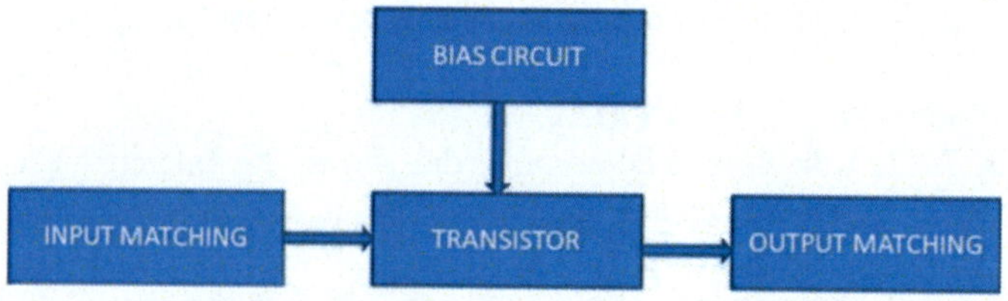

Fig. 2. Basic block diagram of low-noise amplifier

4 Result

In this paper, we have observed various designs of low-noise amplifiers and each design is having its own advantages. The design of low-noise amplifier by using mutual coupling technique can be used in low-power application and by using three stages can be used in noise-sensitive applications. Among all the available designs, the design of low-noise amplifier with interconnecting stages is found to be more advantageous, because this design is working the same for the entire frequency band (Fig. 3).

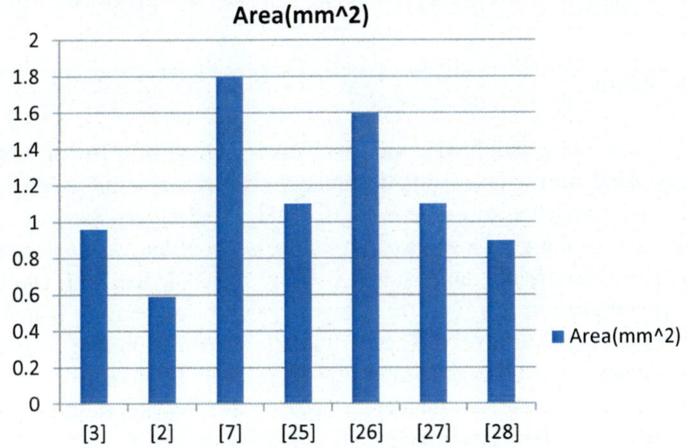

Fig. 3. Comparison of area of CMOS

From the above figure, the area occupied by the CMOS in [2] design is very much less when compared to remaining designs; therefore, when area is the main constraint, the design [2] can be used. In design [7], the area occupied is very much high when compared to other designs, so it cannot be preferred when area is the main constraint (Fig. 4).

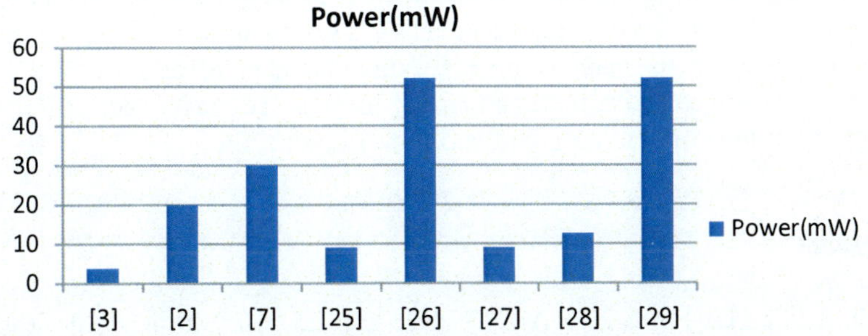

Fig. 4. Comparison of power of CMOS

From the above figure, the power consumed by the design [3] is very much less when compared to other designs; therefore, when power is the main constraint, then the design in [3] can be preferred. In case of area, the design [7] is not preferred; but in case of power, it consumes comparatively less power. When power is the main constraint, then the designs [25] and [26] cannot be preferred.

The bandwidth range of the design [26] is high when compared to other designs; however, the area occupied and power consumed are high. As mentioned above, the area occupied by design [2] is less but the bandwidth range provided by this is less, and the power consumed by the design [3] is less but bandwidth provided this is also less.

5 Applications

The wideband low-noise amplifier which was implemented in 0.18 μm CMOS technology can be used in low-power applications. And it can also be used in the applications where the available area is less [3]. In [22], the designed low-noise amplifier provides high degree flat gain response and this helps in giving constant gain through the entire passband so that the noise figure is being reduced. Hence, it can be used in noise-sensitive applications. In [21], the designed low-noise amplifier is more efficient when compared to previous works. The designed low-noise amplifier is working well for entire frequency band by maintaining low noise. This LNA can be used in applications where the temperature is fluctuating because this low-noise amplifier provides the same results at all the temperatures. In [19], the designed LNA is used in the applications where the power dissipation should be reduced. In [18], the designed low-noise amplifier can be used in applications where the power dissipation should be less and where the linearity should be improved.

6 Conclusion

In [3], a single-stage amplifier is used in order to achieve bandwidth extension, but the single-stage amplifiers efficiency will be affected by atmospheric conditions (temperature) and aging. The design of LNA by interconnecting stages [22] is found to be more advantageous than remaining works because this LNA design is working the same for entire frequency band by maintaining low noise. Number of designs of low-noise amplifiers were evolved and each design having its own advantages such as less power consumption, occupying less area, and less power dissipation.

References

1. Yazdi, A., Lin, D., Heydari, P.: A 1.8 V three-stage 25 GHz 3 dB-BW differential non-uniform downsized distributed amplifier. IEEE ISSCC Technical Digest, pp. 156–158 (2005)
2. Liao, C.-F., Liu, S.-I.: A broadband noise-canceling CMOS LNA for3.1–10.6 GHz UWB receivers. IEEE J. Solid State Circ. **42**(2), 329–339 (2007)
3. Meaamar, A., Chye, B.C., Seng, Y.K.: A 3–8 GHz low-noise CMOS amplifier. IEEE Microwave Wirel. Compon. Lett. **19**(4) (2009)
4. Liao. C.F., Liu, S.I.: A broadband noise-canceling CMOS LNA for 3.1–10.6 GHz UWB receiver. In: IEEE Custom Integrated Circuits Conference, National Taiwan University, Taipei, Taiwan (2012)

5. Ellinger, F., Wickert, M., Eickhoff, R., Mayer U., Hauptmann, S.: Low noise radio frequency integrated circuits in 90 nm SOI CMOS up to 60 GHz. In: Chair for Circuit Design and Network Theory, 01–10 (2010)
6. Mohan, S.S., Hershenson, M.D., Boyd, S.P., Lee, T.H.: Bandwidth extension in CMOS with optimized on-chip inductors. IEEE j. Solid State circ. **35**(3) (2000)
7. Ismail, A., Abidi, A.A.: A 3–10 GHz low-noise Amplifier with wideband LC-ladder matching network. IEEE j. Solid State Circ. **39**(12) (2004)
8. Wei, C.C., Chiu, H.C., Feng, W.S.: An ultra-wideband CMOS VCO with 3–5 GHz tuning range. In: IEEE International Workshop on Radio-Frequency Integration Technology, pp. 05–10 (2005)
9. Lee, H., Mohammadi, S.: A subthreshold low phase noise CMOS LC VCO for ultra low power applications. IEEE Microwave Wirel. Compon. Lett. **17**(11) (2007)
10. Uhrmann, H., Zimmermann, H.: Alow-noise current preamplifier in 120 nm CMOS technology. In: Mixdes 2007 Ciechocinek, Poland, pp. 21–23 (2007)
11. Balemarthy, D.: A 1.8/2.4 GHz dual-band CMOS low noise amplifier using miller capacitance tuning. Indian Institute of Technology, Guwahati, India (2008)
12. Meaamar, A.: A 3–8 GHz low-noise CMOS Amplifier. IEEE Microwave Wirel. Compon. Lett. **19**(4) (2009)
13. Rashid, S.M.S., Ali, S.N.: A 36.1 GHz single stage low noise amplifier using 0.13 μm CMOS process. In: 2009 World Congress on Computer Science and Information Engineering (2008). ISBN: 978-0-7695-3507-4/08
14. Ximenes, A.R.: A wideband noise canceling low-noise amplifier for 50 MHz–5 GHz wireless receivers in CMOS technology (2011). ISBN: 978-1-61284
15. Lim, W.Y., Shi, J., Arasu, M.A., Je ,M.: Geometric scalable 2-port center-tap inductor modeling (2012)
16. Yang, T.: An Ultra-low-power low-noise CMOS bio-potential amplifier for neural recording (2005). https://doi.org/10.1109/tcsii.2015.2457811
17. Liao, W.-R.: A 0.5–3.5 GHz wideband CMOS LNA for LTE application (2016). ISBN: 978-1-5090-1978-6/16
18. Mazhabjafari, B., Yavari, M.: A 2.6–13.7 GHz highly linear CMOS low noise amplifier for UWB applications. In: The 22nd Iranian Conference on Electrical Engineering (ICEE 2014), 20–22 May, Shahid Beheshti University (2014)
19. Zhang, H., Fan, X., Sánchez-Sinencio, E.: A low-power, linearized, ultra-wideband LNA design technique. IEEE J. Solid State Circ. **44**(2), 320–330 (2009)
20. Mehrjoo, M.S., Yavari, M.: A low power UWB very low noise amplifier using an improved noise reduction technique. In: IEEE IEEE International Symposium of Circuits and Systems, pp. 277–280 (2011)
21. Khanapurkar, M.M.: Design of ultra wideband low noise amplifier with the negative feedback using micro strip line matching structure for multiple band application and its Simulation based performance analysis (2016)
22. Singh, V.: Ultra wide band low noise amplifier with self-bias for improved gain and reduced power dissipation (2016). ISBN: 978-1-5090-1666
23. Salama, M., Soliman, A.M.: Low-voltage low-power CMOS RF low noise amplifier. Int. J. Electron. Commun. (AEÜ) **63**(6), 478–482 (2009)
24. Khosravi, H., Zandian, S., Bijari, A.: A low power, high gain 2.4/5.2 GHz concurrent dual-band low noise amplifier (2019)
25. Liu, R., Lin, C., Deng, K., Wang, H.: A 0.5–14 GHz 10.6 dB CMOS cascode distributed amplifier. Symp. VLSI Circ. Dig. **17**, 139–140 (2003)

26. Liu, R.-C., Deng, K.-L., Wang, H.: A 0.6–22 GHz broadband CMOS distributed amplifier. In: IEEE Radio Frequency Integrated Circuits Digest of Technical Papers, pp. 103–106 (2003)
27. Bevilacqua, A., Niknejad, A.: An ultrawideband CMOS low-noise amplifier for 3.1–10.6 GHz wireless receivers. IEEE J. Solid State Circ. **39**(12), 2259–2268 (2004)
28. Bevilacqua, A., Niknejad, A.M.: An ultra-wideband CMOS LNA for 3.1–10.6 GHz wireless receivers. IEEE Int. Solid State Circ. Conf. **XVII**, 382–383 (2004)
29. Kim, C.-W., Kang, M.-S., Anh, P.T., Kim, H.-T., Lee, S.-G.: An Ultra-wideband CMOS low noise amplifier for 3–5 GHz UWB system. IEEE J. Solid State Circ. **40**, 544–547 (2005)

Study on Bicycle-Based Real-Time Information Feedback System by Using IoT

Guthula Hema Mutya Sri[1], Galla Bharggav[1], Rajasekhar Manda[1], and Durgesh Nandan[2(✉)]

[1] Department of ECE, Aditya Engineering College, Surampalem, India
{hemasriguthula, bharggavgalla}@gmail.com
rajasekharm@aec.edu.in

[2] Accendere Knowledge Management Services Pvt. Ltd., CL Educate Ltd., New Delhi, India
durgeshnandano51@gmail.com

Abstract. IoT means connecting, establishing communication between objects by using the Internet. This paper presents a study reports on how bicycling by using IoT becomes an exact health tool and major benefit in terms of health monitor. Nowadays, the bicycle is the most popular exercise in metro cities. At the same time, high-speed Internet and various sensors combination based on IoT devices are widely used. Although, bicycles have all known benefits to health but they fail to provide cyclists person exact health benefits information. If no information, people lose charm to do cycling in the long term. Therefore, this system introduced bicycle-based real-time data feedback system based on combining smartphones and IoT. After completing cycling exercise, the person can see the cycling-related data through the software. This methodology has used various types of sensors for collecting data like handling, orientation, and balancing sensors. This system is to provide real time, correct, and complete data to cyclists for the best experience and to improve their fitness. Data on heartbeat, cycling speed, total time taken to complete, distance traveled, and energy levels are calculated.

Keywords: Arduino board · Cellular phone · Cyclist monitoring · Embedded software · Event detector · Firm-based · IoT

1 Introduction

IoT connects, establishes communication between objects and networks. In recent years, the world is increasingly changing with embedded systems and the Internet of things. There is a need for a real-time data entity. The smartphone is a key component of IoT. It is useful in giving information regarding the

© Springer Nature Singapore Pte Ltd. 2021
P. Deshpande et al. (eds.), *Next Generation Information Processing System*, Advances in Intelligent Systems and Computing 1162,
https://doi.org/10.1007/978-981-15-4851-2_15

Web of things. It calculates distance based on time and physical distances. IoT is used for storing real-time data automatically [5]. We can track the electric bicycle condition based on the running and rest state by using radar. Nowadays, IoT is becoming a most useful application [3]. IoT is possible and more efficient with the use of sensors. We use sensor-based IoT in many places like research center, offices, hospitals, and colleges. Some of IoT-based devices are automatic door detector, operating electronic devices in smartphones [11]. At present, man involvement is less and his time saved is more. IoT is used in measuring energy-calories gained and lost at the time of cycling. IoT is useful for data assurance [13]. It is useful for vehicle over-speed detection. All the information regarding roads, traffic, and vehicles can be given with the use of IoT. All smart electronic devices include the Internet of things. IoT for recording system plays a vital role [15]. With the use of this IoT, queries can be solved easily. Smartphone sensors are responsible for monitoring the conditions in the cycling path [21]. There is a major issue of theft of a bicycle, and these can be overcome by a new technology using IoT based on the current and parking position and decides whether it is low or high risk. The smart module lock is created and works more efficiently in detecting the position of bicycle and deciding whether it is stolen or not [1]. IoT is the reason for secure communication and data assurance. IoT always contains existing, deleted, and any information that is modified. IoT is useful for less power consumption, more efficient working, and storage in local storage devices [22]. IoT is used for big data storage in survey processes in transportation, route planning. It is used to determine cycling person performance based on cloud technologies. IoT systems can better interact with others and convey information directly and it is easy to study, analyze, and understand more clearly and perfectly. The required literature review has been explored further in Sect. 2. Section 3 details the existing methodology to identify the real-time bicycle record system on ground conditions using a smartphone through IoT. The comparison of result which consists of techniques we used and parameter achieved is clearly explained in Sect. 4. Finally, the technique to calculate the health condition and distance travelled are concluded in Sect. 5.

2 Literature Review

The IoT world is becoming integrable with a computer network. This requires technical standards and requires more time [5]. It is providing two data streams—primary streams for road vehicle traffic and secondary for support usage. The location tracking can be found by using radars [3]. IoT is being paid more attention to all the people in the private and public sectors [11]. Automation has the largest attention in this world. All the controlled information can be stored by IoT devices. It can be used for all types of devices which require storing system [13]. Bicycle is a useful sport. At the same time, IoT is used in the cycle for recording ground conditions and total distance travelled, energy lost during cycling, and this information is stored and traced [7, 22]. Even though urban is developing, traffic information is low. By using this IoT device, we can get proper information about traffic and roads [6]. As healthcare costs are increasing, we came up with an IoT device that reduces costs. These devices comprise of many sensors and microprocessors for best health monitoring. The main target is to serve the best developers and researchers [14]. The main idea of this IoT is to address several communication solutions. Identifying and tracking problems are done by IoT technology. This survey is to appear for the best communication solutions [2]. The Web of objects can be applicable to houses, companies, and factories [10, 18]. Wireless sensors networks (WSNs) are mostly employed in object tracking and detection [19].

3 Existing Methodologies

This introduced an embedded system to record the cyclist data to be displayed. The two parts of the data recorder are “Smartphone-based event data recorder” and “bicycle-based real-time information feedback system.” The smartphone-based events information recorder consists of these modules accelerometer module (AM), gyroscope module (GM), system integration module (SIM), electric compass module (ECM), GPS module [22].

Event data recorder using a Smartphone:

(i) **Accelerometer Module**
This module does not require any external device. It examines the cyclist by use of the three-dimensional acceleration of smartphones. It analyses the path conditions, the habit of cycling, and theft preventing [9, 22].

(ii) **Gyroscope module**
This 3D module of smartphones is used to determine cyclist's angle of rotation and data whether the cycle is balanced or fatigue or overturned. Electronic gyroscope also called as a micro-electromechanical gyroscope [22].

(iii) **Electric compass module**
It is used to determine cycling direction to check whether the person is moving in a correct path or not, and differentiates the directions by sensing the magnetic field of the earth [22].

(iv) **GPS module**
It locates the particular point of cyclists and record cyclists coordinates when cyclists are cycling. A cyclist can share the route maps generated from data recorded with another cyclist using the Web. In addition to this module, the system can record the location of defects and vehicle speed [9, 22].

(v) **System Integration Module**
System integration module performs integration work by the accelerometer, gyroscope, electric compass, and GPS. All displayed information has been stored in a smartphone chip for the purpose to calculate various analysis parameters like highest, average, slowest speed, total distance traveled, and calories burned [5, 9, 22].

Communication Medium:
Bluetooth or Wi-Fi can be used for the most relevant communication medium [8] because it is smooth when used with Internet data network communication for uploading databases [8].

Cloud data recorder:
Cloud data recorder is a communication medium to store the data on the Internet through a cloud data recorders provider and manages data storage. Cloud data recorder can access data anywhere and anytime. This is exchanging the date with the communication medium and it pushes the data into the clouds or things speak.

Bicycle-based RTIFS (Real-Time Information Feedback system):
It consists of five modules. They are (1) Control and computing combine Wi-Fi module and Bluetooth module. (2) Reed switch counts the number of rotations, the time required for the wheels, and converts mechanical movement into an electrical signal. (3) Real-time data display using Arduino while displaying the related information. (4) Bluetooth transmission Modules establishes connection with the Bluetooth device which is built in the Smartphone. (5) Cloud will be used to store the data of a particular person in different conditions.

Description of the system:
This cycle real-time-based recorded system consists of three blocks. (1) Bike or cyclist (2) Smartphone event data recorder (3) The real-time information feedback system. The first block consists of a wireless sensor network that is embedded. The bike or cyclist will send the obtained data to the database through coordinator located on the bike or cycle. The server gets the information from WSN that can be accessed by a support user. The nodes of the embedded system contain sensors that are connected to Arduino, each of the nodes gets power from Li batteries. Primary processing is made by Arduino focusing sensor data to send to the coordinator. Wi-Fi and Zigbee protocols are used for communication from WSN to the database and WSN, respectively [4]. IoT architecture has four layers (Fig. 1).

Deep architectures are a source of deep learning that is supervised or unsupervised techniques contain that work upon multi-layers of artificial neural networks. The non-linear response obtained by each layer depends upon the data given from the input layer. The human brain and neurons structure are the fundamental steps toward implementing the DL [17]. In fact, these techniques are developed on top of ANNs and feed-forward neural networks (FNNs) [12,20]. Over a decade, multi-layer perceptions (MLPs) were used. But it has the drawback in implementation and trains the sets for multiple layers [12,16,20].

Smartphone-based event data recorder
1. Accelerometer 2. Gyroscope 3. Electric compass 4. GPS 5. System integration.

Bicycle-based real-time information feedback system
1. Control and Computing 2. Reed switch 3. Real-time data display 4. Bluetooth 5. Wi-Fi.

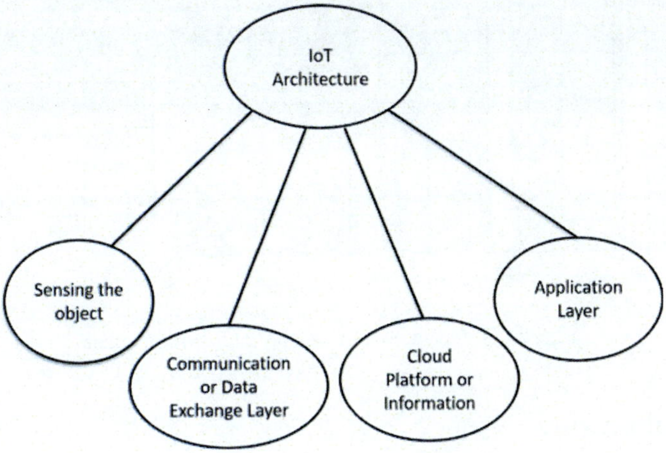

Fig. 1. IoT layer architecture

4 Result

The bicycle recording system consists of an event data recorder and data feedback system. The data is calculated using the path chosen by cyclists, slope, and road conditions. The cycling data consists of calories burned, distance travelled and speed of cyclist. In this system, there is a chance for the sharing of information. The data from these sensors are displayed immediately. Recorded data consists of parameters like heartbeat ranging from 44 to 72 bpm which is 58 bpm on average, speed of cycling ranging from 0 to 22 km per hour which is 11 km per hour on average, maximum focus value of 100 and traveled distance is maximum 3.82 km. The tabular columns explain the different conditions (Table 1).

Table 1. Different conditions for the same cyclist

Parameters	Minimum	Maximum	Average
Heartbeat (BPM)	44	72	58
Cycling speed in km per hour	0	22	11
Focused	96	100	98
Distance travelled (km)	0	3.82	

Figure 2 shows the graphical representation of the tabular values of different conditions for the same cyclist.

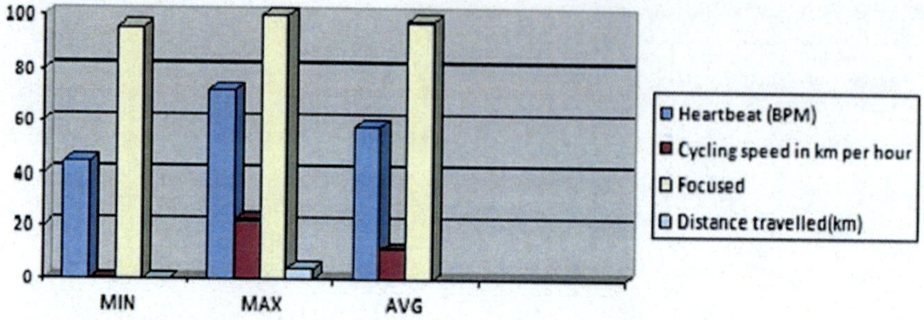

Fig. 2. Different conditions for the same cyclist

5 Applications

IoT provides many numbers of applications in all domains that can build intelligent systems useful for society. Proper communication between the objects and modules opens the path for wide applications in IoT. Database and mobile application This android application simplifies the cyclist trainer to maintain the data sheet for cyclists.

Application in a vehicle sharing system
It is possible to share either a bike or a car in the same city. Of course, this leads to the congestion problem that gives errors in traveling time estimation. Incorporating congestion management can improve performance.

Real-time GPS application
Travelling and waiting time can be estimated. The system can be updated by uploading new data.

Transportation and logistics domain
Nowadays, all the vehicles on roads and tracks are well automated with actuators, sensors, and processing power. This will help to control the road traffic.

Healthcare domain
Patient monitoring systems analyze blood pressure and breathing activity temperature. Emergency situations can be handled easily using wireless devices.

6 Conclusion

This developed system provides complete information for cyclists. All the parameters are calculated, displayed by installing the modules on their smartphones. This cyclist recorded system provides real-time information such as path conditions, geographical location, speed, and kilometers travelled to achieve their goal in burning the calories. The whole system works based on IoT that makes it feasible for cyclists to share their accurate real-time information. The augmented feature in smartphones makes cyclists enjoy their cycling environment.

References

1. Akins, K., Goodson, M., Skeggs, P., Rumbaugh, S., Zyla, C.: Bicycle theft monitoring and recovery devices. US Patent App. 13/712,831 (2013)
2. Atzori, L., Iera, A., Morabito, G.: The internet of things: a survey. Comput. Netw. **54**(15), 2787–2805 (2010)
3. Bahl, P., Padmanabhan, V.N., Bahl, V., Padmanabhan, V.: Radar: An in-building RF-based user location and tracking system (2000)
4. Dias, A.C., Postolache, O.: Cyclist performance assessment based on wsn and cloud technologies. In: 2018 International Conference and Exposition on Electrical And Power Engineering (EPE), pp. 1041–1046. IEEE (2018)
5. Guinard, D., Trifa, V.: Towards the web of things: Web mashups for embedded devices. In: Workshop on Mashups, Enterprise Mashups and Lightweight Composition on the Web (MEM 2009), Proceedings of WWW (International World Wide Web Conferences), Madrid, Spain, vol. 15 (2009)
6. Händel, P., Ohlsson, J., Ohlsson, M., Skog, I., Nygren, E.: Smartphone-based measurement systems for road vehicle traffic monitoring and usage-based insurance. IEEE Syst. J. **8**(4), 1238–1248 (2013)
7. Hao, X., Jin, P., Yue, L.: Efficient storage of multi-sensor object-tracking data. IEEE Trans. Parallel Distrib. Syst. **27**(10), 2881–2894 (2015)
8. Hickman, R.C., Bobbitt, J.E., Tanner, J.C., Lau, P.W.S., Friedman, M.T., Mullally, J.P.: Internet database system. US Patent 6,523,036 (2003)
9. Kumar, G.A., Kumar, A.S., Kumar, A.A., Maharajothi, T.: Road quality management system using mobile sensors. In: 2017 International Conference on Innovations in Information, Embedded and Communication Systems (ICIIECS), pp. 1–6. IEEE (2017)
10. Lee, S., Chong, I.: User-centric intelligence provisioning in web-of-objects based iot service. In: 2013 International Conference on ICT Convergence (ICTC), pp. 44–49. IEEE (2013)
11. Liu, Y., Zhou, G.: Key technologies and applications of internet of things. In: 2012 Fifth International Conference on Intelligent Computation Technology and Automation, pp. 197–200. IEEE (2012)
12. Mohammadi, M., Al-Fuqaha, A., Sorour, S., Guizani, M.: Deep learning for iot big data and streaming analytics: a survey. IEEE Commun. Surv. Tutorials **20**(4), 2923–2960 (2018)
13. Nissanka, S., Senevirathna, M., Dharmawardana, M.: Iot based automatic storing and retrieval system. In: 2016 Manufacturing and Industrial Engineering Symposium (MIES), pp. 1–5. IEEE (2016)
14. Pantelopoulos, A., Bourbakis, N.G.: A survey on wearable sensor-based systems for health monitoring and prognosis. IEEE Trans. Syst. Man Cybern. Part C (Appl. Rev.) **40**(1), 1–12 (2009)
15. Pendor, R.B., Tasgaonkar, P.: An iot framework for intelligent vehicle monitoring system. In: 2016 International Conference on Communication and Signal Processing (ICCSP), pp. 1694–1696. IEEE (2016)
16. Schmidhuber, J.: Deep learning in neural networks: an overview. Neural Netw. **61**, 85–117 (2015)
17. Shahid, N., Aneja, S.: Internet of things: Vision, application areas and research challenges. In: 2017 International Conference on I-SMAC (IoT in Social, Mobile, Analytics and Cloud) (I-SMAC), pp. 583–587. IEEE (2017)

18. Shamszaman, Z.U., Lee, S., Chong, I.: Woo based user centric energy management system in the internet of things. In: The International Conference on Information Networking 2014 (ICOIN2014), pp. 475–480. IEEE (2014)
19. Sheng, X., Tang, J., Zhang, W.: Energy-efficient collaborative sensing with mobile phones. In: 2012 Proceedings IEEE INFOCOM, pp. 1916–1924. IEEE (2012)
20. Svozil, D., Kvasnicka, V., Pospichal, J.: Introduction to multi-layer feed-forward neural networks. Chemometr. Intell. Lab. Syst. **39**(1), 43–62 (1997)
21. Wijerathne, N., Viswanath, S.K., Hasala, M.S., Beltran, V., Yuen, C., Lim, H.B.: Towards comfortable cycling: A practical approach to monitor the conditions in cycling paths. In: 2018 IEEE 4th World Forum on Internet of Things (WF-IoT), pp. 778–783. IEEE (2018)
22. Zhao, Y.X., Su, Y.S., Chang, Y.C.: A real-time bicycle record system of ground conditions based on internet of things. IEEE Access **5**, 17525–17533 (2017)

An Investigation of Different Methodology Used for Achieving Compact Multiband Microstrip Antenna for Wireless Application

Kamireddy Manohar[1](✉), Vijayasri Bolisetti[1], and Sanjeev Kumar[2]

[1] Department of ECE, Aditya College of Engineering and Technology, Surampalem, India
kmanohar.viswa@gmail.com, vijayasri.bolli@acet.ac.in
[2] Accendere Knowledge Management Services Pvt. Ltd, New Delhi, India
sanjeev.kumar@accendere.co.in

Abstract. A microstrip patch antenna with a multiband operation and small size can attract the antenna designers, which can provide overall size reduction in the communication system. In this paper, we mentioned the ways to improve the bandwidth and size reduction of the microstrip patch antenna to operate at different bands of frequencies. The techniques like slots loading, slotted shorted patch, pentagonal fractal patch, defective ground structure, parasitic elements (inverted L- and T-shaped), etc., are explained in this paper. By using these techniques, we can cover Wi-Max, Wi-Fi, 802.119, and C-bands while making it's size compact. All these features allow us to design an antenna that can be suitable for multiband operation in the field of wireless communication, i.e., up coming 5G applications.

Keywords: Compact antenna · Multiband · Slot antenna · Metamaterial

1 Introduction

The salient features of microstrip antenna such as small size, cost-effectiveness, low-profile high operating frequency range made it an excellent one to use it in the handheld devices such as mobile phones [1–4]. And moreover, it allows us to obtain dual or circular polarization and ease of controlling resonance frequency [5]. An antenna with a low profile can play a vital role in the overall size reduction of the wireless communication system [6]. The features of the microstrip antenna made it well suited for wireless communication. However, there exist some drawbacks such as narrow bandwidth, the existence of surface wave modes, and difficulty of controlling radiation patterns for a single element [5]. In general, design of the multiband antenna is the challenge of multiple application systems [7]. The place problem occurs with the usage of antennas which are operating at a single frequency band [4]. The remedy for the above drawbacks is the multiband antenna, which can be operated at Wi-Max (2.3–2.4 GHz), Wi-Fi (2.4–2.45 GHz), and at other frequencies [8]. Different bandwidth enhancement techniques are proposed while keeping the antenna as a low profile one [7, 9].

© Springer Nature Singapore Pte Ltd. 2021
P. Deshpande et al. (eds.), *Next Generation Information Processing System*,
Advances in Intelligent Systems and Computing 1162,
https://doi.org/10.1007/978-981-15-4851-2_16

1.1 Basic Design Equations of the Antenna

Microstrip patch antenna has so many configurations, but the most widely used patch configurations are rectangular patch and circular patch [1].

For rectangular patch

(a) Width of the patch

$$w = \frac{c}{2F\sqrt{\frac{\varepsilon_r + 1}{2}}} \quad (1)$$

(b) Length of the patch

$$L = L_{\text{eff}} = 2\Delta L \quad (2)$$

(c) Extended length

$$\Delta L = 0.412\frac{(\varepsilon_{\text{eff}} + 0.3)\left(\frac{w}{h} + 0.264\right)}{(\varepsilon_{\text{eff}} - 0.264)\left(\frac{w}{h} + 0.8\right)} \quad (3)$$

(d) Effective dielectric constant

$$\varepsilon_{\varepsilon_{\text{eff}}} = \frac{\varepsilon_r + 1}{2} + \frac{\varepsilon_r - 1}{2}[1 + 12\,h/w]^{1/2} \quad (4)$$

(e) Effective length

$$L_{\text{eff}} = \frac{\lambda}{4\sqrt{E_r}} \quad (5)$$

For circular patch
The radius of the patch

$$a = \frac{F}{\left\{1 + \frac{2h}{\Pi a \varepsilon_r\left[\ln\left(\frac{\Pi a}{2h}\right)\right] + 1.7726}\right\}^{1/2}} \quad (6)$$

where c = speed of light, ε_r = dielectric constant of the substrate, F = operating frequency, L_{eff} = effective length of the patch, ΔL = extended length due fringing fields, λ = operating wavelength, h = thickness of substrate, and w = width of patch.

2 Techniques to Obtain Compact Multiband Microstrip Antenna

Due to the rapid increase in miniaturization communication systems and multiple applications, an antenna with a small size and large bandwidth is needed. A compact multiband microstrip antenna can be able to fulfill this requirement. There are some methods to achieve multiband microstrip antenna. They are as follows:

2.1 Compact and Broadband Microstrip Antenna with Shorted and Slot Loading Patch [7, 10]

In this, the antenna's radiating patch is connected to ground via substrate layer using a shorting pin, and the edge of the antenna is wounded with a copper strip to complete the circuit. It can be also done by using a shorting post. The shorting post method is a very easy method when compared with the copper strip wounding around the edges. The type of application will decide the position or place of the shorting pin. The configuration of the compact size microstrip antenna is shown in the [7, 10]. In method slot method, the radiating patch of the microstrip patch antenna can make the antenna to operate at the different frequencies, thereby making its size compact. Two slots are created in the radiating patch in the triangular patch. This configuration can provide 25% of size reduction and multiband operation. This type of antenna configuration is represented in Fig. 1.

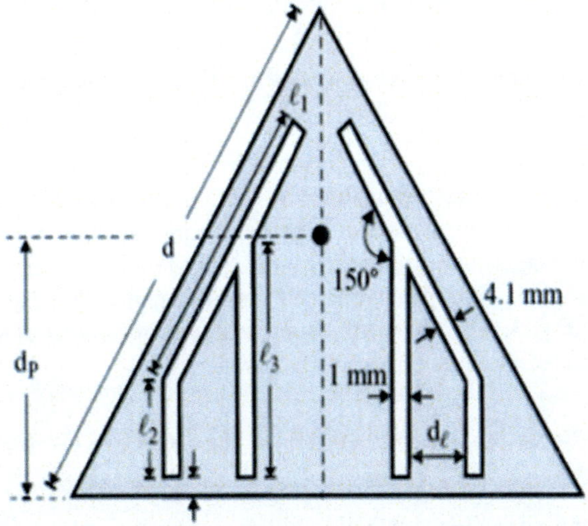

Fig. 1. Triangular patch slotted antenna

2.2 Planar Multiband Antenna for Compact Mobile Transceivers [2]

This type of antenna configuration is fabricated by using FR-4 substrate and consists of two radiating elements of inverted G-shape, which is connected to a feeder (microstrip feeder) symmetrically. A G-shaped radiating element is shorted to the ground plane through a hole. The ground plane slotting is used by the radiator to produce number of resonant frequencies which are essential for modern communication systems. The radiation pattern of this antenna is an omnidirectional pattern, and the size is also compact.

2.3 A Compact Multiband Open-Ended Slot Antenna for Mobile Handsets [11]

This antenna can be formed by making two slots that are open-ended, at the edges of the ground plane, and the slots are made as T-type and E-shape, respectively. This antenna can operate at five bands which can be easily tuned by changing the geometrical parameters of respective slots that are open-ended. Due to this, the size can be optimized and thereby getting multiband operation.

2.4 Compact Multiband Microstrip Antenna Using Inverted L- and T-Shaped Parasitic Elements [12]

A new microstrip antenna that consists of inverted L-shaped and T-shaped parasitic elements is introduced in this technique. This antenna has a ground plane which is made up of aluminum and fabricated on FR-4 substrate. The operating frequencies are independently tuned by varying independent parameters of the antenna.

2.5 Multiband Microstrip Antenna Based on Durer Pentagonal Fractal Patch [4]

The structure of the Durer pentagon is designed by Albert Durer about 500 years ago. This structure is used in microstrip antennas to yield multiband operation and compact size. In this iteration, concept is used for the pentagonal patches. Figure 2 represents the shape of the pentagon and its first iteration. The iteration consists of a pentagon which is divided into six mini pentagons and removed part of five isosceles triangles. This is shown in Fig. 3. The usage of Durer pentagon results in dual-band antenna which can be operated at commonly used bands.

2.6 Compact Microstrip Antenna Based on the Defective Ground Structure [6]

This technique introduces a new method to obtain a miniaturized multiband microstrip antenna. The U-shaped and L-shaped slots help to attain DGS and thereby providing multiband configuration as shown in Fig. 4. The gain of the antenna is reduced significantly by making U-shaped slots on the patch and shorting wall to the ground. The shorting position will decide the operating bands, which means the timing depends on the place of the shorting pin.

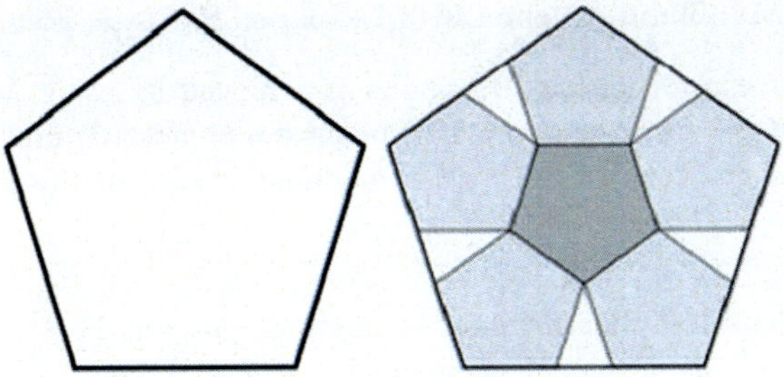

Fig. 2. Pentagon and its first iteration

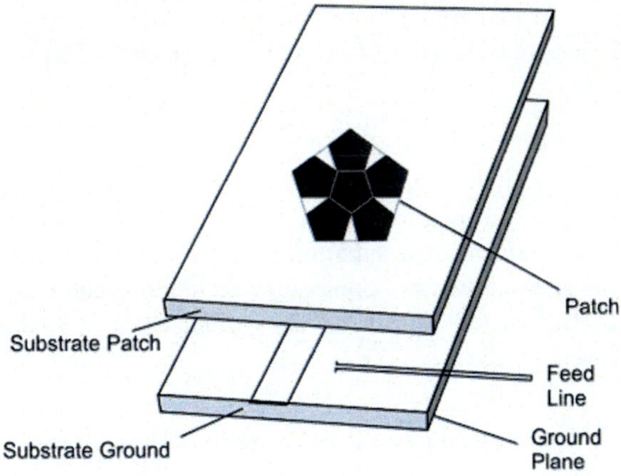

Fig. 3. Durer pentagonal patch

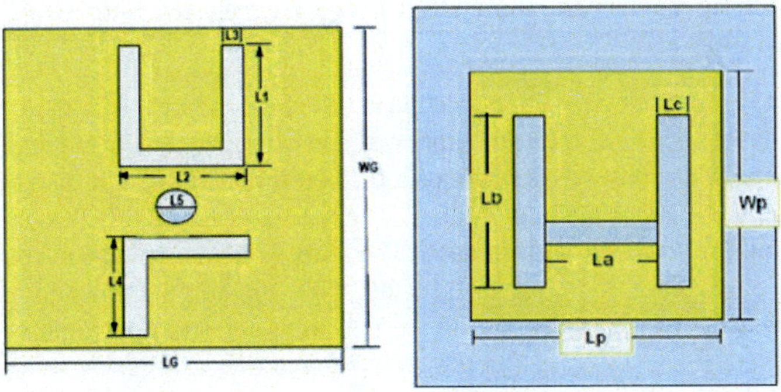

Fig. 4. Defective ground structure

2.7 Compact Multiband Antenna with U-Shaped Parasitic Elements [13]

In this antenna, the resonant band tunability is achieved by using parasitic elements which are in U-shape fabricated on a TMM4 substrate, meaning that the antenna will operate at the single band as well as at multiple bands. In this antenna, a good impedance match at all the bands is observed.

2.8 Compact Multiple Slot Triband Microstrip Antenna [14]

This configuration differs from the normal microstrip antenna in its radiating patch structure, in which an M-slot and two L-slots are etched on the radiating patch. This technique will yield the multiband operation for the required applications.

2.9 Compact Triangular Slotted Microstrip Antenna for Multiband Operation [15]

A rectangular SRR which will act as a partial metamaterial loading for the main patch is used in this technique. Generally, the point to multipoint communication is needed to connect wireless devices. The same thing can be obtained by the trapezoidal monopole antenna that has an omnidirectional radiation pattern. The double-band resonance can be obtained by the trapezoidal monopole in which the rectangular patch is made symmetrically with the transmitting patch. This antenna has a very compact size among all the available antennas.

2.10 Multiband Metamaterial-based Microstrip Antenna [16]

In this technique, two antennas with 2×3 CSSRs are produced behind of the microstrip aerial. In this case, first antenna is denoted as antenna 1, and next antenna is carried out by the way of integrating a CSRR cell into the emission patches, resulting in a triple band tasks for achieving the WLAN and Wi-Max bands. Here, the CSSR acts as a metamaterial configuration to recognize multiband presentation.

2.11 Multiband Patch Antenna Design Using Corrugated Split-ring Resonators [17]

Asymmetrical inspired design of the antenna is introduced in this technique. It is said to be a simple design because it uses a number of non-corrugated SRR materials, and it is in square shape. The substrate is made of a dielectric material like a diamond and the circular metallic parts are made of copper. The spacing between the rings is made as low as possible to get good performance. The narrow spacing between the rings will increase the capacitance, gain, and bandwidth. This antenna will operate at eight resonance frequencies which can be useful for telecommunication networks.

2.12 Dual C-slotted Microstrip MIMO Antenna for Wireless Applications [18]

MIMO means multiple inputs multiple outputs that mean single antenna has the capability to utilize a number of antennas, such type of antenna generally developed by etching the patch onto the substrate of PCB slabs and ground plane. The novel design in this antenna will results in multiresonance broadband frequency. This may be operated for a high data rate of more than 1 GBPS, which is essential for LTE, 4G, and 5G wireless system [19].

2.13 Asymmetrical Parallel Slotted Multiband Microstrip Patch Antenna [20]

A plain compressed design of a parallel slotted asymmetrical patch with a feed stripe of microstrip that operates a number of bands is introduced in this technique. The three layers named substrate, patch, and ground are fabricated on FR-4 substrates. To make this antenna to operate at multiple bands, slots have to be made on the patch with correct dimensions.

2.14 Compact and Broadband Antenna with Stacked Shorted Patch [21]

An increase in bandwidth can be obtained for a constant volume by placing two stacked shorted patches for making both the patches to radiate equally, thereby getting a low radiation quality factor. The multiband operation can be obtained using a stacked configuration or a multiresonator gap coupled configuration. This type of configuration of a compact multiband s-shaped microstrip patch antenna is given in [9].

2.15 60° Sectorial Antenna for Dual-Polarized Multiband and Wideband [22]

A microstrip antenna that can operate at multiple bands is introduced in this antenna. In this, the patch is made with 60° sectorial slots which can change the resonance frequency. The broadside radiation pattern with multiband operation can be obtained or controlled by the corner truncation, which can further reduce patch effective aperture. The orthogonal variations are shown by the current distributions on the surface which are the causes of dual-polarized response.

3 Result Section

Table 1 shows the comparative analysis of the published in term of bandwidth, reduction of size, and methodology applied in the published paper. One more important parameter of antenna is VSWR, i.e., voltage standing wave ratio. From the above table, it is below than 2, which show that antenna is resonating in the desire frequency band.

Table 1. Comparative analysis of the published papers

Broadband and compact methodology used	Alignment	Bandwidth	Reduction size	VSWR	Applications
Stacked shorted patches technique	Square-shape stacked patch	76.25%	24.6%	<2	GPS
	S-shape stacked patch	14%	24.6%	<2	IMT 2000 mobile handset
Slot loading technique	Rectangular slot loaded	Less than v-slot	65%	<2	WLAN, GPS, Wi-MAX
	Triangular patch	3 times that of regular triangular microstrip patch antenna	25%		
	V-slot loaded	Higher than rectangular patch	60%		
Open-ended slot technique	T- and E-shaped slots	850 MHz (wider)	–	<3:1	GSM, WLAN
Origination of parasitic elements	Inverted L- and T-shaped parasitic elements	255 MHz (7.29% max)	–	≤ 2	WLAN, LTE TDD NO. 34, Wi-MAX
Convex pentagon-shaped microstrip patch	Durer pentagon fractal patch	73 MHz	15%	<2	Bluetooth, cellular communication, RFID applications

4 Conclusion

The size and bandwidth of the antenna are the considerable design specifications for any application of microstrip antenna. Sequentially to build the antenna to function at multiple bands, one should follow the proper method to make it as a compact multi-band microstrip antenna, otherwise the size and band of operation are not compromised properly. So, one can follow the specific method for each application. In this paper, the available techniques which are used to yield compact size and multiband operation in microstrip antenna are mentioned. Depending on the application, each technique can obtain the required bandwidth and compact size.

References

1. Balanis, C.A.: Antenna Theory: Analysis and Design, 3rd edn. John Wiley & Sons, Hoboken (NJ) (2005). ISBN: 0-471-66782-X
2. Dadgarpour, A., Abbosh, A., Jolani, F.: Planar Multiband Antenna for Compact Mobile Transceivers. IEEE Antennas Wave Propag. Lett. **10**, 651–654 (2011)
3. Esfahlani, S.H.S., Tavakoli, A., Dehkhoda, P.: A compact single-layer dual-band microstrip antenna for satelite applications. IEEE Antennas Wirels. Propag. Lett. **10**, 931–934 (2011)
4. Iyer, B.: Characterisation of concurrent multiband RF transceiver for WLAN applications. Adv. Intell. Syst. Res. **137**, 834–846 (2017)
5. Anguera, J., Borja, C., Puente. C.: Microstrip fractal—shaped antennas a review. In: 2nd European Conference on Antennas and Propagation, Edinburgh, UK (2007)
6. Shah, S.I.H., Bashir, S., Shah, S.D.H.: Compact multiband microstrip antenna using defected ground structure. In: The 8th European Conference on Antennas and Propagation, Hague, Netherlands (2014)
7. Kakaria, P., Nema, R.: Review and survey of compact and multiband microstrip patch antenna. In: IEEE International Conference on Advances in Engineering and Technology Research, India (2014)
8. Ali, M.M.M., Azmy, A.M., Haraz, O.M.: Design and implementation of reconfigurable quad-band microstrip antenna for MIMO wireless communication applications. In: 31st National Radio Science Conference (NRSC 2014), Ain Shams University, Egypt (2014)
9. Kumar, S., et al.: A multiple band-notched monopole antenna with incorporated GSM and UWB for wireless applications. Inter. J. Adv. Sci. Technol. **28**(16), 362–378 (2019)
10. Mahmoud, S.F.: Compact broadband microstrip patch antennas. In: 4th National Radio Science Conference, Cairo, Egypt (2007)
11. Cao, Y., Yuan, B., Wang, G.F.: A compact multiband open ended slot antenna for mobile handsets. IEEE Antennas Wirel. Propag. Lett. **10**, 911–914 (2011)
12. Kim, J.-W., Jung, T.-H., Ryu, H.-K., Woo, J.-M.: Compact multiband microstrip antenna using—L-shaped and T-shaped parasitic elements. IEEE Antennas Wirel. Propag. Lett. **12**, 1299–1302 (2013)
13. Iyer, B., Abegaonkar, M.P., Koul, S.K.: Reconfigurable inset-fed patch antenna design using DGS for human vital sign detection application. In: Computing, Communication and Signal Processing. Advances in Intelligent Systems and Computing, vol. 810. Springer, Singapore (2019)
14. Rathod, B., Iyer, B.: Concurrent triband filtenna design for WLAN and WiMAX applications. In: Hitendra Sarma, T., Sankar, V., Shaik, R. (eds.) Emerging Trends in Electrical, Communications, and Information Technologies. Lecture Notes in Electrical Engineering, vol. 569. Springer, Singapore (2020)
15. Rahman, S., Kiran, U..: A compact triangular slotted microstrip antenna for multiband operation. In: 2016 International Conference on Wireless Communications, Signal Processing and Networking (WiSPNET), Chennai, India
16. Yu, K., Li, Y.: Multi-band metamaterial-based microstrip antenna for WLAN and Wi-MAX applications. In: 2017 International Applied Computational Electromagnetics Society Symposium—Italy (ACES), Florence, Italy
17. Patel, S.K., Argyropoulos, C., Kosta, Y.P.: Broadband high gain multiband patch antenna design using corrugated split ring resonators. In: 2017 Progress in Electromagnetics Research Syposium—Fall (piers-fall), 19-22 Nov, Singapore (2017)

18. Gupta, R.K, Shanmuganantham, T. Dr., Kiruthika, R.: Dual C-slotted microstrip patch MIMO antenna for multiband wireless applications. In: 2017 International Conference on Intelligent Computing, Instrumentation, and Control Technologies (ICICICT), Kannur, India
19. Kumar, S., Kumar, R., Vishwakarma, R., Srivastava, K.: An improved compact MIMO antenna for wireless applications with band notched characteristics. Int. J. Electron. Commun. (AEÜ) **90**, 20–29 (2018)
20. Mehra, K., Jain, A.: Asymmetrical parallel slotted multiband microstrip antenna. In: International Conference on Information, Communication, Instrumentation and Control (ICICICI 2017), Indore, India (2017)
21. Tian, L., Xue, Z., Li, W., Ren, W.: Stacked triband microstrip patch antenna. In: IEEE 5th International Symposium on Electromagnetic Compatibility, Beijing, China (2017)
22. Ambekar, A.G., Venkata, A.P.C., Deshmukh, A.A., Kadam, A.A..: 60° Sectorial microstrip antenna for dual polarized multiband and wideband response. In: Fourth International Conference on Computing Communication Control and Automation (ICCUBEA), Pune, India (2018)

STP:Suicidal Tendency Prediction Among the Youth Using Social Network Data

Manish Sharma(✉), Bhasker Pant, Vijay Singh, and Santosh Kumar

Graphic Era deemed to be University Dehradun,
566/6 Bell Road Clement Town, Dehradun, Uttarakhand, India
manishsharma.cse@geu.ac.in, http://www.geu.ac.in

Abstract. The suicide tendency is increasing day to day. It is having a negative impact on our youth. Human at possibility of suicide does not want help before trying to attempt and do not feel necessity to any mental health counselling. As a result, suicidal tendency is becoming social challenge. At the same time, social media are becoming popular for communication and exchange emotional expression to the world. Social media like Twitter, Facebook and Instagram play major role for emotion sharing. Therefore, huge number of people publishes their emotion of depression and happiness within notes in these social media like Twitter. Social platform like Twitter has a large number of collection of emotional notes. In this situation, machine learning helps in early prediction of the depression and suicidal tendency. Therefore, in this paper, we develop a soft solution which is able to early detect suicidal tendency among the youth. In this approach, we train a model with genuine suicidal notes and post collected from different sources and generate score for input social media tweet high or low for social tendency prediction. With the help of social media postings like Twitter, we are able to identify risk of suicide.

Keywords: Social media · Social network analysis · Twitter · Computational social science · Suicide

1 Introduction

Suicide is a major social problem in the world, and suicide creates a big issue to community mental health. In this area of research, it is very difficult due to the lack of useful information special about suicide attempts and is also difficult to know the information after the death of person. Problem is that only some individual share the emotion in form of notes and posts. So, gathering suicide data for research is barriers [3].

Although social media are growing tool that may help in research, social media influence subsequently large number of individual to share their emotion

© Springer Nature Singapore Pte Ltd. 2021
P. Deshpande et al. (eds.), *Next Generation Information Processing System*,
Advances in Intelligent Systems and Computing 1162,
https://doi.org/10.1007/978-981-15-4851-2_17

in real time. So, the large amount of data generated by social media using Twitter platform, where user know as Twitter post status and update the tweets frequently in real time [1], and these tweets broadcast publicly and are openly available to all [4,5]. Twitter is publicly available in interest of followers.

The purpose of this suicidal tendency prediction model is to determine based on Twitter conversation whether Twitter user is at risk of suicide or not. In this paper, we employed it in two phase process firstly we identify keywords that represent high risk factor of suicide tendency through natural language processing of suicide notes and posts and generate word cloud of keywords as shown in Fig. 1. In second phase, we pilot tested each keywords appeared in individual tweets in context of suicide risk. Finally, based on appearance of keywords, generate score of tweet for prediction [6].

1.1 Motivation

In our society, suicide behaviour is a challenging community issue, and it is very difficult to research in this area. Many problems we face to suicidal prediction are due to lack of surveillance. And at the same time, it is difficult to collect information after the death of individual. This problem motivates us to establish the level of concern for Twitter users and individual after analysis of his or her tweets which made direct or indirect textual relationship [8]. With the help of this process, our aim is to design and implement a system to compute and predict the tendency of suicidal. The framework of this system is very help full to new data scientist.

2 Methodology

Machine learning (ML) and natural language processing (NLP) have most frequently used to establish and identify correlation between social media data and suicide notes and posts. We are developing a model using ML and NLP to detect individual tendency of suicide through social media like Twitter. Using this model, we are able to predict the high or low tendency of suicide as shown in Fig. 2. First phase is original data collection of suicidal notes and post and then for prediction for pilot test data collection of Twitter user. In second phase, it established the relationship between it and generated prediction.

2.1 Suicidal Notes and Posts Collection

Dataset of genuine suicide note taken from published in a blog [16] http://russelljohn.net/journal/2008/03/a-collection-of-suicide-notes. Some other source of suicide notes publish [15] in http://suicideproject.org. In this project website, there are two type suicide notes which published those who died and those who survived form their suicide attempt. In this project website, individual are free to share their thoughts and feeling in form of blog. And all blogs are freely published online as anonymous. In some blog, personal detail of individuals were revealed like age, gender and demographic [10] information [15]. Some genuine suicide notes sample [7,9,11].

2.2 Twitter Data Collection

Using programming language R and library tweet [17], we collect the twitter real-time data. rtweet design provide specific filter criteria such as keyword, time and individual attributes. Twitter developer account is also available for researcher for collecting Twitter data. Some open-source application-programming interface API is also provided by Twitter. In Twitter, tweets keywords include warning signs like drug abuse, family violence and previous suicide attempts [1]. Twitter archiver is also used for data collection based upon defining specific rules [21].

2.3 NLP and ML Computational Process

The machine learning algorithms is used to train the model from the given processed data from the data passed through NLP phase. This data can generalize unseen random data of suicidal notes and posts. Machine learning performs three-way, supervised, semi-supervised and unsupervised learning. In first supervised learing, training data are already well defined and labelled, in semi-supervised learning some part of data are defined and labelled, and in last one unsupervised learning, main challenge is to define the structure of data [12,13].

3 Suicidal Tendency Prediction Analysis Model

There are five phases of this model, which start with data cleaning, feature engineering, training, score validate model and evaluate model.

Fig. 1. Word cloud of suicide notes

3.1 Phase 1: Suicide Notes Text Dataset Cleaning and Pre-processing

In phase 1, we perform text cleaning and pre-processing module of natural language processing, to reduce the noise. Text cleaning is also used to find the important features of dataset and improve the accuracy of the STP model. In cleaning phase, we remove the stopwords like "a", "an", or "the" and duplicated characters, numbers and special characters. And also change the dataset in lowercase. For that, we use Python library "nltk".

3.2 Phase 2: Feature Engineering from Pre-processed Clean Dataset

In this phase, we extract numeric feature like N-Gram from pre-processed and clean dataset. In this module, we remove words and arrange in column by using whitespaces between words and generate a dictionary of unique words from pre-processed and clean dataset. Count the frequency or N-gram of appearance in dataset and create a vectors of features from those frequency counts. Figure 1 shows word cloud of suicide notes.

3.3 Phase 3: Training Model and Learning Algorithm

In this phase, dataset classification using learning algorithm is performed. Classified text is transformed in to numeric value based on N-Gram values. Now, we perform two-class logistic regression to predict score and review it. So that, text problem is converted into regular classification of dataset in this phase.

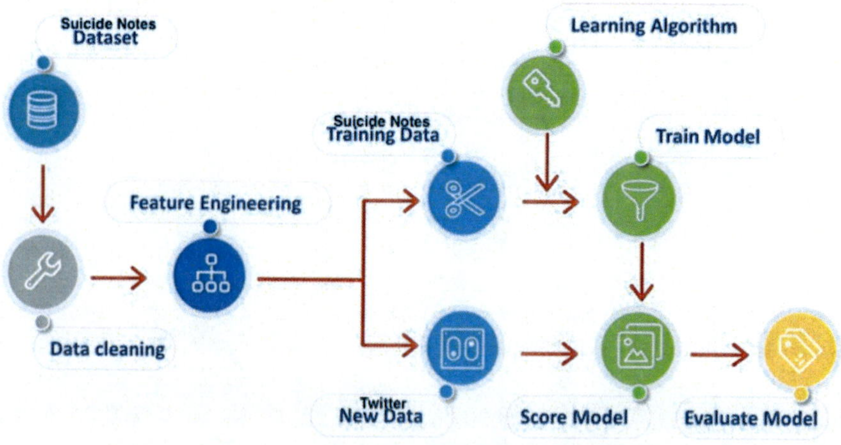

Fig. 2. Suicide tendency score prediction process. *Online Image Source* https://towardsdatascience.com/data-preprocessing-3cd01eefd438

3.4 Phase 4: Score Modelling for Validation

In [18] this phase, we validate trained model using score against of test dataset and calculate accuracy. Model learned the vocabulary from the training dataset using N-Gram and frequency of words. So that, we should use the weight and vocabulary when features extract from test data [21].

3.5 Phase 5: Evaluate Model

In [19,20] this phase, model is ready for evaluating test data and also trained for deploying. Based upon the score of test data, model returns the prediction result as high or low tendency of suicide. It also trains the model for making prediction accurate.

Fig. 3. Word cloud of suicide notes after removing stopwords

4 Implementation

4.1 Dataset Cleaning and Pre-processing and Feature Extraction

Using Python open source library nltk, string, re, collections, pandas, nltk.corpus stopwords, we pre-process suicide notes and post. In pre-process, we remove stops words and symbols as shown in Fig. 3. In feature extraction process, the

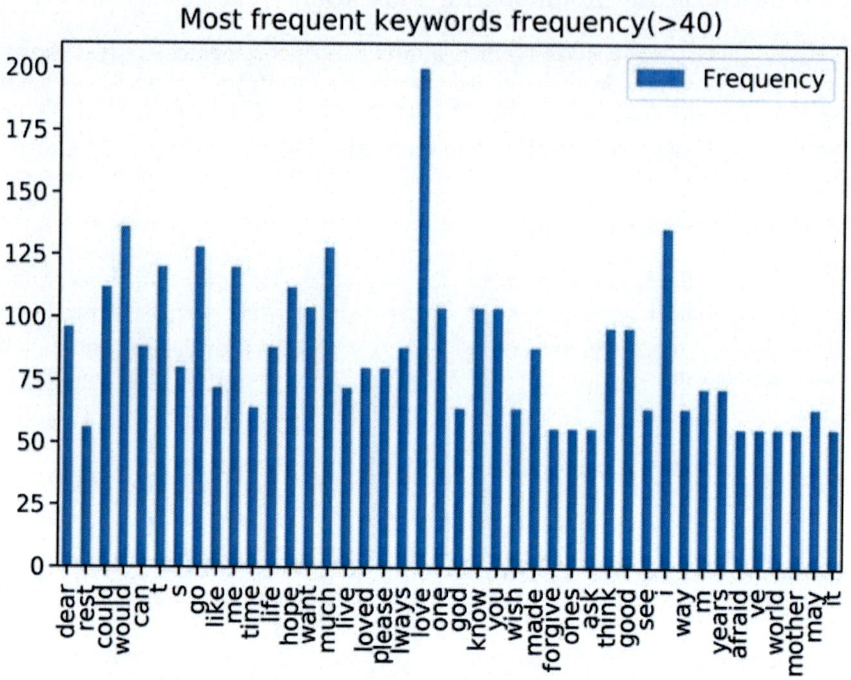

Fig. 4. Keywords frequency graph

$$TFIDF\ score\ for\ term\ i\ in\ document\ j = TF(i,j) * IDF(i)$$

where

$$IDF = Inverse\ Document\ Frequency$$

$$TF = Term\ Frequency$$

$$TF(i,j) = \frac{\text{Term i frequency in document j}}{\text{Total words in document j}}$$

$$IDF(i) = \log_2\left(\frac{\text{Total documents}}{\text{documents with term i}}\right)$$

and

$$t = Term$$

$$j = Document$$

Fig. 5. Term Frequency

frequency of unique words. Most frequent keywords and phrases appear in suicide notes and posts like: "want love, no hope, known, think, life, always, kill myself, end, life never, can't go, die, without me, better off dead, tired of living, don't want to be, die alone, go to sleep forever". Frequence graph generated using matplotlib of Python library as shown in Fig. 4.

4.2 Training Model and Machine Learning Algorithm

ML is used to train the computational model. In our framework, we trained our model by using genuine suicide dataset collected from online [15,16]. And it can be generalized to unseen training dataset.

Term frequency and inverse document frequency (TF-IDF) : it is used for information retrieval and also used for text mining. TF-IDF used the scheme of assign weight to each term in a document, and this weight assigns based on frequency of the particular term. The weight of higher term scores more important as compare to other term. There are two types of frequency, one is term frequency and other is inverse document frequency. Term frequency defines frequency of term in document, and inverse document frequency measures term importance of term in document. Model is defined in Fig. 5.

Bag Of Words Model: The bag of words model is extracting features from text. A bag of words is a representation of text that describes the occurrence of words within a document. It involves two things:

1. A vocabulary words.
2. A calculate the presence of words.

5 Results

We implement both classifier in Python and perform machine learning training and testing on dataset https://www.kaggle.com/kazanova/sentiment140. It contains 1,600,000 tweets extracted. From this dataset, 20% suicide notes are used for training, and 80% are used for testing the classifier performance. Compersion results are given in Table 1. As per result, TF-IDF is better performed than BOW.

Table 1. TweetClassifier prediction compersion

	TF-IDF	BOW
Precision	0.9230769230769231	0.8666666666666667
Recall	0.5106382978723404	0.2765957446808511
F-score	0.6575342465753424	0.4193548387096774
Accuracy	0.8820754716981132	0.8301886792452831

6 Conclusion

In this paper, we demonstrate that Twitter tweet can be used to calculate the risk factor and suicidal tendency prediction. These platform initially provides the validation for Twitter dataset for future suicidal tendency research group. And also provide new dimension for social service and public health. Score finding defines that there is a association between suicidal notes and tweets and risk for actual suicide. In future, this model can be improved based on rich training dataset.

References

1. Marres, N., Gerlitz, C.: Interface methods: renegotiating relations between digital social research. STS and sociology. Sociol. Rev. **64**(1), 21–46 (2016). https://doi.org/10.1111/1467-954x.12314
2. Borra, E., Rieder, B.: Programmed method: developing a toolset for capturing and analyzing tweets. Aslib J. Inf. Manage. **66**(3), 262–278 (2014). https://doi.org/10.1108/ajim-09-2013-0094
3. Marres, N., Moats, D.: Mapping controversies with social media: the case for symmetry. SSRN Electron. J. (2015). https://doi.org/10.2139/ssrn.2567929
4. Christensen, H., Batterham, P., O'Dea, B.: E-health interventions for suicide prevention. Int. J. Environ. Res. Pub. Health **11**(8), 8193–212 (2014). https://doi.org/10.3390/ijerph110808193
5. Namratha, P., Kishor, M., Sathyanarayana Rao, T.S., Raman, R.: Mysore study: a study of suicide notes. Indian J. Psychiatry **57**(4), 379–382 (2015)
6. Synnott, J., Ioannou, M., Coyne, A., Hemingway, S.: A content analysis of online suicide notes: attempted suicide versus attempt resulting in suicide. Suicide Life Threat. Behav. **48**(6), 767–778 (2017). https://doi.org/10.1111/sltb.12398
7. Prokofyeva, T. (2013). Language use in two types of suicide texts. Published MA thesis. Linkoping University
8. Robinson, J., Cox, G., Bailey, E., Hetrick, S., Rodrigues, M., Fisher, S., Herrman, H.: Social media and suicide prevention: a systematic review. Early Intervention Psychiatry **10**(2), 103–121 (2015). https://doi.org/10.1111/eip.12229
9. Hamilton, D.R.: Suicide as an escape from pain: an analysis of suicide notes and case files. Browse all Theses and Dissertations. 667.https://corescholar.libraries.wright.edu/etd_all/667 (2012)
10. Berti, D.: Suicide notes under judicial scrutiny in India. South Asia Multidisc. Acad. J. [Online], 17 — 2018, Online since 19 February 2018, connection on 30 April 2019. http://journals.openedition.org/samaj/4481; https://doi.org/10.4000/samaj.4481
11. Rani, M., Girdhar, S., Murty, O.: Suicide note: the last words. J. Forensic Med. Toxicol. **32**(2), 35–41 (2015)
12. Pestian, J.P., et al.: A machine learning approach to identifying the thought markers of suicidal subjects: a prospective multicenter trial. Suicide Life Threat. Behav. **47**(1), 112–121 (2017)
13. Gomez, J.M.: Language technologies for suicide prevention in social media. In: Workshop on Natural Language Processing in the5th Information Systems Research Working Days, pp. 21-29, Quito, Ecuador (2014)

14. Jashinsky, J., Burton, S.H., Hanson, C.L., West, J., Giraud-Carrier, C., Barnes, M.D., et al.: Tracking suicide risk factors through Twitter in the US. Crisis **35**(1), 51–9 (2014)
15. The Suicide Project—Share Your Suicide Stories. Suicideproject.org. Retrieved Oct, 2019 from http://suicideproject.org (2004)
16. A Collection of Suicide Notes & Letters. Published in a weblog: http://russelljohn.net/journal/2008/03/a-collection-of-suicide-notes
17. R API for Collecting tweets of individuals. https://github.com/ropensci/rtweet
18. Briscoe, T., Medlock, B., Andersen, Ø.E.: Automated assessment of ESOL free text examinations. Technical Report UCAM-CL-TR- 790, University of Cambridge, Computer Laboratory (2010)
19. Shermis, Mark D.: Contrasting state-of-the-art in the machine scoring of short-for constructed responses. Educ. Assess. **20**(1), 46–65 (2015)
20. Crossley, S., Allen, L.K., Snow, E.L., McNamara, D.S.: Pssst... textual fea- tures... there is more to automatic essay scoring than just you! In: Proceedings of the fifth international conference on learning analytics and knowledge, pp. 203–207. ACM (2015)
21. Zhang, X., Fuehres, H., Gloor, P.A.: Predicting stock market indicators through Twitter "I hope it is not as bad as I fear." Retrieved from http://dx.doi.org/10.1016/j.sb-spro.2011.10.562

Check for updates

Feature Reduction-Based DoS Attack Detection System

Jahed Momin Shaikh(✉) and Deepak Kshirsagar

Department of Computer Engineering & IT, College of Engineering Pune (COEP), Pune, India
jahed22shaikh@gmail.com; mominsjl8.comp@coep.ac.in
ddk.comp@coep.ac.in

Abstract. Denial-of-service (DoS) has still been popularly used by attackers. It can be seen in China and the USA who are major victims of DoS attacks in recent years. For this reason, the development of an intelligent intrusion detection system (IDS) remains a challenging task. This study proposes a system for the detection of DoS attacks with feature reduction using a rule-based PART classifier. The reduced feature set is identified based on the combination of information gain and correlation attribute evaluation methods. The system is implemented and tested on CICIDS 2017 dataset. Finally, the proposed system provides an accuracy of 99.9871% for the detection of DoS attacks with 56 reduced features.

Keywords: Rule-based classifiers · Feature reduction · Information gain · Correlation · CICIDS 2017

1 Introduction

In the present digital era, DoS attack is a major concern for cybersecurity. DoS attacks are directed [1] towards web applications and IoT infrastructure. This attack is generated at different layers of TCP/IP model with the help of different tools by the attacker. The behaviour of an application layer DoS attack is different from a network and transport layer.

Intrusion detection system (IDS) is one of the security layers for the network to defend attacks. Generally, IDSs are classified into host, network and distributed depending upon architecture. IDSs are classified into misuse and anomaly [2] based on detection strategy. In misuse detection, the system captures [3] network traffic, search for the pattern, and based on rules, it identifies malicious traffic. The anomaly IDS based on the profile generates alarms for malicious traffic. The benefit of anomaly IDS is that it detects unknown attacks with a high detection rate, but it creates a high false alarm rate.

Machine learning techniques are currently used by many researchers for the identification of relevant and irrelevant features present in captured network traffic. Most of the researchers use KDDcup99, NSL-KDD, and UNSW-NB15 datasets for the evaluation of the proposed method. This study mainly concentrates on the application layer DoS attack, namely SlowHTTPTest, Hulk, Slowloris, GoldenEye, and

© Springer Nature Singapore Pte Ltd. 2021
P. Deshpande et al. (eds.), *Next Generation Information Processing System*,
Advances in Intelligent Systems and Computing 1162,
https://doi.org/10.1007/978-981-15-4851-2_18

Heartbleed. The above traditional datasets do not consist of the application layer DoS attacks mentioned above. This motivates us to use CICIDS 2017 DoS dataset for the identification of relevant and irrelevant features for DoS.

The major contributions of this study are as follows:

1. The proposed feature reduction method is based on information gain (IG) and correlation attribute evaluation.
2. The proposed system is implemented and tested on CICIDS 2017 dataset with PART rule-based machine learning classifier.

The rest of the manuscript is organized as follows: the literature survey is provided in Sect. 2. Section 3 describes the proposed system with feature reduction. Details of implementation and result analysis are given in Sect. 4. Conclusion and future scope are provided in Sect. 5.

2 Literature Review

Pullagura et al. [4] propose an intrusion detection system with robust feature selection. The system uses an ensemble feature selection technique based on Chi-square, Euclidean distance, and correlation. The system is evaluated on the KDDcup99 dataset with support vector machine (SVM). The system produces accuracy, precision, and recall of 96.25, 80.20, and 78.96%, respectively, with SVM using reduced features.

Patgiri et al. [5] propose recursive feature elimination for intrusion detection. The system uses random forest (RF) and SVM. The system is tested and evaluated on NSL-KDD dataset. The experimental result shows that SVM performs better compared to RF in terms of accuracy.

Ployphan et al. [6] propose a cybersecurity threat detection system using hybrid machine learning on adaptive boost algorithm. The system uses correlation feature selection method and hybrid classifier uses a combination of k-NN, C4.5, MLP, LDA, and SVM based on adaptive boosting. The system is tested on UNSW-NB15, NSL-KDD, and KDDcup99 datasets.

Azar et al. [7] propose an intrusion detection system based on combine feature selection techniques. The combine feature selection is based on correlation, gain ratio, and information gain. The system uses multilayer perceptron (MLP), k-NN, and Naïve Bayes classifiers for the classification. The system is tested on KDDcup99 dataset. k-NN produces the highest accuracy and precision for the classification with six reduced features.

Manzoor et al. [8] propose an intrusion detection system with reduced features using artificial neural network (ANN). The system uses information gain and correlation with ranker. The sets are formed based on the rank of the feature selection techniques with the help of union and intersection operation. The system produces an accuracy of 93.8% for DoS with 25 reduced features using ANN on the KDDcup99 dataset.

Lu et al. [9] propose a hybrid network intrusion detection model with ANN. The extracted features are divided into different feature subsets, and then, these feature subsets are used to train the ANN model. The system uses backpropagation with genetic algorithm for the classification. The system produces the highest precision of 99.9% with reduced features for DoS on the KDDcup99 dataset.

Velliangiri [10] proposes a hybrid feature reduction method for intrusion detection. The hybrid feature reduction method is based on the binary grey wolf optimization (BGWO) and kernel principal component analysis (KPCA). The system produces an accuracy of 95.381% for DoS with SVM on the KDDcup99 dataset.

Wankhede et al. [11] put forward the use of machine learning and neural network for the detection of DoS attack. The system does not provide any feature reduction technique. The system provides an analysis of RF and multilayer perceptron (MLP) on CICIDS 2017 dataset. The experimentation result shows RF produce improved results compared to MLP.

Aljawarneh et al. [12] suggest a new hybrid model for intrusion detection using feature selection analysis. The feature selection analysis is based on information gain and vote scheme. The system uses a hybrid classifier based on J48, random tree, REEP tree, AdaBoostM1, decision stump, Naive Bayes, and Meta Paging classifiers. The hybrid model produces 99.9% for DoS on the NSL-KDD dataset.

Patil et al. [13] recommend distributed DoS attack detection system with the feature selection technique. The system uses IG with ranker for feature selection. The original features are divided into primary and secondary categories based on the value of IG. The system is tested on CICIDS 2017 DDoS dataset with RF, J48, and logistic model tree (LMT). The experimentation result shows J48 which produces 0.43% improvement in detection rate for DDoS attack with 69 reduced features.

This literature review motivates to develop a feature reduction technique using a combination of the information gain and correlation for the detection of an application layer DoS attack.

3 The Proposed System

Figure 1 shows the proposed system for intrusion detection with feature reduction. The first step consists of data pre-processing of the selected dataset for the system. The dataset consists of duplicate features as well as some missing and infinite values. The duplicate features are removed and missing values are replaced with zeros. Infinite values present in the dataset are also replaced with zeros. Data pre-processing results in a compact dataset.

Information gain and correlation attribute evaluation methods with ranker are performed on a compact dataset. Information gain attribute evaluation (IGAE) calculates the weight for each feature based on entropy. IGAE with ranker produces the list of features in a descending order based on the weight. Similarly, correlation attribute evaluation (CAE) calculates the weight of each feature based on Pearson's coefficient and produces a list of features in a descending order based on the weight using ranker. The features that are associated with zero weight based on the information gain and correlation are discarded.

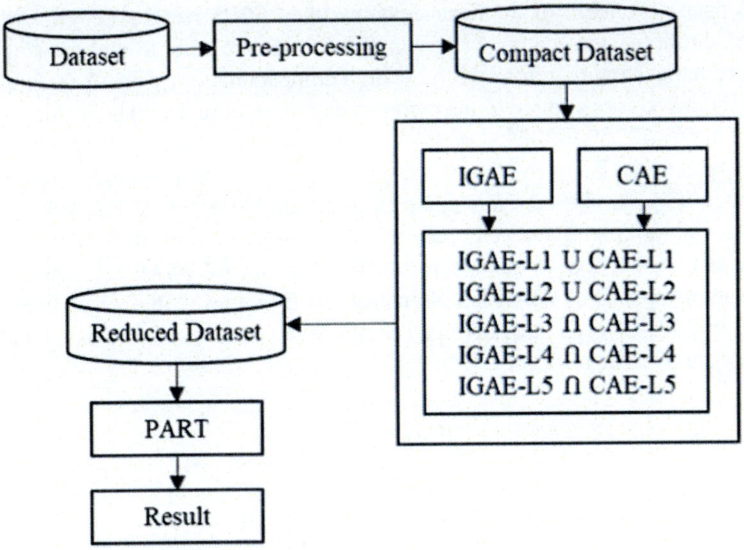

Fig. 1. Proposed DoS attack detection system

The remaining IGAE features are divided into five levels, namely from IGAE-L1 to IGAE-L5, and CAE features are divided into five levels from CAE-L1 to CAE-L5 based on rank. IGAE-L1 to IGAE-L4 and CAE-L1 to CAE-L4 consist of 15 features in each level, and IGAE-L5 and CAE-L5 contain the remaining features.

In the next step, union operation is performed on IGAE-L1 with CAE-L1 and IGAE-L2 with CAE-L2. Intersection operation is performed on IGAE-L3 with CAE-L3, IGAE-L4 with CAE-L4, and IGAE-L5 with CAE-L5. These operations on levels obtain 56 reduced features from the original 77 features.

The reduced dataset obtained with 56 features is given to the rule-based PART classifier for training and testing. PART classifier trains and tests the model for the classification. The proposed system is tested on the dataset and evaluated in terms of performance.

4 Implementation and Result Analysis

The proposed system is implemented with the help of open-source Weka [14] 3.8.3 tool. The pre-processing of the dataset is carried out through the Scikit-learn library of Python programming. The proposed system is tested on Intel(R) Xeon(R) CPU E3-1271 v3 coupled with 32 GB of RAM.

The experimentation and result analysis of the proposed system are based on CICIDS DoS 2017 [15] dataset. The dataset consists of 440031, 5499, 231073, 5796, 10293, and 11 records of Benign, SlowHTTPTest, Hulk, Slowloris, GoldenEye, and Heartbleed, respectively, with 78 features as mentioned in [16].

The dataset consists of duplicate features like "FWD Header Length" which are removed. Some of the feature values in the dataset are infinite and missing. These values are replaced with zeros and to obtain a compact dataset with 77 features. IGAE and CAE methods are applied to the compact dataset with ranker. The features <32, 33, 34, 50, 56, 57, 58, 59, 60, 61> are associated with zero weights which are discarded from IGAE and CAE.

Further, IGAE divides the remaining features into five levels as IGAE-L1 to IGAE-L5 based on ranking. IGAE-L1 consists of the top 15 ranked features. IGAE-L2 consists of the next top 15 ranked features and so on till IGAE-L4. The remaining seven features consist of IGAE-L5. Similarly, CAE divides features into five levels as CAE-L1 to CAE-L5 based on ranking. The five levels of features associated with IGAE and CAE are as shown in Table 1.

Table 1. IGAE and CAE feature levels

Level	IGAE	CAE
L1	66, 2, 6, 13, 16, 17, 19, 37, 40, 41, 42, 43, 53, 55, 65	13, 11, 14, 19, 23, 24, 40, 41, 42, 43, 53, 55, 74, 76, 77
L2	1, 5, 7, 11, 14, 15, 18, 21, 22, 23, 24, 36, 38, 63, 67	1, 2, 8, 12, 17, 18, 21, 22, 28, 29, 39, 44, 48, 49, 52
L3	4, 9, 10, 26, 27, 29, 35, 54, 64, 70, 72, 73, 74, 76, 77	9, 16, 26, 27, 31, 37, 38, 45, 47, 54, 66, 67, 69, 70, 75
L4	3, 8, 12, 20, 25, 28, 30, 39, 44, 48, 52, 62, 68, 69, 75	4, 5, 10, 15, 20, 25, 30, 46, 51, 63, 64, 68, 71, 72, 73
L5	31, 45, 46, 47, 49, 51, 71	3, 6, 7, 35, 36, 62, 65

Union operation is performed on IGAE-L1 with CAE-L1 and IGAE-L2 with CAE-L2 that results into <6, 11, 13, 14, 16, 19, 23, 24, 37, 40, 41, 42, 43, 53, 55, 65, 66, 74, 76, 77> and < 1, 2, 5, 7, 8, 12, 15, 17, 18, 21, 22, 28, 29, 36, 38, 39, 44, 48, 49, 52, 63, 67>, respectively. Intersection operation is performed on IGAE-L3 with CAE-L3, IGAE-L4 with CAE-L4, and IGAE-L5 with CAE-L5 that results into <9, 26, 27, 54, 70>, <3, 20, 25, 30, 62, 68>, and <46, 51, 71>, respectively.

The result of union and intersection on different levels as mentioned in the above implementation finally obtains a reduced feature set of 56 features as shown in Table 2. The reduced dataset with 56 features is used for further experimentation.

Table 2. Reduced feature set

# Features	Feature numbers
56	13, 55, 14, 11, 23, 42, 76, 74, 24, 19, 77, 40, 41, 43, 53, 66, 6, 65, 16, 37, 18, 21, 2, 12, 22, 48, 39, 17, 44, 28, 1, 29, 52, 8, 49, 7, 38, 15, 5, 63, 36, 67, 9, 54, 26, 27, 70, 30, 25, 3, 62, 20, 68, 71, 51, 46

From the suite of machine learning, rule-based classifiers named as decision table, PART, JRip, and OneR are tested on the dataset with original features. PART outperforms in terms of accuracy with 50% training and testing using original features. Therefore, the system uses PART for further experimentation and result analysis with 50% training and testing.

PART uses the reduced dataset with 56 features, and the performance is measured in terms of precision, recall, and false positive rate (FPR) as mentioned in [9, 16]. The performance of the PART on test dataset with and without feature reduction is shown in Table 3.

Table 3. Comparison of PART classifier with and without feature reduction

Proposed method	Class	Precision (%)	Recall (%)	FPR
Without feature reduction (77 features)	Benign	99.9827	99.9686	0.0301
	Slowloris	99.3437	99.0017	0.0055
	SlowHTTPTest	99.2698	98.9087	0.0058
	Hulk	99.9463	99.9783	0.0269
	GoldenEye	99.478	99.7673	0.0079
	Heartbleed	100	85.7143	0
With feature reduction (56 features)	Benign	99.9795	99.965	0.0356
	Slowloris	99.3793	99.2083	0.0052
	SlowHTTPTest	99.1985	99.0542	0.0064
	Hulk	99.9498	99.9775	0.0251
	GoldenEye	99.4194	99.6122	0.0088
	Heartbleed	100	85.7143	0

Table 3 shows the PART provides an improvement in precision for Slowloris and Hulk with reduced 56 features. However, precision for Heartbleed remains the same and slightly decreases for Benign, SlowHTTPTest, and GoldenEye.

The analysis of PART in terms of precision with and without feature reduction is depicted in Fig. 2. It shows the PART provides the precision of 99.3793 and 99.9498% for Slowloris and Hulk DoS attacks, respectively, with 56 reduced features that are improved as compared to original 77 features.

Table 4 presents a comparison of the proposed system with current state-of-the-art systems. The NA stands for not applicable (NA) in the table.

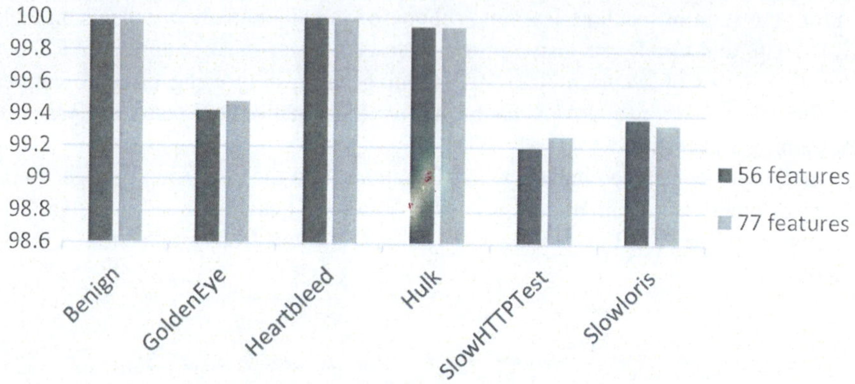

Fig. 2. Precision analysis of PART

Table 4. Comparison with current state-of-the-art systems

Work	Feature Reduction techniques	Classifier	Dataset	DoS accuracy (%)	DoS precision (%)
[8]	IG and CR	ANN	KDDcup99	NA	99.9
[10]	BGWO and KPC	SVM	KDDcup99	95.381	NA
[12]	IG and vote	Hybrid	NSL-KDD	99.9	NA
[16]	CSE with Naive Bayes	Decision Tree	CICIDS 2017	89.8427	82.8
Proposed	IG and CR	PART	CICIDS 2017	99.9871	99.5894

5 Conclusion

In this study, the reduced feature set is obtained with the help of information gain and correlation attribute evaluation methods with ranker. The proposed feature reduction method reduces 56 features from the original 78 features. The experimentation and result analysis using PART produce improved precision of 99.3793 and 99.9498% for Slowloris and Hulk DoS attacks, respectively. Finally, the proposed system is compared with current state-of-the-art systems.

In future, the method can be extended to find an optimal feature set using a combination of different feature reduction techniques.

References

1. Zong, Y., Huang G.: A feature dimension reduction technology for predicting DDoS intrusion behavior in multimedia internet of things. Multimedia Tools Appl. 1–14 (2019)
2. Selvakumar, K., Karuppiah, M., SaiRamesh, L., Hafizul Islam, S.K., Hassan, M.M., Fortino, G., Raymond Choo, K.-K.: Intelligent temporal classification and fuzzy rough set-based feature selection algorithm for intrusion detection system in WSNs. Inform. Sci. **497**, 77–90 (2019)
3. Kshirsagar, D., Sale, S.S., Tagad, D.K., Ganpat, K.: Network intrusion detection based on attack pattern. In: 2011 3rd International Conference on Electronics Computer Technology, vol. 5, pp. 283–286. IEEE (2011)
4. Priyadarsini, P.I., Sai, M.S.S., Suneetha, A., Santhi, M.V.B.T.: Robust feature selection technique for intrusion detection system. Inter. J. Control Automation. **11**(2), 33–44 (2018)
5. Patgiri, R., Varshney, U., Akutota, T., Kunde, R.: An investigation on intrusion detection system using machine learning. In: 2018 IEEE Symposium Series on Computational Intelligence (SSCI), pp. 1684–1691. IEEE (2018)
6. Sornsuwit, P., Jaiyen, S.: A new hybrid machine learning for cybersecurity threat detection based on adaptive boosting. Appl. Artif. Intell. **33**(5), 462–482 (2019)
7. Salih, A.A., Abdulrazaq, M.B.: Combining best features selection using three classifiers in intrusion detection system. In: 2019 International Conference on Advanced Science and Engineering (ICOASE), pp. 94–99. IEEE (2019)
8. Manzoor, I., Kumar, N.: A feature reduced intrusion detection system using ANN classifier. Expert Syst. Appl. **88**, 249–257 (2017)
9. Pandey, V.C., Peddoju, S.K., Deshpande, P.S.: A statistical and distributed packet filter against DDoS attacks in cloud environment. Sādhanā **43**(3), 32 (2018)
10. Velliangiri, S.: A hybrid BGWO with KPCA for intrusion detection. J. Exp. Theor. Artif. Intell., 1–16 (2019)
11. Wankhede, S., Kshirsagar, D.: DoS attack detection using machine learning and neural network. In 2018 Fourth International Conference on Computing Communication Control and Automation (ICCUBEA), pp. 1–5. IEEE (2018)
12. Aljawarneh, S., Aldwairi, M., Yassein, M.B.: Anomaly-based intrusion detection system through feature selection analysis and building hybrid efficient model. J. Comput. Sci. **25**, 152–160 (2018)
13. Patil, A., Kshirsagar, D.: Towards feature selection for detection of DDoS attack. In: Computing in Engineering and Technology, pp. 215–223. Springer, Singapore (2020)
14. Shailesh, S.P., Raiwani, Y.P., Panwar, L.S.: Evaluation of network intrusion detection with features selection and machine learning algorithms on CICIDS-2017 dataset. Available at SSRN 3394103 (2019)
15. Frank, E., Witten, I.H.: Generating accurate rule sets without global optimization. In: Fifteenth International Conference on Machine Learning, pp. 144–151 (1998)
16. Sharafaldin, I., Lashkari, A.H., Ghorbani, A.A.: Toward generating a new intrusion detection dataset and intrusion traffic characterization. In: ICISSP, pp. 108–116 (2018)

Vehicle Number Plate Recognition for Toll System

Satishkumar S. Chavan[1(✉)] and Satishkumar L. Varma[2]

[1] Don Bosco Institute of Technology, Kurla West, Mumbai, Maharashtra 400070, India
satyachavan@yahoo.co.in
[2] Pillai College of Engineering, New Panvel, Navi Mumbai, Maharashtra 410206, India
varmasl@yahoo.com

Abstract. Vehicle number plate detection and recognition (VNPR) is a pioneering methodology which has a large impact on the development of road safety, automation in toll collection, transportation efficacy, and support to the traffic authorities. In this paper, a number plate detection and recognition system for toll collection is presented. Use of large sample data sets has made the system efficient and robust enough. Contrast enhancement is a preprocessing used followed by conventional techniques to locate the number plate. The percentage accuracy in locating the number plate in a given image is 94.87%. Horizontal and vertical profiles with a ratio of 1:2 are used to separate characters in the detected number plate. Backpropagation neural network is applied on the extracted characters to recognize them for authentication of license plate. The presented system is compared with other conventional methods for evaluating its effectiveness and efficiency. The average number plate recognition accuracy of the proposed system is 90.21%.

Keywords: Number plate detection · Number plate recognition · Vehicle identification · License plate recognition · Vehicle authentication · Character segmentation · Neural network · Toll collection system

1 Introduction

Due to improved lifestyle and affordability to buy own vehicle, traffic on the road has been increased. This resulted in many fold issues like traffic congestion, road safety, air pollution, increased use of fuel, etc. It also affected toll collection centers which are manually operated in India. In manual toll collection, cash is accepted by the operator and receipt is issued. However, this process delays the traffic flow due to human intervention and problem related to change of cash. This issue is resolved to some extent by keeping separate lane exclusively for daily commuters and also by increasing the number of lanes. These daily commuters use season passes or pay using electronic transaction. Automatic toll collection through number plate detection and recognition may reduce traffic congestion and also may help in identifying illegal vehicles.

© Springer Nature Singapore Pte Ltd. 2021
P. Deshpande et al. (eds.), *Next Generation Information Processing System*, Advances in Intelligent Systems and Computing 1162,
https://doi.org/10.1007/978-981-15-4851-2_19

Number plate detection and license recognition have a wide range of applications, viz. monitoring of traffic [1], vehicle identification [2, 3], tracking suspicious vehicles [1], etc. It can be also used for collecting parking charges automatically, recording and verifying the legal vehicles at petrol pumps, malls, cinema halls, etc. It is also useful for storing vehicle information, accessing owners' information, and collecting tolls electronically without affecting the traffic flow. To enable the use of an automatic system for such applications, the accurate detection and recognition of the vehicle license is necessary. It is also expected to have a system which is robust and efficient enough for number plate detection and recognition. Although the use of electronic paying systems and rapid growth in the development of Internet of Things (IoT), Bluetooth, and Wi-Fi is rising, a large number of constraints still exist to make a number plate recognition robust system. The major challenge is the accuracy of locating number plate and character recognition. The effect of illumination, background changes, and environmental conditions also plays a significant role in the efficiency of the recognition system.

The paper is organized as follows: Sect. 2 presents an overview of the existing techniques of the number plate recognition. Section 3 describes the vehicle detection and recognition system (VNPR). Section 4 gives result analysis for the detection and recognition accuracy of the presented system. The conclusion is presented in Sect. 5.

2 Related Work

The transportation system has been evolving with rapid growth in technology, new devices, and algorithms. Automatic detection of the number plate and its recognition (VNPR) has many applications such as automatic toll collections, verifying authentic vehicles, and restricting illegal vehicle. VNPR may be useful at petrol stations, parking areas, malls, cinema halls, etc. for surveillance. Automatic toll collection systems are being nowadays preferred globally. National Highways Authority of India (NHAI) has also deployed FASTag electronic system in many cities which uses radio frequency identification (RFID) technology for direct collection of toll [4].

A large number of methodologies have been proposed in the literature for number plate recognition and character recognition for traffic and security purposes. Agarwal et al. have developed a license plate recognition system for traffic violation detection [5]. The edge detection followed by morphological operations is used by Patel et al. to detect number plate. They identified a number plate using a template matching algorithm for character recognition [6]. Anagnostopoulos et al. have proposed sliding concentric windows method of segmentation along with connected components. It is followed by character recognition using two-layer probabilistic neural networks which provided 89.1% accuracy of plate recognition [7]. Web-based Colombian license plate identification and recognition using Rasp Raspberry PI 3 has been demonstrated by Arrieta-Rodríguez et al. [8]. Number plate detection using images with rainy or foggy environmental distortions, low illumination and contrast, and other similar and titled number plates have experimented on 850 car images with Bangla script [9]. Real-time car license plate detection to cater rotational inaccuracies using CNN-based MD-YOLO framework is proposed by Xie et al. [10].

The work presented in this paper involves number plate detection, character recognition, and identifying license plate. Once the number plate is recognized, the relevant record of the owner is accessed for identifying the legal vehicle. Then, the vehicle owner is charged toll fees allowing the vehicle to pass by the toll booth. It will reduce the human resources and waiting time in queue. This will also help in avoiding vehicle congestion on the road/toll booth resulting in smooth traffic flow.

3 Proposed Vehicle Detection and Recognition System (VNPR)

3.1 System Architecture

The number plate detection and recognition system involves locating license plate, extraction and identification of characters, recognition of the license plate, and accessing the owner's detail for collecting toll charges automatically. It involves many image preprocessing steps like normalizing the image, applying thresholding, edge detection, and finding connecting components for locating the number plate. Once the license number is extracted and identified, it is validated using the RTO database for authentication of legal vehicle. The legal vehicles are charged toll fees electronically. If the license number is not matching with the database, the vehicle is treated as illegal and concerned authorities are alerted.

Figure 1 presents the VNPR system. The camera captures the vehicle's frontal or rear image which will have a number plate. The captured image was filtered for removal of noise using Gaussian and median filters. Some enhancement techniques like sharpening, histogram equalization, smoothing, etc. were used to improve the quality of the image. Then, binarization (thresholding) and edge extraction were used. The morphological operations like thinning, dilation, boundary extraction, connectivity trace, etc. were used to locate the number plate. The next step was to apply character segmentation to separate alphanumeric characters. These characters were recognized using a neural network backpropagation algorithm for verifying license plate authorization. This resulted in getting the record of the vehicle owner, transaction history, and registered details. Then, the toll is collected through electronic payment.

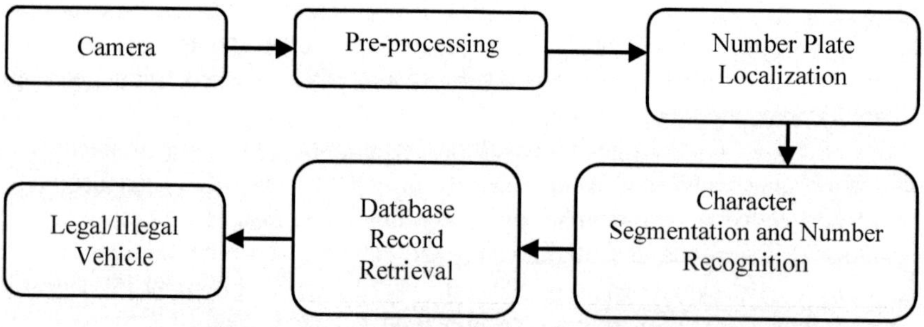

Fig. 1. Vehicle number plate detection and recognition system (VNPR)

3.2 Preprocessing

Noise Removal The image was processed for removal of the noise. In this filtering, 3 × 3 average mask was used to remove the Gaussian noise, and median filtering was used to attenuate the impulse noise.

Histogram Equalization It was used for improving contrasts automatically for poor illumination images. This step also helped in the proper selection of the threshold for binarization.

Binarization It was a useful preprocessing technique for number plate extraction as shown in Fig. 2b. It provided a distinct vehicle number plate which was extracted using a few morphological operations.

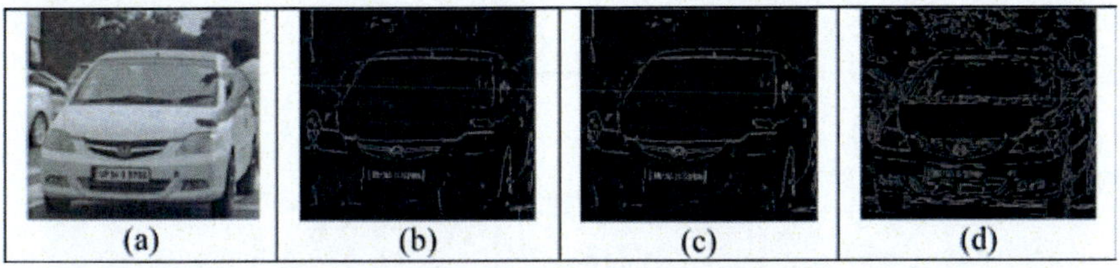

Fig. 2. Edge detection **a** original image, **b** Prewitt, **c** Sobel, **d** Canny

3.3 Edge Detection

The edge detection was another important step in the preprocessing which helped in deciding the boundaries of various objects. One of them was the number plate. In this work, various edge detection methods, namely Prewitt, Sobel, and Canny were used for boundary detection as given in Fig. 2b–d, respectively. Sobel edge detection has given better result compared to other techniques in most of the test cases. The techniques commonly used for localization of number plate were Sobel edge detection and edge linking using the Hough transform. Few morphology operations such as dilation, thinning, and connecting components were also used to make the localization efficient.

3.4 Number Plate Localization

The steps involved in the number plate localization were feature extraction, number plate region extraction, and morphological operations.

Feature Extraction To extract the number plate region, the features like aspect ratio (R), area of bounding box (A), and density (D) were used. These features were computed using Eqs. (1–3). Here, R is the aspect ratio, i.e., height (H) and width (W) ratio. These features helped in discarding most of the irrelevant objects while retaining relevant (number plate) region.

$$R = \frac{H}{W} \tag{1}$$

$$A = H \times W \tag{2}$$

$$D = \frac{n}{\text{Area}} \tag{3}$$

Number Plate Region Extraction In this step, the smearing algorithm was used to retain few regions which satisfy the features obtained using Eqs. (1–3). One of these regions was the number plate region. The process is illustrated with the help of Fig. 3. In the smearing method, the image was scanned along horizontally as well as vertically. The number of white pixels, along scan line which does not satisfy the selected thresholds (Th1 and Th2) criteria, was converted into black pixels.

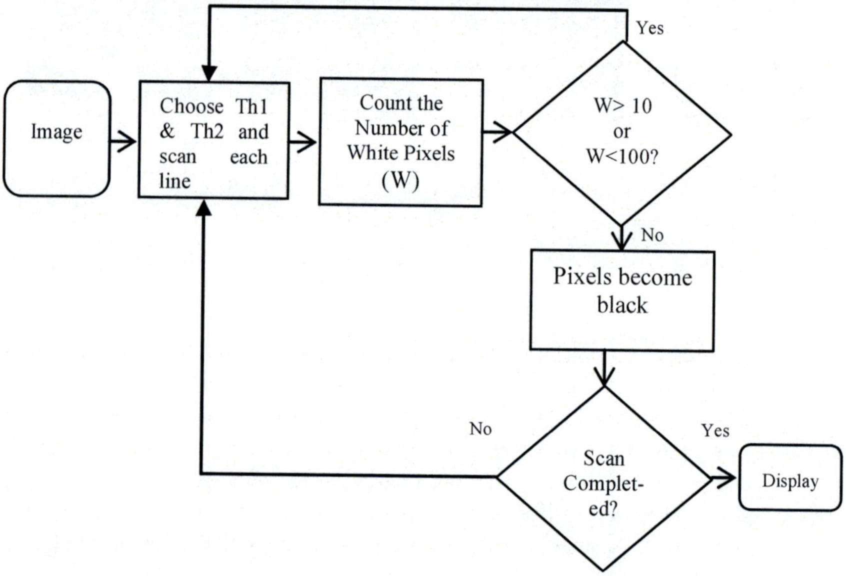

Fig. 3. Number plate extraction algorithm

Using the algorithm given in Fig. 3, approximately 2–3 candidate regions were retained, while most of the objects were discarded. If more than one region gets extracted, then a new set of rules were applied to extract an accurate number plate region.

Rule 1: The image was divided into three horizontal regions. The region in the lower two-third part was selected as the region of number plate localization.
Rule 2: The farthest location of the object from the image border was considered as a candidate object.

Morphological operations Morphological operations like dilation, thinning, connectivity trace, and closing were used in this process. Dilation resulted in gap filling and connecting boundaries of the edge image. Thinning operation provided single-pixel representation of edge. Connectivity trace was used to instituting the boundaries which helped in identifying position of number plate. The number plate extraction steps are shown in Fig. 4.

Fig. 4. Number plate localization: **a** original color image, **b** binarized image, **c** region of interest after morphological operations, and **d** license plate detection

Character segmentation and identification Characters, namely letters and numbers, were extracted using horizontal and vertical profiling. The characters on the license plate were segmented after delineating the number plate. Region growing approach was preferred to segment bright characters from dark background. A priori information about ratio of height to width and character spacing is used in character segmentation. This approach is simple, easy to implement and faster. Then, these extracted numbers and letters were normalized for equal size with the width–height ratio of 1:2. The backpropagation neural network (BPNN) was designed for training and testing using variety of character patterns from number plates. The result of test images for character recognition is shown in Figs. 5 and 6.

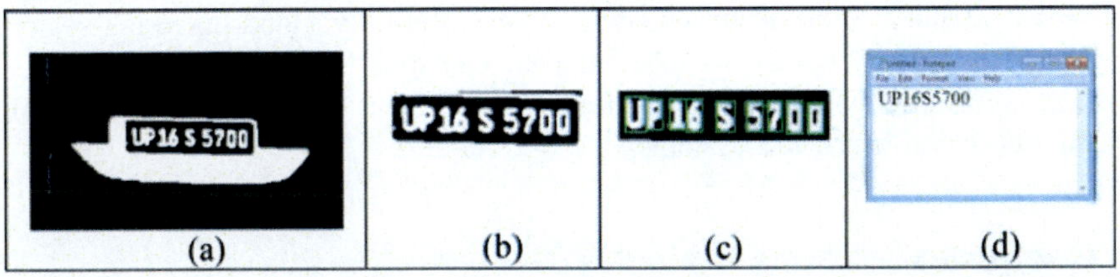

Fig. 5. Number plate recognition: **a** region of interest (ROI), **b** number plate extraction, **c** character segmentation, and **d** license plate recognition

4 Result and Discussion

4.1 Number Plate Localization and Character Segmentation

The total 1,590 images of cars including frontal and rear view were used for locating the number plate in the image. Most of the images carry number plates in the bottom

Fig. 6. Sample result images of number plate detection and recognition using BPNN

one-third portion of the image whereas few images consist of the number plate in the middle region of the image. Table 1 presents the percentage accuracy with which localization of number plate and character segmentation was achieved using the presented work.

Table 1. Accuracy of number plate extraction and character segmentation

Parameter	Accuracy (%)
Number plate detection	94.87
Character segmentation	93.23

4.2 Character Recognition

The automatic VNPR system was tested using data sets for English 26 letters and 10 digits. The data set consisted of a total 720 images comprising 20 images of each character. The BPNN was trained for 36 classes. Total 576 images (80%) were used for training, and 144 images (20%) were used for testing. Table 2 shows the classification accuracy for each character. The comparative analysis of the presented technique with other conventional methods is presented in Table 3. Recently, cloud-based IoT systems are gaining popularity due to its inherent characteristics like remote operations and seamless connectivity [11, 12]. In the future, the present technology will be implemented with cloud-based IoT systems.

5 Conclusion

As a part of the traffic monitoring or smart toll collection system with authentication of legal vehicles for automation, the number plate detection and recognition plays a significant role. This paper presents a system for collecting toll charges from legal vehicles through automatic detection and localization of number plate in the frontal or rear view of vehicle and character recognition of license plate. The system has two front end stages, viz. number plate localization and license plate recognition. Third stage is accessing the database of legal vehicle and charging the owner for usage of the

Table 2. Classification accuracy for English characters

Character	Accuracy	Character	Accuracy	Character	Accuracy	Character	Accuracy
0	87.2	A	95.2	K	90.4	U	86.7
1	90.4	B	93.4	L	92.6	V	89.5
2	85.5	C	94.7	M	93.1	W	88.6
3	92.1	D	93.3	N	92.7	X	93.4
4	90.8	E	91.1	O	81.5	Y	94.8
5	91.6	F	89.2	P	86.2	Z	89.9
6	96.5	G	86.7	Q	86.3	–	–
7	93.5	H	91.6	R	91.6	–	–
8	94.6	I	84.1	S	87.4	–	–
9	91.8	J	81.2	T	88.4	**Avg.**	**90.21**

Table 3. Comparative analysis of the presented system with other techniques

S. No.	Method	Recognition accuracy (%)
1.	License plate and character recognition [8]	81.00
2.	Optical character recognition [5]	93.00
3.	Probabilistic neural network (PNN) [7]	89.10
4.	Feature extraction model and BPNN [10]	97.70
5.	Presented system	90.21

road. The presented system provides 94.87% of accuracy for number plate detection and 93.23% accuracy for character segmentation. The segmented characters are recognized with 90.21% of accuracy.

References

1. Shreyas, R., Kumar, B.V.P., Adithya, H.B., Padmaja, B., Sunil, M.P.: Dynamic traffic rule violation monitoring system using automatic number plate recognition with SMS feedback. In: Second International Conference on Telecommunication and Networks (TEL-NET), Noida, pp. 1–5 (2017)
2. Hofman, Y: License Plate Recognition—A Tutorial. Hi-Tech Solutions (2008). URL: http://www.licenseplaterecognition.com/ Accessed on 19 Nov 2019
3. Ozbay, S., Ercelebi, E.: Automatic vehicle identification by plate recognition. Inter. J. Comput. Inform. Eng. **1**(9), 1418–1421 (2007)
4. National Electronic Toll Collection (NETC). URL: https://www.npci.org.in/netc. Accessed on 30 Nov 2019
5. Agarwal, P., Chopra, K., Kashif, M., Kumari, V.: Implementing ALPR for detection of traffic violations: a step towards sustainability. Procedia Comput. Sci. **132**, 738–743 (2018). International Conference on Computational Intelligence and Data Science (ICCIDS 2018)
6. Patel, F., Solanki, J., Rajguru, V., Saxena, A.: Recognition of vehicle number plate using image processing technique. Control Syst. Eng. **2**(1), 1–7 (2018)

7. Anagnostopoulos, C.N.E., Anagnostopoulos, I.E., Loumos, V., Kayafas, E.: A license plate-recognition algorithm for intelligent transportation system applications. IEEE Trans. Intell. Transportation Syst. **7**(3), 377–392 (2006)
8. Arrieta-Rodríguez, E., Murillo, L.F., Arnedo, M., Caicedo, A., Fuentes, M.A.: Prototype for identification of vehicle plates and character recognition implemented in raspberry pi. In: IOP Conference Series, Materials Science and Engineering, vol. 519(012028), pp. 1–5 (2019)
9. Azam, S., Islam, M.M.: Automatic license plate detection in hazardous condition. J. Vis. Commun. Image Rep. **36**, 172–186 (2016)
10. Xie, L., Ahmad, T., Jin, L., Liu, Y., Zhang, S.: A new CNN-based method for multi-directional car license plate detection. IEEE Trans. Intell. Transp. Syst. **19**(2), 507–517 (2018)
11. Deshpande, P., Sharma, S.C., Peddoju, S.K.: Implementation of a private cloud: a case study. In: Pant, M., Deep, K., Nagar, A., Bansal, J. (eds.) Proceedings of the Third International Conference on Soft Computing for Problem Solving. Advances in Intelligent Systems and Computing, vol. 259. Springer, New Delhi (2014)
12. Deshpande, P.: Cloud of everything (CLeT): the next-generation computing paradigm. In: Iyer, B., Deshpande, P., Sharma, S., Shiurkar, U. (eds.) Computing in Engineering and Technology. Advances in Intelligent Systems and Computing, vol. 1025. Springer, Singapore (2020)

Prediction of High Recommendation Mobile Brands Using Sentiment Analysis

Smita Bhanap(✉) and Seema Kawthekar

Dr. Babasaheb Ambedkar Marathwada University, Aurangabad, India
srbhanap@gmail.com, seema_babrekar@yahoo.co.in

Abstract. Digital world is growing very fact every day. It is the very important resource for generation of digital data. Technology is taking us to more sophisticated and user-centric applications in order to perform our daily requirements. Mainly, usage of social networks and online transactions is more preferred by people which generates huge data in terms of useful insights for future references. People are more comfortable giving their views, ideas, and reviews online instead of offline. These reviews are the source which if used in appropriate manner can help to understand needs of customers and help them in recommending some brands as per their requirements and help them in decision making. In this paper, we are using tweets of various mobile brands as our input. We then use naïve Bayes and support-vector machine supervised machine learning algorithms for prediction of sentiments and prediction of highly recommended brands. In our experiment, we got more accurate results for support-vector machine than naïve Bayes.

Keywords: Sentiment analysis · Naïve Bayes · Support-vector machine · Prediction

1 Introduction

Nowadays, with availability of technical support along with access to social sites, people are more prone to share their views on such sites. These reviews or comments are good source of information about events, products, social aspects which can be used by other people in community for better decision making. The usefulness of these reviews is the motivation for many researchers to work on this data and help to derive new conclusions. The reviews generated from social media are useful to derive some hidden patterns for decision making, but these reviews may not be in proper format. To refine these unstructured reviews into structured format and to be used further, some data models are required to be designed. In this paper, we focus to process and analyze the reviews or the sentiments of the social media by applying some machine learning algorithms.

Sentiment analysis can also be referred to as a kind of text classification. It categorizes texts based on the opinions of people. It plays an important part of natural language processing. Process of detecting the contextual polarity of the text as positive, negative, or neutral is referred to as sentiment analysis. Sentiment classification approaches are named as lexicon-based approach and machine learning approach [1].

© Springer Nature Singapore Pte Ltd. 2021
P. Deshpande et al. (eds.), *Next Generation Information Processing System*,
Advances in Intelligent Systems and Computing 1162,
https://doi.org/10.1007/978-981-15-4851-2_20

Lexicon-based approach calculates sentiment score using a lexicon of negative and positive words. The positivity or the negativity of that word is determined by a value assigned to it. The sentiment of the text is sum or average of all the words in given sentence.

The semantic analysis of the sentence helps to increase the meaning and accuracy of the result [2, 3]. Analysis of customer reviews is important to maintain product quality and satisfy customer expectations. This helps the organization to increase sales by matching customer requirements [4]. Sentiment analysis considers a training set for its performance and gives accurate evaluation of the text [3]. Currently, many researchers are working in the field of sentiment analysis as it is very challenging to deal with type of reviews given by customers.

As shown in Fig. 1, the process of sentiment analysis can be described in three levels:

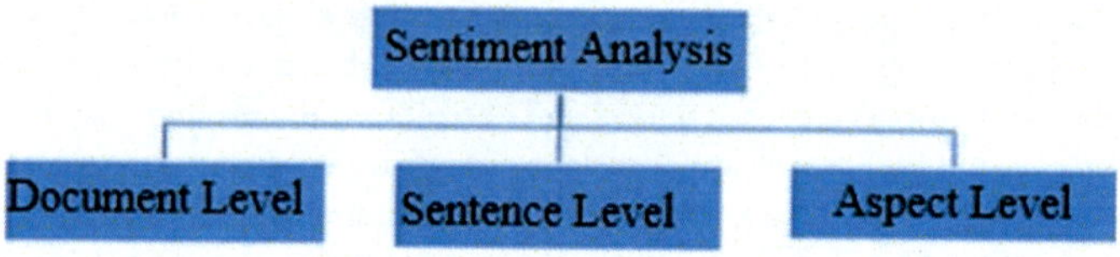

Fig. 1. Levels of sentiment analysis

- **Document-level** considers entire document at a time in order to find overall opinion of the document [5].
- **In Sentence-level approach**, the document is braked into sentences, and polarity of each sentence is calculated separately [3].
- **Aspect-level handles** each feature term of a product as main target [3].

The supervised machine learning algorithms like naïve Bayes, support-vector machine (SVM), maximum entropy, decision tree use labeled datasets in which words or phrases are features. Various feature selection mechanisms can be used to select features hidden in the text and perform detailed analysis of the sentences as a part of sentiment classification.

In this proposed research, we have used two classifiers named naïve Bayes and SVM for feature classification. We have used feature selection method to classify polarity as positive and negative. Also, we have used n-gram approach for better results. The results show that SVM classifier performs much better with accuracy of 97%. In case of naïve Bayes algorithm, we have used performance measures like precision, recall, and F-measure so as to compare accuracy.

2 Literature Review

Researchers are working in the field of sentiment mining on various aspects. Sentiment classification task is mainly divided into three approaches named machine learning approach, lexicon approach, or the hybrid approach [1]. Nowadays, people express their views using different platforms like blogs, tweets, and discussion forum of various social sites. These methods have opened possibilities in the field of research as the contents written by people are heterogeneous in nature. It is required to focus on selecting important features from these user contents. For sentiment classification of movie reviews, authors of [2] have used feature selection method and a feature optimization technique based on genetic algorithm. E commerce-based companies make use of user-generated content in order to serve customers better and increase their sales [6]. Sentiment analysis poses several challenges related to aspect detection and mapping aspect words [6]. Different machine learning algorithms are used by researchers to define polarity and classify them. The authors of [7] shows that naive Bayes with binary features work better for several text classification tasks. The authors of [8] analyze tweets based on aspect approach. In [9], authors propose an enhanced method for classifying the feedbacks using support-vector machine. Authors of [10] present a Web product ranking system by using sentiment analysis in which product ranking results are shown after firing a query for required product. In [11], authors have used SentiWordNet for calculating polarity of products and decide which is highly rated.

3 Proposed Methodology

Proposed work focus on predicting highly recommended brands based on user reviews of different mobile brand like Iphone, Vivo, and Oppo posted on Twitter. The process flow is shown in Fig. 2.

These reviews are collected and preprocessed after which stored in repository. By comparing their results, we want to find the most popular brand. For this purpose, machine learning classifiers like naïve Bayes and support-vector machine are used to build a model. We have used open-source tool R for our experimental work as it is available freely. Various packages required for this process are tm, ggplot, Twitter, and e1071. The methodology is shown in Fig. 2 and described as per sections given below.

3.1 Data Collection and Cleaning

Twitter API's are used to download tweets of different mobile brands. For this, we used Twitter Outh keys. Once the reviews are downloaded, we then store them in csv files into a repository. These csv files need to be preprocessed as it may contain unwanted text. For preprocessing, tm package is used to get text from tweets. In preprocessing, removal of punctuations, some specific stop words, removing special characters is done.

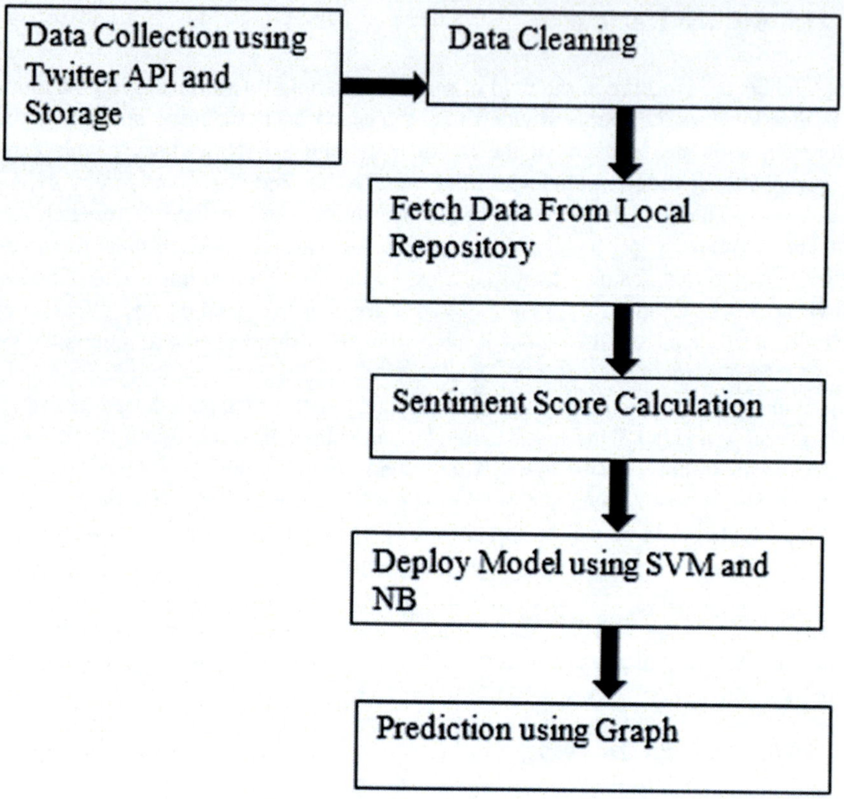

Fig. 2. Proposed methodology

3.2 Sentiment Score Calculation

After preprocessing, reviews are in only text form. We use dictionary approach to compare words in the review with preloaded positive and negative dictionaries. Here, in this work, we have used n-gram approach so as to find accurate polarity of the review as positive and negative. We are mainly interested in positive and negative tweets only as it is going to decide most popular mobile brand. For sentiment classification, we have used two naïve Bayes and support-vector machine classifiers.

- Naïve Bayes classification technique calculates probability of each aspect of the given text [4]. In our reviews, polarity for each word of sentence is calculated.
- Support-vector machine classifier is used to divide the dataset into different classes. Support vectors are the data points which are near to the hyper plane which classifies a set of data [4]. In other words to classify a new data correctly, we use hyper plane which is obtained from a greatest possible margin [4].

By applying these two classifiers, we have calculated accuracy for each brand so as to compare results. The results show that SVM gives higher accuracy as compared to naïve Bayes for the datasets which we have used. The performance of these two algorithms is compared on the basis of accuracy, precision, recall, and F-measure. These parameters can be calculated as follows:

Precision is defined as number of true responses retrieved against total number of true and false both positives. **Recall** is the count of number of instances in the test set which is correctly labeled by the classifier out of total instances in the test set that is actually labeled for a particular class [10]. **F-measure** is the weighted harmonic mean of precision and recall for a particular class [10].

4 Experimental Results

See Table 1 and Fig. 3.

Table 1. Accuracy of mobile brands using NB and SVM

	NB				SVM
	Precision	Recall	F-measure	Accuracy	
OPPO	72	66	69	76	79
IPHONE	71	65	68	75	66
ONE PLUS	84	83	85	85	87
VIVO	64	62	58	69	80

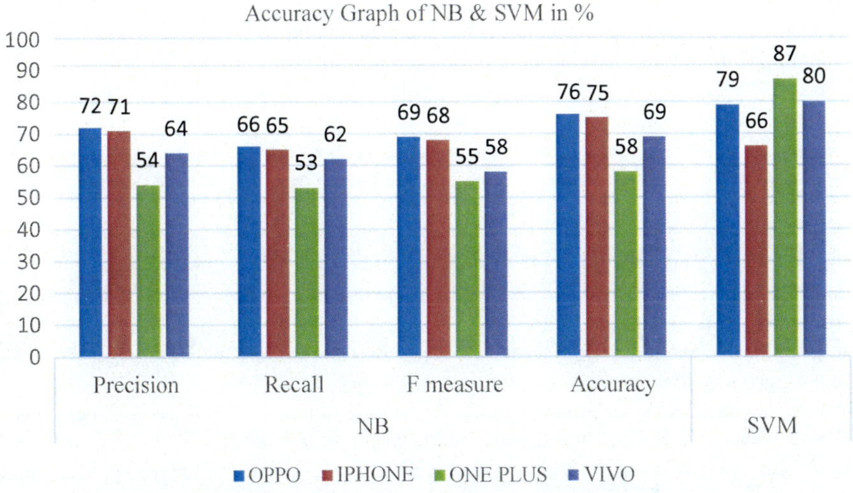

Fig. 3. Graphical representation of accuracy of NB and SVM classifiers

5 Conclusion

In this paper, we have proposed a methodology which helps to predict highly recommended mobile brands based on Twitter reviews. Analysis of reviews in terms of sentiments or polarity helps to find popular brands among customers and increases quality by focusing on customer's comments or suggestions or negative reviews. We have used naïve Bayes and support-vector machine algorithms for the sentiment classification, and the performance is compared based on Precision, Recall, and F-measure. Results show that support-vector machine algorithm gives more accuracy than naïve Bayes algorithm. In future, our work will focus on increasing dataset size and also implementing more algorithms like decision tree, random forest, and perform accuracy comparisons.

References

1. Hu, M., Liu, B.: Mining and summarizing customer reviews. In: Proceedings ACM SIGKDD, pp. 168–177 (2004)
2. Sangam, S., Shinde, S.: A novel feature selection method based on genetic algorithm for opinion mining of social media reviews In: ICICCT 2018, CCIS, vol. 835, pp. 167–175 (2019)
3. Sarawgi, K., Pathak, V.: Opinion mining: aspect level sentiment analysis using SentiWordNet and Amazon web services. Inter. J. Comput. Appl., 3136 (2015)
4. Vanaja, S., Belwal, M.: Aspect-level sentiment analysis on e-commerce data. In: International Conference on Inventive Research in Computing Applications (ICIRCA). IEEE Xplore Compliant Part Number: CFP18N67-ART; ISBN: 978-1-5386-2456-2 (2018)
5. Sadegh, M., Ibrahim, R., Othman, Z.A.: Opinion mining and sentiment analysis. Inter. J. Comput. Technol. **2** (2012)
6. Anand, D., Naorem, D.: Semi-supervised aspect based sentiment analysis for movies using review filtering. Procedia Comput. Sci. **84**, 86–93 (2016)
7. Wang, S., Manning, C.D.: Baselines and bigrams: simple, good sentiment and topic classification. In: Proceedings of the 50th Annual Meeting of the Association for Computational Linguistics, Jeju, Republic of Korea, pp. 90–94 (2012)
8. Lek, H.H., Poo, D.C.C.: Aspect-based twitter sentiment classification. In: 2013 IEEE 25th International Conference on Tools and Artificial Intelligence, pp. 366–373 (2013)
9. Bhaskar, J., Sruthi, K., Nedungadi, P.: Enhanced sentiment analysis of informal textual communication in social media by considering objective word and intensifiers. In: International Conference on Recent Advances and Innovations in Engineering (ICRAIE-2014), pp. 1–6 (2014)
10. Swati, A., Kale, N.D.: Sentimate analysis for web product ranking. Inter. J. Recent Innov. Trends Comput Commun. **2**(11), 3602–3604. ISSN: 2321-8169
11. Chothani, H.D., Menaria, S.: Opinion mining of customer review for Amazon product. Inter. J. Trend Sci. Res. Dev. **2**(3), 406–410 (2018)

Dynamic Virtual Machine Provisioning in Cloud Computing Using Knowledge-Based Reduction Method

R. Bhaskar[1(✉)] and B. S. Shylaja[2]

[1] Department of CSE, DBIT, Bengaluru, India
bhaskardbit@gmail.com
[2] Department of ISE, Dr. AIT, Bengaluru, India

Abstract. Cloud infrastructure performance extremely depends ahead on the task scheduling and load balancing. The recent growth of cloud computing and service provider's key challenge is scheming proficient mechanism for managing the restricted resources shared by different applications. Resource administration method has to do efficient assignment of resources for virtual machines by ensuring optimal resource exploitation of available physical machines. This paper proposes the application of rough-set model for provisioning of virtual machines. The proposed method uses knowledge-based reduction technique, and it generates the rules to reduce unnecessary attributes of the virtual machines. These rules help virtual machine managers for making effective administration of restricted resources.

Keywords: Cloud computing · Data center · Physical machine · Rough-set model · Virtual machine

1 Introduction

The current rush forward in the fame of infrastructure as a service cloud systems can be credited to the on-demand accessibility of computing resources such as processor cores, memory, and disk. Virtual machines are hosted on all physical servers. Incoming request or application may claim virtual machines of different capacities. On-demand request allows the cloud users to hourly purchase the computing facility to attain cost-effective service for the unpredictable workloads and packages of virtual machines are billed based on the usage [1, 2].

Various instances of operating system can run on single physical machine is the key benefit of virtualization, and each individual instances of operating system is called virtual machines (VM). As a result making at most utilization of the accessible resource capacities of the physical machine which facilitates enormously savings in energy and infrastructure costs, the recent growth of cloud computing, service providers' key challenging problem is scheming competent mechanism for administering the restricted resources shared by diverse applications. Resource administration mechanism has to do competent sharing of resources for virtual machines by guaranteeing optimal resource exploitation of available physical machines. Resource administration mechanism

© Springer Nature Singapore Pte Ltd. 2021
P. Deshpande et al. (eds.), *Next Generation Information Processing System*,
Advances in Intelligent Systems and Computing 1162,
https://doi.org/10.1007/978-981-15-4851-2_21

allows cloud users and also service provider to the effective utilization of their available resources.

This paper addresses the problem of resource provisioning. Resource provisioning covenant with mapping and scheduling of cloudlets to virtual machines also mapping and scheduling of virtual machines onto physical machines, and once virtual machine is created in cloud, the objective is to find the best physical machine to run the virtual machine. Virtual machine provisioning includes accessibility of one or more virtual machines each one with different capabilities of processor cores, memory, and disk, with the objective of ensuring optimal resource utilization of the available physical machines to reduce the costs. This paper proposes the application of rough-set model for provisioning of virtual machines. The proposed method uses knowledge-based reduction technique, and it generates the rules to reduce unnecessary attribute of the virtual machines. These rules help virtual machine manager for making the efficient selection of virtual machine.

2 Related Work

Researchers approached the problem of virtual machine provisioning in cloud, and various resource allocation methods have been proposed for cloud computing to provide cost-effective service.

Sambit Kumar Mishra and et al. proposed energy-aware task-based virtual machine consolidation algorithm [3], for execution of new task, and then new task is in the local queue of that virtual machine. When there are no tasks in the local queue of a virtual machine, then the virtual machine is de-allocated from the host and the resources are free. The main goal of this method is to reduce the makespan time along with the power frenzied by cloud system.

Xiuchen Qie and et al. proposed energy-efficient VM allocation strategy for cloud data centers [4]; virtual machine is an independent server, and an available virtual machine is be assigned to the first task queuing in the system buffer. The proposed strategy as a type of novel queuing model with partial asynchronous multiple vacations with an objective of reducing the energy frenzied by the cloud system.

B. Muthulakshmi and et al. proposed artificial bee colony-based hybrid method for virtual machine allocation [5]; the proposed algorithm integrates the functionality of simulated annealing into artificial bee colony algorithm to perform efficient scheduling based on the size, priority of the request, and the closest distance between client node to server. This method improves the makespan time of the host and virtual machine.

Parvathy S. Pillai and et al. [6] presented a new methodology for dynamic resource allocation using the uncertainty principle of game theory. In this method, the game is planned to be played flanked by each broker and the manager. Each broker manager permutation is able to outline their payoff matrix. Each broker attempts to make the manager pay and forms its own list of combinations.

Zhe Hao [7] presented resource allocation based on the improved ant colony algorithm. In this technique, ant is in g_i at moment t_i, as soon as incisive the resource to satisfy the prerequisite in the network; the primary step is to inspect the region that is adjacent to g_i. The ant on the node chooses the next hop based on the pheromone of its adjacent node and selects the node that has the max pheromone in the path.

Yanzhi Wong and et al. [8] projected a novel method of resource allocation based on game-theoretic framework. In this method, the pool of service request consists of single kind of application service requests that are created from all the clients. A service demand is free to be dispatch to any server belonging to any cloud service provider for the reason that all the servers in the cloud can sustain such application type. At the same time, service request is posted to server created a dedicated virtual machine for that service request heaps the application executable and starts execution. The total profit of each cloud service provider for maximization is the total income attained from servicing the request, which depends on the typical request-response time as specified in the service level agreement subtracted by the energy cost of the servers.

Hossein Morshedlou and et al. [9] proposed a proactive resource provisioning method. In this approach, as soon as service provider countenance risk of resource outflowing or its resource provisioning policy is not succeed; it finds which virtual machines are better contender to release their resources. Releasing resources of these virtual machines make it feasible to use the released resources for administering vital requests of the queue. Service provider endeavors to discard the requests which are insignificant. Based on the significance of users and reliability level of virtual machines, total of n pairwises can be constituted.

Jinhhao Liu and et al. [10] proposed aggressive resource provisioning method called SPRNT, which does planning flanked by workload metrics and actions and is accumulated in cerebellar model articulation (CMAC) table. The CMAC table is updated by a belligerent reward strategy to formulate the decision engine proficient to regulate the resource provisioning effectively. With the employment of the belligerent reward strategy, the actions that considerably increase resource provisioning are optimistic to be preferred by the decision engine as soon as the resources are not adequately provisioned.

Kastas Katsails and et al. [11] proposed dynamic weighted round-robin scheduling algorithm to achieve service-level agreements. In this technique, as an alternative of formulating one decision leading each service achievement, it packs collectively numeral of decisions and fix them in a vector. When servers turn out to be vacant, the subsequently unemployed decision in the vector is elected. This method has drawback that few decisions are prepared earlier with usual a lesser amount of information about system state.

3 Rough-Set Model

Rough-set model is introduced by Z. Pawlak in the early 1980s, is a mathematical technique for imprecise information system, and has developed into a key tool for data investigation of soft computing. Rough-set model has well-built qualitative investigation qualifications to articulate efficiently imprecise information. It has been broadly used in decision analysis, rule-making, machine learning, smart control, etc. The key characteristic of rough-set model is firm mathematical definitions and toughness [12]. Rough-set model can be distinct in general by the way of topological operations, and rough-set model investigation is based on two approximations; inferior approximation and superior approximations are defined as follows: given a information base $l = (S, R)$ where S is the space and R is the equivalence relation. Let x be a subset of S about information k are defined as,

$$k_{-}(x) = \{v|(\forall v \in S) \wedge ([v]k \subseteq v\} \tag{1}$$

$$k-(x) = \{v|(\forall v \in S) \wedge ([v]k \neq \Phi\} \tag{2}$$

where $[v]k$ specifies a correspondence class of object v about information k.

The inferior approximation (1) of a v with admiration to k is the set of all substances, which surely fit into the set v on the space S based on information k. The superior approximation (2) of set v with admiration to k is the set of all substances, which possibly belong to the set $v$on the space S based on information k [13]. Knowledge reduction is the key crisis in the imprecise information processing. Majorly, two ways of reduction in imprecise information systems, viz. the information system which does not comprise assessment attributes, and information system which comprise conditions and assessment attributes.

This work considers the information table of virtual machine mapping and scheduling onto cloudlets as a knowledge expression system. Each row represents an object (status of virtual machine). Each column represents an attribute that can be measured for an object (properties of virtual machine) [14, 15].

4 System Architecture

The cloud data center has unconstrained homogeneous/heterogeneous physical machines and offers virtual machines renting service. When consumer put forward their virtual machine requests, physical machine is identified and selected to create the required virtual machine in the cloud data center; this process is called as virtual machine assignment. When there are many virtual machines in cloud, virtual machine manager has quickly to select the virtual machine from the resultant sets. Figure 1 shows system architecture, and virtual machine requests from different consumers are often handled and which are set up over numerous physical machines.

The virtual machine assignment problem is a multi-objective optimization problem. Consider a system with "m" physical host machines and "n" virtual machines. Each virtual machine is represented as vi, and each host is represented as pj. Each virtual machine runs on exactly one physical machine, although one physical machine can host an arbitrary number of virtual machines. Once initial assignment of virtual machines by approximation of resource demanded for every application, the proposed method of resource provisioning continuously monitors the resources used and re-assigns the resources based on demands of other applications; aim is to find an optimal assignment of virtual machines to physical machines.

This paper proposes rough-set model for making efficient decision on the selection of virtual machine. In rough-set model, virtual machines and their attributes can be

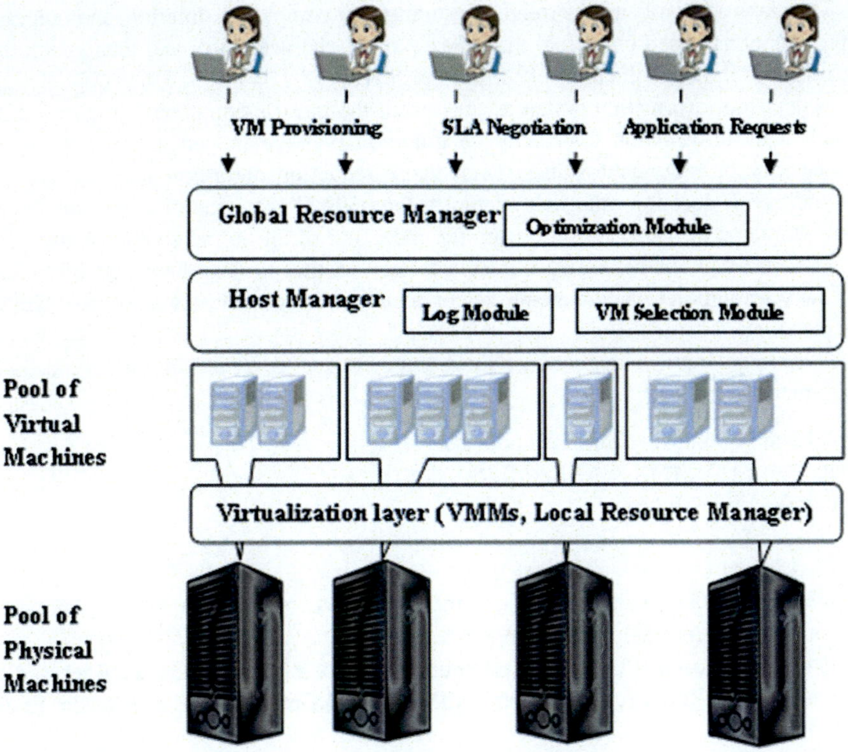

Fig. 1. System architecture

represented in a tabular structure. The rows of the table represent list of physical machines and columns represent properties of the respective physical machine. Table 1 represents the sample decision information system of cloud computing. S = {þ$_M$1, þ$_M$2, þ$_M$3, þ$_M$4, þ$_M$n} represents the physical machines and v = {ƛ1, ƛ2, ƛ3, ƛ4, ƛ5, ƛ6, ƛ7, ƛ8} is the set of attributes of virtual machine, where ƛ1, ƛ2, ƛ3, ƛ4, ƛ5, ƛ6, ƛ7, and ƛ8, respectively, the HostCPU, HostRam, HostDisk, HostBandwidth, UCPU, URam, UDisk, UBandwidth. $D = \{\delta\}$ is the decision attribute.

Table 1. Sample decision information system

Name	ƛ1	ƛ2	ƛ3	ƛ4	ƛ5	ƛ6	ƛ7	ƛ8	δ
þ$_M$1	4	100	1024	1024	4	100	650	512	NA
þ$_M$2	8	100	1024	1024	4	98	850	626	A
þ$_M$3	16	100	1024	1024	5	95	340	626	A
þ$_M$4	16	100	1024	1024	16	65	999	512	NA
þ$_M$n	8	100	1024	1024	4	40	100	512	A

The key idea of using rough-set model is computing inferior and superior approximations based on virtual machines explicit characteristics and then generating the rules for the consolidation of virtual machines.

A decision information system expresses all the information about virtual machine model. This information table may be gratuitously large in part since it is having unnecessary or redundant values. Knowledge reduction algorithm helps to achieve minimal set of decision rules that guarantee the optimal provision of virtual machines. The δ contains decision attribute from the reduction of all the relative attributes.

The virtual machine manager is able to rely on these rules; these rule offers significant information to formulate proficient decision for optimal provisioning of Virtual machine.

A multi-resource rough-set based technique is used to set off the virtual machines assignment.

Algorithm Virtual machine assignment
Input: PMList, VMList, VMOnPMList
Output: VM is assigned

1. PMList ← GetTotalNoOfHost();{þᴍ = {þᴍ1, þᴍ2, . . ., þᴍn} ⊆ Uhs }
2. VMList ← GetTotalNoOfVM(); {ʋmList = {ʋm1, ʋm2, . . ., ʋmj, . . ., ʋmm}}
3. for each ʋm in VMList Util(ʋm$_i$)
4. if(ʋm.GetUnusedCPU > = ċĺ.GetRequiredCPU && ʋm.GetUnuesdRam > = ċĺ.GetRequiredRam&& ʋm.GetUnusedDisk > = ċĺ.GetRequiredDisk) {Add ʋm← ċĺ}
5. VMOnPMList ←GetCurrentNoOfVMOnHost(); {Utili[k] = Σj = 1,m utili[j][k]}
6. foreach ʋm in VMList do {
7. if (ʋmOnPMList! = null) {
8. for each Host in ʋmOnPMList do {
9. if(Host.GetUnusedCPU > = ʋm.GetRequiredCPU && Host.GetUnuesdRam > = ʋm.GetRequiredRam&& Host.GetUnusedDisk > = ʋm.GetRequiredDisk) {

   ULHostList.add(Host);}
10. End

5 Experiments and Results

It is exceptionally intricate to carry out reiterated bulky amount of experiments on a real infrastructure. The CloudSim toolkit [2] has been preferred as a simulation platform. CloudSim toolkit is developed by Melbourne University and is an extensible simulator, whose objective is to facilitate the simulation and modeling of cloud computing environment. It allows simulating virtualized resources and cloud computing entities like data centers, virtual machines, and physical hosts. Thus, CloudSim toolkit facilitates to implement diverse resource assignment policies and estimate the policy performance.

This work is carried out a simulation of data center, physical machine, virtual machine, data center broker, and virtual machine assignment policy. Simulation conducted for diverse numeral of task with arbitrary length and by considering three virtual machines of different processing power, and 10 cloudlets initially for conduction of experiments.

Once cloudlets are assigned to all the virtual machines as per assignment policy; the proposed knowledge-based reduction method monitors the resources of each physical machine which is computed for efficient exploitation of resources. The proposed knowledge-based reduction method minimizes the makespan time and also increases the resource exploitation ratio.

Performance metrics are used to evaluate the algorithm efficiency. It is essential to describe the execution time and makespan time. Execution time is the ratio of million instruction of cloudlets to the million instruction per second of the assigned virtual machine.

$$\mathrm{ET}_i = \frac{C_{ri}}{C_{vj}} \tag{3}$$

Cvj is the figure that demonstrates the total amount of CPU competence of virtual machine VM*j*, and *Cri* is the figure describes length of the task.

Similarly, completion time is the sum of execution time and the completion time of the previously assigned cloudlet for the same virtual machine

$$\mathrm{CT}i = \mathrm{ET}i + \mathrm{CT}(i-1) \tag{4}$$

Makespan time of virtual machine is the sum of time to complete all cloudlets assigned to the same virtual machine.

$$\mathrm{MS}t(vm) = \sum_{i=0}^{m} vmj \tag{5}$$

Makespan time of host is the maximum makespan time of the virtual machine from the assigned virtual machines in that host.

$$MSt(host) = \max_{0 \le m \le n} MSt(vm) \tag{6}$$

Simulation is carried out to find out the result using the space shared policy and time shared policy, shortest job first, conductance dynamic assistance, and compared with the knowledge-based reduction method. Throughout the experimentation, each virtual machine is arbitrarily assigned and each virtual machine is monitored every 60 s. The proposed method is simulated 1 host, 3 virtual machines, and 10 cloudlets. The proposed knowledge-based reduction method improves the makespan time and to prove the efficiency; the proposed method is compared with the different VM policies existed. Tables 2 and 3 show the comparison of makespan time of the proposed method with different VM policies, inference, and results are drawn and shown in Figs. 2 and 3.

Table 2. Comparison of host makespan time

S. No.	VM policy	Host makespan time
1	Space shared	3744.09
2	Time shared	3743.96
3	PBSA	3014.37
4	SJF	480.63
5	CODA	418.57
6	KBR	366.67

Table 3. Comparison of VM makespan time

S. No.	VM policy	VM makespan time		
		VM0	VM1	VM2
1	Space shared	3744	228.24	338.14
2	Time Shared	3743.96	228.24	338.14
3	PBSA	3014.37	145.63	280.25
4	SJF	0	480.63	178.22
5	CODA	0	136	418.57
6	KBR	0	381.76	56.82

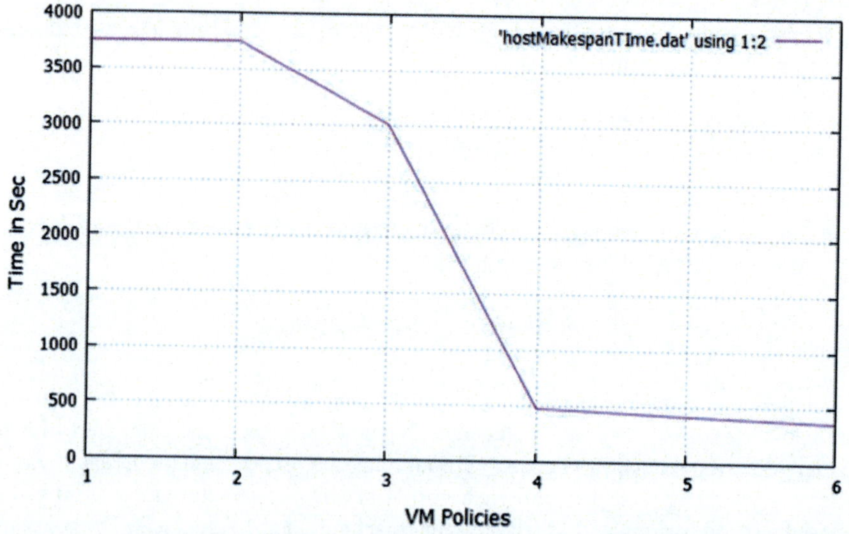

Fig. 2. Comparison of physical machine makespan time

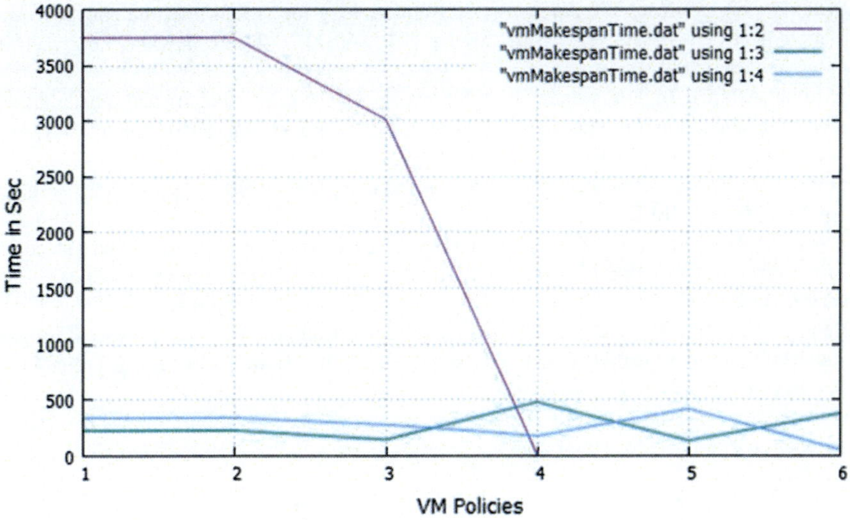

Fig. 3. Comparison of virtual machine makespan time

6 Conclusion

This paper addressed the difficulty of provisioning the on-demand virtual machine. The proposed knowledge-based reduction method for solving virtual machine provisioning problem considers the existence of resources of multiple types. The proposed method generates rules for making efficient decision in selection and mapping of cloudlets to virtual machines for the virtual machine manager. The major goal of the proposed system is to supervise all the virtual machines constantly and improves the utilization ratio of available resources. Hence, this algorithm minimizes the makespan time, and execution speed of applications is increased. The algorithm usefulness and competence is validated by conducting experiments.

The proposed method requires further improvements, i.e., identify the unused switched ON physical machine and convincing the constraint that left over physical machine will not get over-exploited and then turn to standby the physical machine with no resource exploitation.

References

1. Bhaskar, R., Deepu, S.R., Shylaja, B.S.: Dynamic allocation method for efficient load balancing in virtual machines for cloud computing. Adv. Comput. Int. J. (ACIJ) **3**(5), (2012)
2. Calheiros, R.N., Ranjan, R., Beloglazov, A., Rose, C.A.F.D., Buyya, R.: Cloudsim: a toolkit for modeling and simulation of cloud computing environments and evaluation of resource provisioning algorithms: Softw.: Pract. Exp. **41**(1), 23–50 (2011)
3. Mishra SK, Puthal D, Sahoo B, Jayaraman PP, Jun S, Zomaya AY, Ranjan R.: Energy-efficient VM-placement in cloud data center. Sustain. Comput. Inform. Syst. **20**, 48–55 (2018). https://doi.org/10.1016/j.suscom.2018.01.002

4. Qie, X., Jin, S., Yue, W.: An energy-efficient strategy for virtual machine allocation over cloud data centers. J. Netw. Syst. Manag. **27**, 860–882 (2019). https://doi.org/10.1007/s10922-019-09489-w
5. Muthulakshmi1, B., Somasundaram, K.: A hybrid ABC-SA based optimized scheduling and resource allocation for cloud environment. Clust. Comput. **22**, S10769–S10777 (2019). https://doi.org/10.1007/s10586-017-1174-z
6. Pillai P.S., Rao, S.: Resource allocation in cloud computing using the uncertainity principle of game theory: IEEE Syst. J. (2014)
7. Gao, Z.: The allocation of cloud computing resource based on the improved ant colony algorithm. In: Sixth IEEE International Conference on Intelligent Human Machine System and Cybernetics
8. Wang, Y., Lin, X., Pedram, M.: Game theoritic framework of SLA—based resource allocation for competitive cloud service providers. In: Sixth IEEE Green Technologies Conference (2014)
9. Morshedlou, H., Meybodi, M.R.: Decreasing impact of SLA violations: a proactive resource allocation approach for cloud computing environments. IEEE Trans. Cloud Comput. **2**(2) (2014)
10. Liu, J., Zhang, Y., Zhou, Y., Zhang, D., Liu, H.: Aggressive resource provisioning for ensuring QoS in virtualized environments. IEEE Trans. Cloud Comput. **3**(2) (2015)
11. Katsalis, K., Paschos, G.S., Viniotis, Y., Tassiulas, L.: CPU provisioning algorithms for service differentiation in cloud—based environments. IEEE Trans. Netw. Serv. Manag. 12 (1) (2015)
12. Zdzisław, P.: Rough-set theory and its applications. J. Telecommun. Inf. Technol. (2012)
13. Rissino, S., Lambort-Torres, G.: Rough-set theory-fundamental concepts, principles, data extraction and applications. In: Data Mining and Knowledge Discovery in Real Life Applications, pp. 293–299 (2010)
14. Bhaskar, R., Shylaja, B.S.: Knowledge based reduction for virtual machine provisioning in cloud computing. Int. J. Comput. Sci. Inf. Secur. (IJCSIS) **14**(7) (2016)
15. Liu, Y., Esseghir, M., Boulahia, L.M.: Cloud service selection based on rough-set theory. IEEE (2015)

Early Detection of Grape Stem Borer Using IoT

Kainjan Sanghavi[1(✉)] and A. M. Rajurkar[2]

[1] SNJB COE Chandwad, Research Scholar SGGS, Nanded, Maharastra, India
kainjan@gmail.com
[2] MGM COE, Nanded, Maharashtra, India
rajurkar_am@mgmcen.ac.in

Abstract. Grape stem borer is a serious threat to grapes due to its severe symptoms and loss of production. Traditional diagnosis of grape stem borer depends upon symptom identification, due to sensitivity limits of identification tools in vineyards. Grape stem borer prime indications are parching and sneering of affected branches. Recognition of the borer in early stages is a most challenging chore. This paper presents a novel system, utilizing sound sensor for detection of stem borer in grape vineyard using Internet of things. Foremost contribution of this work is a technique for early detection of stem borer pest based on IoT through an handheld device. The analytic solution detailed in this paper does not necessitate the farmer or any user to be an IoT expert in order to use it. The accuracy achieved for the identification of grape stem borer is higher than 90%. The system is envisioned to incorporate the significant advancements in communication technologies and wireless sensor networks.

Keywords: Grape stem borer · Grape vineyard · Internet of things (IoT) · Early detection · Grape diseases

1 Introduction

Grape is a vital ephemeral, climatic produce, enormously grown in India. Maharashtra accounts for 70% of India's total grape acreage and 63% of production. It is confronted by approximately 100 insect pests, which advances different types of damages to the grape vineyards. Amongst them, grape stem borer is a stern pest that is becoming one of the restraining causes in grape farming, mostly in Nashik district [1]. Recently, IoT-based systems are used for remote monitoring of objects [2].

Grape stem borer was previously considered to be a problem only in longstanding and deserted vineyards. Nevertheless, extreme occurrence of this pest is observed in even one-year-old grape grounds in topical years. The disease caused by this pest causes damage which is initially invisible to the naked eye. If there is 1 ha of land, then the general farmer obtains 10 lac income for export. Nevertheless, if the farm is contaminated with this pest, there is a loss of approximately 3.75 lacs per hectare in the exportation of grapes.

Extreme boring can eradicate the plant through the casing. The damage can lessen the amount of rudders for reproduction. When the contamination is late, the white

© Springer Nature Singapore Pte Ltd. 2021
P. Deshpande et al. (eds.), *Next Generation Information Processing System*,
Advances in Intelligent Systems and Computing 1162,
https://doi.org/10.1007/978-981-15-4851-2_22

pustules grow. Yield losses of approximately 20% on early plants and 80% in late-planting crops result due to stem borer.

There are several ways and approaches to diagnose these pests that affect the grape plant [3].

- Visual examination of the grape plant to check for deficit emblems: Using this, only vital insufficiencies are reported, but the symptoms of damage often detected may be inaccurate.
- Soil exploration: Meet the requirements of topsoil nutrient and other characteristics.
- Vegetal matter analysis: Nutrient level measurements in plant tissue. It can find flaws in soil testing that could not be identified.
- Bioassays: Methods for detecting nutrient deficiencies that incorporate tissue analysis methods and plant screening.
- Ground research: This experiment is the oldest and most accurate way for diagnosing nutrient deficiencies, but it is an expensive procedure.

These techniques are used as a first step in the discovery of the plant because the field tests are quite expensive and difficult to handle. However, they can typically not be performed on the plant but only in the laboratory.

There have been significant high-tech expansions in agriculture over the years. They can be categorized according to their existence. Identification of natural disasters, the monitoring and regulation of crops, biodiversity and geophysical measurement systems, organic importance of ecological monitoring and forest fire detection are certain applications based on the information technology [4, 5]. Furthermore, it is tranquil to have arrayed sensor structures for observing all environmental meteorological parameters such as temperature, humidity and air [6] or distributed monitoring systems based on the Internet of things (IoT) technology [7]. Such technology will enhance the existing methods along with the working conditions of the farmers and also improve the identification of diseases to ensure the final product's quality.

The proposed system uses sensors in IoT device to determine the existence of stem borer. The system generates an alert and sends a message to notify the farmer when it senses any abnormal status in the grape stem. The reimbursements of this arrangement are enormous, and the farmer will also need very less cost to invest. The focal contributions of this paper are as follows:

- A comprehensive design of the handheld device comprises sensors to detect the pest, i.e., stem borer.
- An innovative way of estimating the number of nodes needed for the region to be sensed.
- Use of a lucrative sensors that are freely available in the arcade.

The rest of the paper is organized as follows. Section 2 discusses about the life cycle of stem borer and traditional ways to control the development of stem borer. Section 3 describes the layout of the wireless sensor network, the installation of the wireless sensor node and its procedure. Section 4 demonstrates the results of vineyard control system by using the proposed system. Finally, Sect. 5 draws the conclusions and proposes future work.

2 Related Work

2.1 Life Cycle of Stem Borer

Four stages, embryo, larva, pupa and adult are part of the life cycle of stem borer [8] as depicted in Fig. 1. **Eggs**: Pocket-sized eggs are set in silts, which in turn are enclosed with a rubbery surface. These silts are positioned individually. **Larvae**: Anew hatched flat-headed cream burst into the trunk or arms and fed into them, thus making them hollow. **Pupa**: The larvae pupate to dark brown color in next three to five days. **Adult**: The adult beetle lives in the wine until May. The insect is a borer that drills into stalks which twigs and causes the affected branches to dry and wither. During July to September, the grown beetle instigates to appear from the vines with a round opening. Female beetles make prominent slits on the bark of the trunk and the arms of the seed.

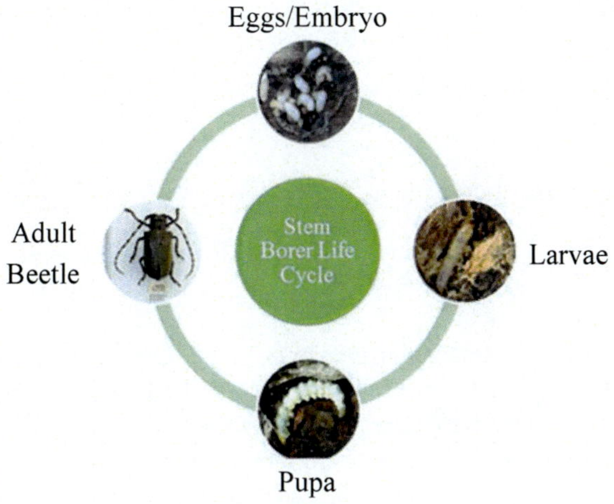

Fig. 1. Life cycle of stem borer

Larvae stage is the most treacherous stage of damage to grape plant because it consumes the innermost tissues of the paddy plant.

2.2 Traditional Ways to Control Stem Borer [9]

Though above practices as discussed in Table 1 are used to detect and prevent the growth of stem borer, it is challenging to control the damage caused by the stem borers as they are interior feeders. Farmers get to know about the growth of this disease at a very latter stage, i.e., adult stage, thereby bearing heavy economical and quality production loss.

Farmers then attempt to use dichlorvos in the holes and cover the holes in mud or place a half tablet of aluminum phosphide in the hole and close it. The farmers then spray the pesticiside prepared by combining 0.07% of Endosulfan in 2 ml water, in the event of extreme infestation to kill adults [2].

Table 1. Traditional ways to control stem borer

Technique	Description of control
Power-driven control	The ovum frames are congregated from the pitch and destroyed
Growth destruction	The stubbles of plant are evacuated and scorched after garnering
Sunlit conning	The stem borers get easily fascinated to brightness. Hence, sunlit conning is commonly used to draw and exterminate the pest
Natural control	Use of marauders and raiders or former vermin to reduce the growth
Organic control	Use of appropriate chemicals to protect the damage instigated by stem borer

Nevertheless, the residue of these chemicals stays on the plant through the application of all these insecticides. As a result, the grapes do not pass export checks. For 1 ha of land in the grapes field, the general farmers can earn 10 lac revenue from exports. However, if that insect is infectious, the export of grapes causes almost 3.75 lac per hectare loss.

2.3 Motivation to Use IoT

The focus has recently been on Internet of things (IoT). IoT is a wide and far-reaching technology that offers potential applications in all manufacturing, automation and supply chain services sectors.

In order to implement intelligent networks, IoT proves to be an economical and reliable technology [10]. IoT can be an effective tool for pest disease diagnosis and hence more successful than the related work described.

Various sensors can be used to detect different types of pests. Study in [11] shows the use of these sensors for the growth and tracking of several pests. Stem borer produces a sound while boring the stem [12], and hence, as discussed in [11], an acoustic sensor can be used, thereby integrating IoT with agriculture for the stem borer detection

2.4 Findings

So far, farmers are using traditional methods or techniques that do not regulate these stem borers effectively because they bore within the stem. Monitoring of stem borer in the field needs a huge human effort which is not accurate but instead time consuming, and hence, automation in agriculture is a big need of the day. Currently, existence of an early detection system is unobtainable.

This paper hence discusses a grape stem borer detection system based on the Internet of things to cover problems arising in traditional methods, such as chemical residues, the time needed to work on any tree and the need for several manual work.

3 Methodology

In the plantations where the infestation is serious, the rattling sound is heard [12], i.e., the grape stem borer generates sound while boring the stem. We therefore use the sound sensor to detect the stem borer when it develops inside the stem. The acoustic ability for larvae in situ detection was assessed by this analysis.

The system architecture of the proposed system is depicted in Fig. 2. The grape stem borer detector node is structured using Arduino UNO R3 having the microcontroller ATMEGA328, sound sensor (VMA309), buzzer (>85 db), LCD (16 Characters * 2 lines) and power supply (+5 V) (Fig. 3).

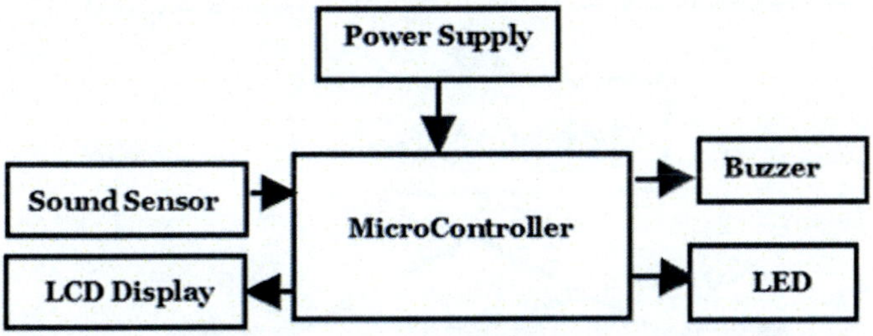

Fig. 2. Proposed system architecture

Fig. 3. Working model of proposed system

By using the proposed system, one can detect and inform the farmer at the precise time about the occurrence of the stem borer so that he or she can take appropriate action to save the grape trees. Therefore, the losses due to stem borer can be minimized to about 3.75 lacs per hectare, which is a boon to an ordinary farmer of India.

The general overview of the entire platform is shown in Fig. 4. Each of them will then be described in detail.

Processing: Process starts by scanning the grape tree with the device designed as shown in Fig. 2. While scanning the tree if a stem borer at larva stage is developed in the stem of the grape tree, the stem borer detector checks its presence by examining the

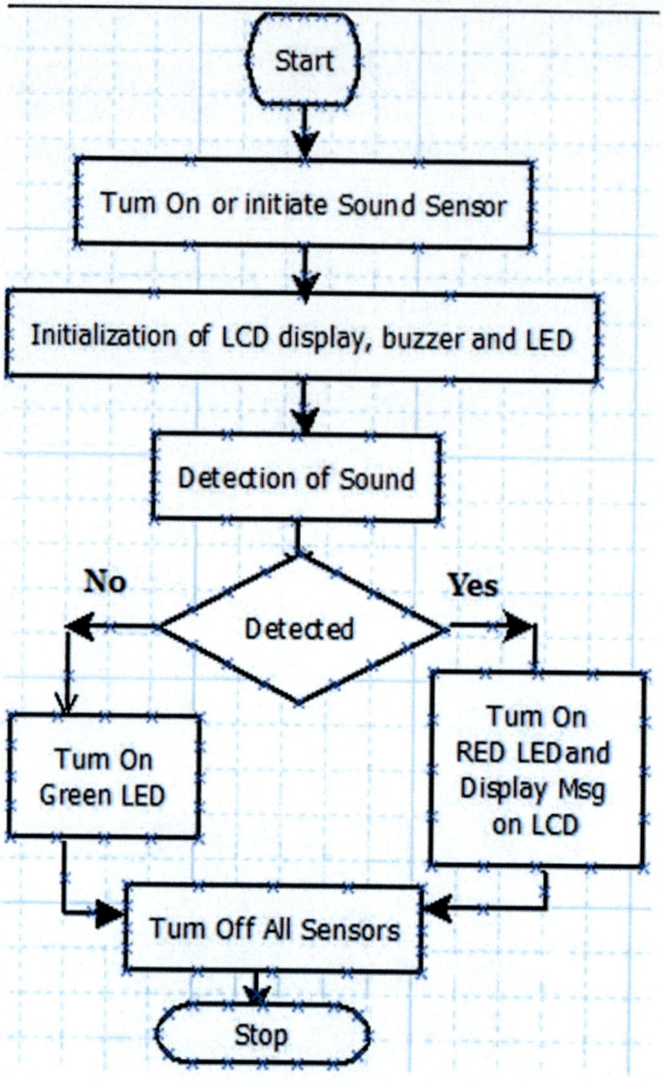

Fig. 4. System flow

sound intensity produced in the stem by this insect. If the sound intensity matches to 55 Hz or more, an alert will be generated by stem borer detector through a buzzer. This indicates the presence of stem borer in the stem of grape tree. Along with this, stem borer detector turns on the green LED as well as display the notification on the LCD available on this device. However, if the sound intensity is not matching the threshold, the Red LED will glow, and an appropriate message will be displayed on the LCD.

Much of an experimental research is done in order to identify the frequency of sound generated by the grape stem borer. As a result, it is deduced that if the sound frequency generated by the stem borer, at larvae stage, goes beyond 55 Hz the borer has started damaging the grape tree.

4 Results and Discussions

4.1 Experimental Setup

The acoustic records of 750 Root systems were collected in the Materwadi, Nashik (owner Prashant Agro Farm), a commercial viñeyard, and 750 Root Systems were recorded in the Sakura (Boraste Agro Farm), Nashik commercial viñeyard. Recordings were taken during July 28 to August 26, 2019. During the recordings, air temperatures ranged from 27 to 38 °C. Of the 1500 plants, 520 have been infected, and 230 have been not inoculated. We used the grape stem borer detector to test the strength for correct identification as depicted in Fig. 5. We performed 1000 cross-validation trials and evaluated the detection using sound sensors. Total samples selected for validation are 150 out of which 45 plants are randomly selected from the 520 diseased plants. Hence for testing, out of the sample of 150, forty five (45) trees which are diseased, and rest 105 are not inoculated.

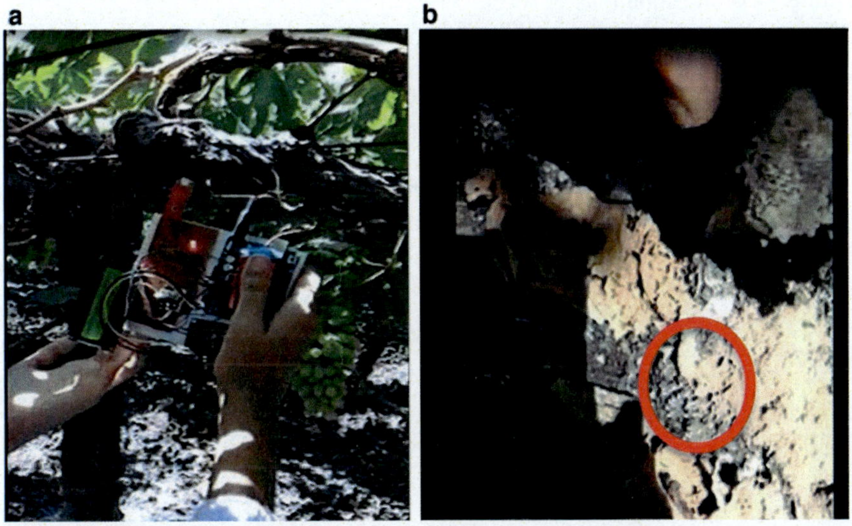

Fig. 5. **a** Testing done at Prashant agro farm. **b** Grape stem borer found in Materwadi stem detected in Fig. 54a

4.2 Evaluation

The evaluation uses confusion matrix as its respective classifier. Figure 6 shows the confusion matrix of samples taken for testing along with resultant values of accuracy, recall and precision.

Total Samples = 150	Predicted No	Predicted Yes	
Actual No	Tn = 45	Fp = 0	$Precision = \frac{Tp}{Tp + Fp} = 100\%$
Actual Yes	Fn = 3	Tp=42	45
	$Recall = \frac{Tp}{Tp + Fn} = 93\%$		

Fig. 6. Confusion matrix (testing)

When grape trees are given as input, out of the 45 diseased samples, 42 are correctly classified, whereas three are misclassified. Hence, the correct classification rate for 45 samples is 93%. However, for trial, when pomegranate tree was given as on input, 45 were correctly classified as invaild input and 0 misclassified, thereby giving the correct classification rate as 100%.

From above information, accuracy is computed using Eq. 1.

$$\text{Accuracy} = \frac{\text{Tp} + \text{Tn}}{\text{Tp} + \text{Tn} + \text{Fp} + \text{Fn}} = 0.97 = 97\% \quad (1)$$

So, overall system performance of classification is 97%. Table 2 represents all the values of performance metrics. Suggested scheme delivers encouraging results according to the principles. Figure 7 depicts the graphical representation of Table 2.

Table 2. Results of testing

Phase	Total samples	Infected by stem borer	Accuracy (%)	Precision (%)	Recall (%)
Training	1500	520	100	100	100
Testing	150	45	97	100	93

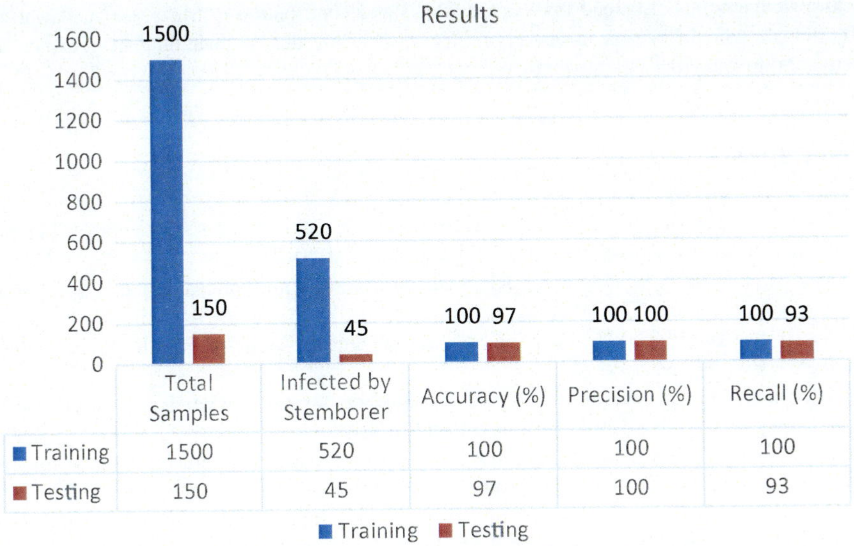

Fig. 7. Performance analysis chart

5 Conclusion

Grape stem borer vermin hover the commercial crops, leading to a decline in grape production and severe global economic losses. Different technologies and methods have been developed and used to monitor and mitigate the damage caused by pests. While traditional methods, such as power-driven control and organic control technologies, are commercially available and commonly used, their field detection applications are limited because of the need for a laboratory setup, time for testing and cost constraints. Recent and newly developed innovations including IoT have been highly appreciated but still need development. It is therefore necessary to develop advanced, in-time, cost-effective and portable early stage detection devices or technologies.

An innovative method for fulfilling these requirements, a handy grape stem borer system with sensitive sound sensor can provide advantages over conventional technologies, addressing this challenge. This paper discussed the grape stem borer detector processing.

The experimental results show that the approach proposed is a valuable approach which can contribute significantly to an accurate detection of grape stem borer present in the stem of the grape plant in a small measurement effort and at a much early stage, thereby reducing the economic losses incurred by the farmers.

The work will be extended to detect various other diseases like Downey Mildew and Powdery Mildew again at an early stage. Another, scope of this work can be extended by using other sensors for much accurate identification in more early stage to identify the stem borer.

Acknowledgements. I extend my sincere thanks and contribution to Mr. Prashant Pawar and Mr. Boraste who helped us in this effort. It was possible due to their diligent guidance and motivation to carry out this research.

References

1. https://www.midh.gov.in/VCS%20Reports/7-Value%20chain%20for%20Grapes%20crop % 20in %20Nasik%20district%20of%20Maharashtra.pdf(2017)
2. Iyer, B., Pathak, N.P., Ghosh, D.: RF sensor for smart home application. Int. J. Syst. Assur. Eng. Manag. **9**, 52–57 (2018). https://doi.org/10.1007/s13198-016-0468-5
3. Van Den Driessche, R.N.: Prediction of mineral nutrient status of trees by foliar analysis. Botan. Rev. **40**, 347–394 (1974)
4. Garcia, M., Bri, D., Sendra, S., Lloret, J.: Practical deployments of wireless sensor networks: a survey. Int. J. Adv. Netw. Serv. **3**, 136–178 (2010)
5. Lloret, J., Garcia, M., Bri, D., Sendra, S.: A wireless sensor network deployment for rural and forest fire detection and verification. Sensors **9**, 8722–8747 (2009)
6. Anand, C., Sadistap, S., Bindal, S., Botre, B.A., Rao, K.S.N.: Wireless multi-sensor embedded system for Agro-industrial monitoring and control. Int. J. Adv. Netw. Serv. **3**, 1–10 (2010)
7. Di Palma, D., Bencini, L., Collodi, G., Manes, G., Chiti, F., Fantacci, R., Manes, A.: Distributed monitoring systems for agriculture based on wireless sensor network technology. Int. J. Adv. Netw. Serv. **3**, 11–21 (2010)
8. Raypuriya, N.: Insect Pest Management of Grape Vine Stem Borer and Stem Girdler, BioTech Articles (2016)
9. Lakshmi, K., Gayathri, S.: Implementation of IoT with image processing in plant growth monitoring system. J. Sci. Innov. Res. **6**(2), 80–83 (2017)
10. Deshpande, P.: Cloud of everything (CLeT): the next-generation computing paradigm. In: Iyer, B., Deshpande, P., Sharma, S., Shiurkar, U. (eds.) Computing in Engineering and Technology. Advances in Intelligent Systems and Computing, vol. 1025. Springer, Singapore (2020)
11. Biz4Intellia homepage.: A complete guide for IoT based pest detection with its benefits. All Rights Reserved Biz4intellia Inc (2019). Last seen on 13/11/2019
12. Salini, S., Yadav, D.S.: Occurrence of stromatium barbatum (Fabr.) (Coleoptera: Cerambycidae) on grapevine in Maharashtra, India. Pest Manag. Hortic. Ecosyst. **17**(1), 48–50 (2011)

Controlled Privacy-Aware (CPA) Protocol for Machine-to-Machine Communication in Internet of Things

Poonam Ninad Railkar(✉), Parikshit Narendra Mahalle, and Gitanjali Rahul Shinde

Computer Engineering Department, Smt. Kashibai Navale College of Engineering, Savitribai Phule Pune University, Pune, India
{poonamrailkar,aalborg.pnm,gr83gita}@gmail.com

Abstract. Internet of things (IoT) is buzzword nowadays, and rapid growth of IoT in the market makes the impact in private and public domains. It is a giant network of devices, people, and services. The devices and things are integrated with a variety of sensors generate heterogeneous data, and these devices are connected to the IoT platform. This data is a very important asset of many organizations. There are a variety of applications that carry sensitive data. The security of such applications becomes a challenging responsibility for security researchers due to the resource constraint environment of IoT. There are various aspects of security, but confidentiality of data is one the major aspect that needs to be addressed. In IoT communication, the data traveling from one to another node, there are chances of a replay attack, Man-in-Middle attack, eavesdropping, etc. To overcome these attacks and challenges, there is a need for efficient lightweight algorithms that maintain confidentiality and privacy of data. At the same time, algorithms must efficiently work in a resource constraint environment. This paper presents Controlled Privacy-Aware (CPA) Protocol using elliptic curve cryptography Diffie–Hellman algorithm to exchange key, as well as for encryption and decryption of data to maintain the privacy. This paper presents formal security analysis against the above-mentioned attacks and gives a comparison of performance in terms of computational time with state of art.

Keywords: Elliptic curve cryptography · Internet of things · Diffie–Hellman · Privacy aware · Security · Confidentiality · Machine-to-machine communication

1 Introduction

Internet of things (IoT) is a collection of interlinked devices which can be mechanical, electronics, computing devices, living things, and object which have unique identification in network and have the capacity to communicate data with other things over the network with or without human interference [1]. IoT is a seamless integration of a variety of technologies and domains including RFID, sensors, smart devices, and actuators. Nowadays, IoT attracting people from academia, research, and industry as it is providing a small communication network and allowing a human being to

© Springer Nature Singapore Pte Ltd. 2021
P. Deshpande et al. (eds.), *Next Generation Information Processing System*,
Advances in Intelligent Systems and Computing 1162,
https://doi.org/10.1007/978-981-15-4851-2_23

communicate with things. IoT is helping businesses in all dimensions. The seamless integration of resource constraint devices and communication technologies with the current Internet possesses some critical problems like privacy and confidentially of the IoT ecosystem. The threats are varying from application to application and network to network. The IoT devices are open to the Internet so these devices are always on risk. IoT devices are having low memory, low computation and communication ability so these devices are vulnerable to attacks. We cannot use existing security algorithms because these algorithms include heavy mathematical computation which may drain a major portion of the device battery, and this can affect the lifetime of devices. In the IoT ecosystem, there is a need for a lightweight algorithm that can protect against a variety of network attacks.

Machine-to-Machine (M2M) communication requires a direct communication channel to interact with after devices [2]. M2M communication is attracted to the automation industry and the academic sector. Unfortunately, with the growth of Internet-based technologies, security challenges also increase. To prevent a malicious attack, illegal access and trust management in businesses, governments, and individual security experts need to overcome challenges for the network.

Elliptic curve cryptography (ECC) is public key cryptography used for digital signatures, encryption, and decryption of data [3]. To get same security, small key size is required compared to other non-ECC algorithms.

Elliptic curve cryptography Diffie–Hellman (ECCDH) is a modification of the Diffie–Hellman algorithm for elliptic curves. It is used for key exchange as well as encryption and decryption of data.

Privacy is nothing but confidentiality. When two devices are communicating and that communication is confidential, then it is called privacy. But actually, when two devices are communicating with each other and they are transferring their credentials, then the first step is authentication and then access control. Access control means, who can access what information. Authentication means X device wants to communicate with Y device then Y device ensuring device X identity. Privacy is used to keeps the information secure from an unauthorized person. Privacy can be achieved in different ways; it can use physical security or mathematical algorithms to encrypt data. In the era of IoT, digitally connected devices provide a comfortable, convenient, and enjoyable life to the end-user. Everyone observing the exponential growth of IoT devices; on the other hand, IoT devices also becoming an attractive target to the hackers. These hackers targeting end-users with their smart devices, so for the researchers, it becomes a challenging task to provide secure IoT solutions in the minimum period. Delay in providing security solution leads to inevitable disaster. Proper access control and privacy will keep the system secure [4].

This paper is organized as follows: Sect. 2 evaluates the related work. Section 3 presents proposed CPA Protocol for M2M communication in IoT. Formal security analysis for CPA protocol presents in Sect. 4. Section 5 presents protocol evaluation in terms of computational and setup time with other existing solutions. Finally, Sect. 6 concludes paper.

2 Related Work

Tianyi et al. [5] proposed an analytical study of issues related to security and privacy of smart applications which includes smart home, grid, and WSN. The authors proposed a novel communication protocol in which a device-to-device communication is carried out with the help of agents via a central controller. The author ensures integrity by using MAC for data transmission. Boussada et al. present a new scheme for the E-health IoT system for maintaining security and privacy by using a hybrid scheme which mainly uses two stages. First, system initialization and registration are carried out, and then at the second stage, actual patient's data is transmitted. They have performed simulation using identity-based symmetric cryptography. The system gives more rapid data transmission and found robust against attacks like eavesdropping attack, replay attack, known key security, and time correlation attack [6]. Heisari et al. [7] proposed a method for accurate and lossless sensitive data for smart city applications by proposing a software defined network.

Mohanty et al. [8] suggested a new framework by incorporating consensus algorithm, certificate less cryptography, and distributed throughput management for ensuring security in smart city application. This system is compared with different existing methods and found improved processing time with less energy utilization. However, in proposed system, constrained devices depend on the federal management system for generating a private key.

Ciphertext-policy attribute-based encryption method is proposed in [9]. User revocation is achieved by implementing a new Walsh–Hadamard transform algorithm. This lightweight system is able to handle two attacks successfully with minimum computation overhead. Blockchain-connected gateway method is proposed by Shi-Cho et al. for protecting the personal information of the user. The purpose of this method is not to reveal private information without the permission of owner [10]. Double-layered architecture having a command node and a clustered node is implemented based on Rubik's cube concept for private key management in WSN. This method is proposed by Naik et al. Dynamic key generation for every subtransaction is implemented using SPF algorithm. This system is secured as the key is not shared over the network. ID-based signature scheme has been proposed in [11] to prevent IP spoofing and Man-in-the-Middle Attack across a network. This system used unidirectional proxy re-encryption which has been experimented to prove that it is secure against chosen cipher text attack. Mahalle et al. experimented IECAC protocol using ECC on AVISPA tool. This protocol worked successfully against three types of attacks. For key generation and exchange, it uses key distribution center (KDC). Han, et al. [12] contributed an identity-based technique to store data. This technique is implemented for both intra and inter domain by considering user-dependent access grants. Also, the technique is effective against collision attack. Chakravorty et al. introduce a novel privacy-aware technique for a cluster of the smartphone in order to submit their data to analytics in [13]. In this, the partial dictionary has been prepared in which key is considered as a user in smart home application and value is considered as data associated with each user and this data is hashed and stored. According to [14–18], the requirement of key size 1024 bits is needed in RSA to achieve 80 bit of security level, however, ECC needs only 160 bits key size to achieve the same security level. Hence, they concluded that ECC is preferred for space-constrained devices.

Table 1 gives a detailed survey and gap analysis of state of art. This table tells that which system is lightweight, attack resistant, and supports distributed algorithm. In IoT communication, the data traveling from one to another node, there are chances of attacks like, replay, the man in the middle, eavesdropping, DoS, etc. To overcome these attacks and challenges, there is a need for an efficient lightweight algorithm that maintains the confidentiality and privacy of data. At the same time, algorithms must efficiently work in a resource constraint environment.

Table 1. Survey and gap analysis of state of art

Paper	Attack resistant	Distributed	Algorithm used	Lightweight
[19]	NC, RE, MiM, Tm	No	Rubik cube-based key management	Moderate
[20]	MiM, Re	Yes	Privacy preference preservation using blockchain	No
[21]	MiM	No	Enhanced security algorithm using hybrid encryption and ECC	No
[5]	MiM	No	A privacy-preserving communication protocol	Yes
[20]	MiM, Re, DOS	Yes	Identity establishment scheme, ECCDH	Yes
[11]	DoS, MiM	Yes	Identity-based cryptography	No
[21]	MiM, EvsD and Re	No	ICAC scheme	Yes
[22]	MiM and Re	Yes	Revocable IBE scheme	No

MiM Man-in-Middle attack, *DoS* denial of service attack, *Re* replay attack, *Tm* tamper attack, *NC* node compromise attack, *EvsD* eavesdropping

3 Proposed CPA Protocol for M2M Communication in IoT

Proposed Controlled Privacy-Aware Protocol for machine–to–machine communication basically divides into three parts as shown in Fig. 1. Domain parameter decision, key generation and exchange, and encryption and decryption

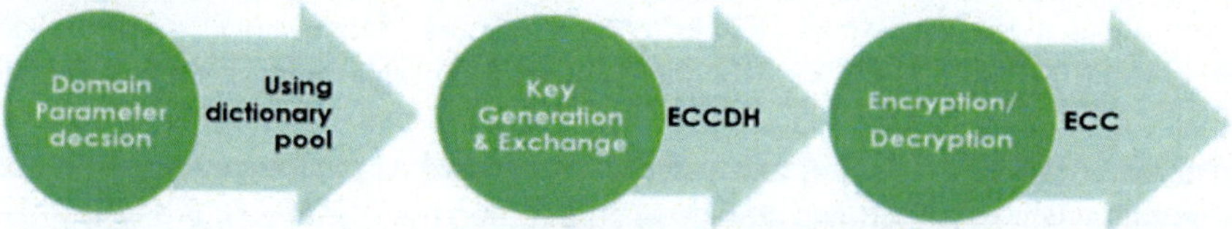

Fig. 1. Flow of CPA Protocol

(A) Domain Parameter decision Devices who want to communicate will decide which domain parameters are to be considered and which elliptic curve is to select over GF(p) which is finite field. p and q are a prime numbers and P is base point with large order of q. Dictionary of elliptic curve and domain parameter is stored and ID of these data is decided by devices at runtime as shown in Fig. 2. Constants a and b are used in defining an equation for the elliptic curve. To maintain the freshness of the key, the timestamp T_s is also checked during the decision of ID for the domain parameter.

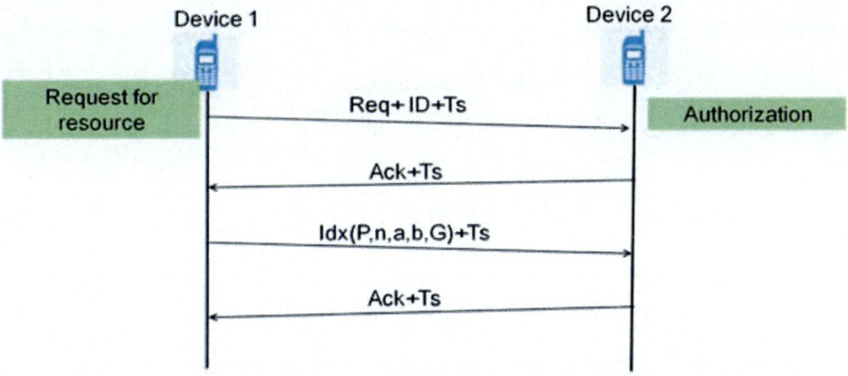

Fig. 2. Domain parameter decision

(B) Key generation and exchange Key generation is process to generate public key and private key. Using ECCDH Protocol, key generation and exchange are done as shown in Fig. 3. G is point generator and n is order of G. D_{pr1} and D_{pr2} are private keys and D_{pb1} and D_{pb2} are public keys of device 1 and device 2, respectively. Private keys are actually random numbers in interval [1, n] and public keys are calculated as $D_{\text{pb1}} = D_{\text{pr1}} * G$ and $D_{\text{pb2}} = D_{\text{pr2}} * G$. Device 1 share D_{pb1} to device 2 and device 2 share D_{pb2} to device 1. Both devices calculate secret key X_s. Timestamp, T_s is encrypted with that secret key and sent to second device to check the valid session. Current session is valid or not is decided upon current time T_{current}, T_s and threshold value for time.

(c) Encryption and Decryption Figure 4 shows how encryption and decryption happen for the message. In this step, ECC algorithm is used for encryption and decryption of message. When device 1 wants to send a message to device 2, then device 1 selects a random value K between interval (1, ($n-1$)). C_1 is calculated as K * G. C_2 is calculated as $M * K * D_{\text{pb2}}$ where M is the message. C_1 and C_2 are sending to another device 2. Now, device 2 decrypts the message $D(M)$ by calculating $C_2 * D_{\text{pr2}} * C_1$. Timestamp T_{xs} is also sent by encrypting it with a generated secret key to keep the track of the session.

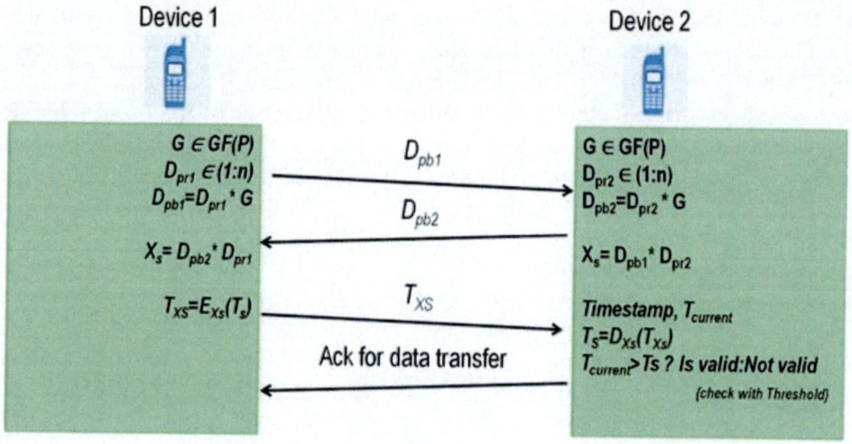

Fig. 3. Key generation and exchange

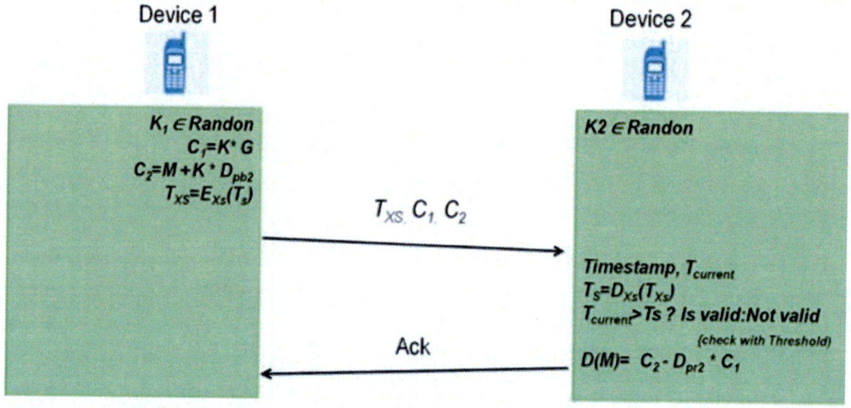

Fig. 4. Encryption and decryption

4 Formal Security Analysis

By considering CPA Protocol, formal security analysis is done against DoS, Man-in-Middle, and replay attacks.

(1) DoS

In DoS attack, the attacker generates a large number of requests and sends all requests to the destination node. Then, the destination node is unable to handle these huge numbers of requests and the targeted device gets flooded. In the proposed algorithm, Token T_s is introduced which will keep the track of the session. In communication between D_i and D_j devices, when D_i receives a message from D_j, first, timestamp validity will check. If the timestamp exceeds or wrong, then the message will discard by D_i else, a message will decrypt. For every session, one ID will be used, and so from the same ID, multiple requests will be restricted.

(2) Man-in-Middle attack
There are various types of Man-in-Middle attacks like eavesdropping and Masquerade attack. Traditionally, in ECC, the generator G is fixed and public. In the proposed system, the generator is hidden, so there is no way to know which point is generator; therefore, the attacker cannot make the Man-in-Middle attack. In the proposed system, all domain parameters are hidden. Device 1 sends public key D_{pb1} to Device 2 and attacker will be able to intercept it, but it is extremely difficult to find out its private key, as all domain parameters are hidden, and also ECC uses a one-way function.

(3) Replay attack
In replay attack, intruder can intercept the message sent by from D_i. But, it is not possible in CPA Protocol because it can simply notice by checking timestamp T_s. If T_s is older than predefined threshold value, it is invalid. In proposed system, change in T_s is not possible because its encrypted using secrete key X_s. At the time of communication between devices D_i and D_j, when D_i receives a message from D_j first, timestamp validity will check. If the timestamp exceeds to current timestamp, then the message will discard by D_i else, a message will decrypt.

5 Results and Discussion

This proposed CPA Protocol is implemented on a laptop device to check the time required for set up, key generation, encryption, and decryption of information. Following three schemes are used for comparison.

Scheme 1: IBE with revocation [22]
Scheme 2: IBE without revocation [13]
Scheme 3: Identity-based data storage in cloud computing [12]
Scheme 4: Ciphertext-policy attribute-based encryption [9]
Proposed System: Controlled Privacy-Aware Protocol.

A graph in Fig. 5 shows a pictorial representation of required execution time in controlled Privacy-Awareness Protocol and other schemes. Table 2 shows execution time required by set up, key generation, encrypt, and decrypt algorithms in different schemes and also in the proposed system. For communication between two devices, approximately 1 MB data is used for encryption and decryption. The time required for encryption and decryption is calculated by computing the average of running the operations 10 times. The time required for key generation is 14 ms. To encrypt message, it takes 5 ms and to decrypt ciphertext, it takes 3 ms. In the proposed system, ECC is used for key generation, encryption, and decryption and to exchange the key, ECDH mechanism is used. ECC is used for ease of key management. For the same level of security, very short keys are required, thus it takes less time than other schemes.

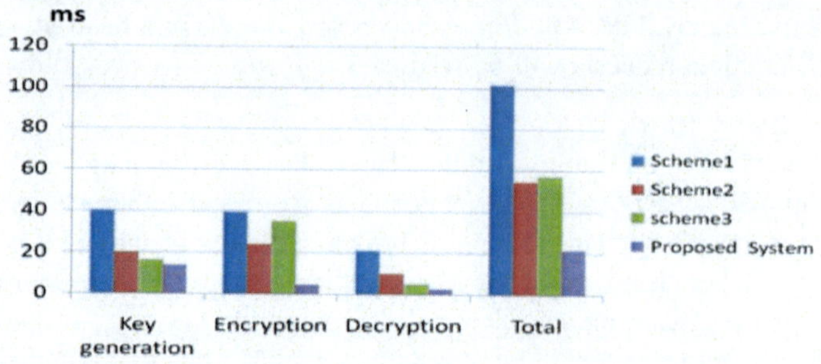

Fig. 5. Evaluation graph for previous scheme and proposed scheme

Table 2. Execution time required for different schemes

Paper	Key generation	Encryption	Decryption	Total time (ms)
Scheme 1	40.369	39.84	21.278	101.487
Scheme 2	20.121	24.595	10.285	55.001
Scheme 3	16.501	35.509	5.13	57.14
Scheme 4	8	12	16	40 (considering minimum attributes
Proposed system	14	5	3	22

6 Conclusions

This paper discusses Controlled Privacy-Aware Protocol for machine-to-machine communication and also gives a comparative analysis of various techniques which are used to secure IoT communication. This protocol ensures the privacy and security of the data generated by IoT devices. This generated data is not exposed as this data is encrypted and it is decrypted by a key which is generated at that session. The proposed work is also implemented and findings have been validated by comparing results with state of art. In this proposed system, domain parameters are hidden, so it is not easy to find out the private key and also ECC uses a one-way function. Distributed, light-weight, and attack resistant solution is supported by this CPA protocol. This protocol used ECCDH for key exchange and generation. For encryption and decryption also it uses ECC algorithm. ECC provides equal security with smaller key size compared to RSA, AES. To check a valid session, a timestamp is also maintained. Comparison in terms of computational time shows that CPA protocol scheme is efficient as compared to other solutions. According to formal security analysis, this protocol is attack resistant to DoS, Man-in-Middle, and replay attack.

References

1. Rouse, M.: Internet of things (IoT). IOT Agenda. Retrieved 14 Aug 2019 (2019)
2. Barki, A., et al.: M2M security: challenges and solutions. IEEE Commun. Surv. Tutor. **18** (2), 1241–1254 (2016)
3. https://en.wikipedia.org/wiki/Elliptic-curve_cryptography
4. Railkar, P.N., Mahalle, P.N., Shinde, G.R.: Access control schemes for machine to machine communication in IoT: comparative analysis and discussion. In: 2018 IEEE Global Conference on Wireless Computing and Networking (GCWCN), IEEE, pp. 59–63 (2018)
5. Song, T., et al.: A privacy preserving communication protocol for IoT applications in smart homes. IEEE Internet Things J. **4**(6), 1844–1852 (2017)
6. Boussada, R., Hamdane, B., Elhdhili, M.E., Saidane, L.A.: PP-NDNoT: on preserving privacy in IoT-based E-health systems over NDN. In: 2019 IEEE Wireless Communications and Networking Conference (WCNC), IEEE, pp. 1–6 (2019)
7. Gheisariy, M., Wang, G., Khanz, W.Z., Fernández-Campusano, C.: A context-aware privacy-preserving method for IoT-based smart city using software defined networking. Comput. Secur. (2019)
8. Mohanty, S.N., Ramya, K.C., Rani, S.S., Gupta, D., Shankar, K., Lakshmanaprabu, S.K., Khanna, A.: An efficient lightweight integrated blockchain (ELIB) model for IoT security and privacy. Futur. Gener. Comput. Syst. **102**, 1027–1037 (2020)
9. Pasupuleti, S.K., Varma, D.: Lightweight ciphertext-policy attribute-based encryption scheme for data privacy and security in cloud-assisted IoT. In: Real-Time Data Analytics for Large Scale Sensor Data, pp. 97–114. Academic Press, Cambridge (2020)
10. Cha, S.-C., et al.: Privacy-aware and blockchain connected gateways for users to access legacy IoT devices. In: 2017 IEEE 6th Global Conference on Consumer Electronics (GCCE), IEEE (2017)
11. Radhakishan, V., Selvakumar, S.: Prevention of man-in-the-middle attacks using ID based signatures. In: 2011 Second International Conference on Networking and Distributed Computing (ICNDC), IEEE (2011)
12. Han, J., Susilo, W., Yi, M.: Identity-based data storage in cloud computing. Futur. Gener. Comput. Syst. **29**(3), 673–681 (2013)
13. Boneh, D., Matt F.: Identity-based encryption from the weil pairing. Annual International Cryptology Conference. Springer, Heidelberg (2001)
14. Huang X., Shah, P.G., Sharma, D.: Protecting from attacking the Man-in-Middle in wireless sensor networks with elliptic curve cryptography key exchange. In: 2010 Fourth International Conference on Network and System Security. IEEE (2010)
15. Vasundhara, K.L., et al.: A comparative study of RSA and ECC. Int. J. Eng. Res. Appl. **8**(1), 49–52 (2018). www.ijera.com. ISSN: 2248-9622
16. Mahto, D., Yadav, D.K.: RSA and ECC: a comparative analysis. Int. J. Appl. Eng. Res. **12** (19), 9053–9061 (2017). ISSN: 0973-4562
17. Pharkkavi, D., Maruthanayagam. D.: Time complexity analysis of RSA and ECC based security algorithms in cloud data. Int. J. Adv. Res. Comput. Sci. **9**(3) (2018)
18. Shaikh, P., Kaul, V.: Enhanced security algorithm using hybrid encryption and ECC. IOSR J. Comput. Eng. (IOSR-JCE) **16**(3), 80–85 (2014)
19. Naik, S.C., Mahalle, P.N.: Rubik's cube based private key management in wireless networks. In: 2013 15th International Conference on Advanced Computing Technologies (ICACT), IEEE (2013)
20. Mahalle, P., et al.: Identity establishment and capability based access control (IECAC) scheme for internet of things.In: WPMC (2012)

21. Mahalle, P.N., et al.: Identity driven capability based access control (ICAC) scheme for the internet of things. In: 2012 IEEE International Conference on Advanced Networks and Telecommunications Systems (ANTS), IEEE (2012)
22. Li, J., et al.: Identity-based encryption with outsourced revocation in cloud computing. IEEE Trans. Comput. **64**(2), 425–437 (2015)

Performance Analysis of Wireless Sensor Network (WSN)

Chevvuri Naga Sridevi[1], Murrey Neeladri[1], and Durgesh Nandan[2(✉)]

[1] Department of ECE, Aditya Engineering College, Surampalem, India
sridevi.chevvuri9866@gmail.com, neeladri.m@aec.edu.in

[2] Accendere Knowledge Management Services Pvt. Ltd., CL Educate Ltd., New Delhi, India
durgeshnandano51@gmail.com

Abstract. In wireless environments, latency and noise must be included in the system plan for continuous control of dispatch vehicles. A compact, wearable, lovable, ergonomically, and at low-cost WLAN node, suitable for detecting a variety of different physical phenomena was the main focus of the research work. It is important to build a large-scale wireless sensor network through an efficient network connection mechanism. For example, ZigBee can assemble a WSN dependent on the bunch tree utilizing a straight-forward system association, address, and steering component. In any case, it cannot offer alluring availability for the hub principally due to the misuse of system profundity and systems administration contortion when applied to huge-scale WSN. In this paper, we consider organization of WSN and try to explained utility and applications of wireless network in innovative and creative manner which shows remarkable performance improvements in large-scale WSN compared to conventional ZigBee.

Keywords: Network design · Wireless sensor networks(WSN) · Delay control · Performance evaluation · Mobility · Harsh environments · Network connection · Scalability · Network depth · Parent/child selection

1 Introduction

As, we move toward deep-seated missions, financially savvy improvement of cutting-edge space innovations and dispatch frameworks is developing. To expand the current scope of rocket, profoundly viable dispatch frameworks are required. Remote wireless sensor systems (WSN) can be utilized so as to improve the eco-friendliness of the dispatch vehicles and to decrease the expense and weight of long wires, links, and connectors [2]. When planning a remote system, the topology of these hubs and the related system parameters are significant elements. In a recent report, data from multiple sensors were used to estimate the

© Springer Nature Singapore Pte Ltd. 2021
P. Deshpande et al. (eds.), *Next Generation Information Processing System*, Advances in Intelligent Systems and Computing 1162,
https://doi.org/10.1007/978-981-15-4851-2_24

landing point of the rocket projectile. A seismic sensor can measure the tremor caused by a rocket landing [2]. When planning a remote system, the topology of these hubs and the related system parameters are significant elements. In a recent report, data from multiple sensors were used to estimate the landing point of the rocket projectile. A seismic sensor can measure the tremor caused by a rocket landing [1]. WSN can be employed in space in different areas, such as low-earth orbit (LEO) weather missions, single-sample structural health monitoring of spacecraft, little satellite hubs fit as a fiddle, and the physical/compound detecting of different planets airs, surfaces, and soils [19]. They were helpful in space applications on account of the convenience and the minimal effort [3]. This work plans to give a superior outline of the system execution and its effect on the future rocket structure of sensors that gather increasing speed data from the strong rocket body [15]. The levels of vibration are crucial as even the smallest quantity of vibration may shed a rocket off its intended trajectory and reduce motor power. The overall sensor node network performance was investigated in this work. WSN including environmental monitoring, health applications, home automation, traffic control, logistics automation, and managerial systems is used in numerous fields of technology. In situations that cannot use cables or sensors are mobile, they are particularly efficient. The monitoring of physical phenomena through small, non-invasive devices is a challenge for the future among all possible applications. The paper presents the implementation of a mobile wireless sensor network (MWSN), which monitors the security conditions of building workers, especially the exposure of UV and micro-dust particles. These individuals are at risk of developing various environmental pathologies. Overexposure can lead to mutations in DNA that could lead to cancer in the skin or to other cellular diseases [17]. Exposure to dust particles also leads to "dangerous," lung loss as a result of cumulative exposure to airworthy powder [13], and autoimmune conditions such as silica-related scleroderma and arthritis [13]. The deployment of low cost, a small area WSN grids was facilitated by recent developments in wireless communication and semiconductor technology [4]. Be that, as it may given the quick development sought after for enormous-scale WSN development (e.g., the shrewd framework, catastrophe readiness and building/industrial facility robotization) [14]. The utilization of ordinary WSNs presents basic specialized difficulties, for example, the versatility of the system and vitality consumption [10]. In specific, a system association component is basic, empowering every one of the hubs to be associated with an enormous WSN proficiently while limiting vitality use [20]. There have been various system joining instruments intended to give solid access to internet applications (for example the DHCP and DAD) [7,8,18]. Be that as it may, for the most part as a result of the communicate storm issue, they may not be vitality effective, so they may not be applied to far reaching WSN [12]. ZigBee uses disseminated cluster-tree topology network protocols [5] to confine vitality consumption. In expansion to the system profundity [6,11], ongoing examination has considered the traffic trouble or potentially signal quality pointer (RSSI). In the following sections, the location of the sensors and the network topology were shown. The aim is to improve performance in terms of delay and network performance. Section 2 shows the topology of the network. The existing scheme has been discussed in Sect. 3. The comparative results are provided in Sect. 4. Section 5 discuss applicability of this field. Pros and corns have been discuss in Sect. 6.

2 Network Topology

The OPNET organize test system was utilized to dissect arrange execution. It is an amazing programming that gives a graphical interface. The ZigBee standard is furnished with a rich library. ZigBee has a minimal effort WLAN standard, which is low-fueled. It depends on the standard IEEE 802.15.4. The minimal effort makes it conceivable to utilize the innovation broadly in remote control and checking applications. It can work inside three distinctive recurrence ranges: Europe with 868 MHz, America and Australia with 915 MHz and worldwide with 2.4 GHz. The gadget ranges from 20 kbps in the recurrence band 868 MHz to 250 kbps in the recurrence band 2, 4 GHz [16]. Native to star and tree typical grids and general mesh networks, the ZigBee network layer supports both. We used the tree network in this work. The three-axis accelerometer provides 12-bit resolution outputs. This device transmits and receives the information of the accelerometer. It also has an 8-bit mode to quickly transfer a single byte which is shown in Fig. 1.

Describes the set up of the network. Three settings are taken into account. Setup-1 is made up of tandem sensors that are suitable for narrow rocket systems with space available in the middle. Similar to the first configuration, 2 where mid-section space is limited. The last set up place's zigzag sensors in bigger rocket diameter. A versatile detecting gadget, a passage, and a remote unit structure organize engineering. The sensor unit (SN) must go with the laborer the whole day, so it should be wearable, light-weight, close to nothing, and powerful. The SN location, handling, and transmission to the GW are utilized to identify the physical wonder. Contingent upon the application, the connection between the SNs and the GW is made utilizing various techniques.

A ZigBee standard of 245 GHz is connected to SNs for workers working in closed and indoor environments. A ZigBee standard of 868 MHz is preferred for large outdoor yards. The ad hoc solutions with an unlicensed bandwidth of 433 MHz are used for larger environments. The system with ZigBee is robust but the distance that can be covered is not that large and the power required to transmit is too high. Long distances without power can be reached with an ad hoc standard. The GW receives data from all sensor systems and transmits it to a central server that collects all data relating to all employees via standard connections like 3 G, Hyper LAN, and Wi-Fi via the Internet. Wireless sensor networking system: devices are used by manufacturers and collect information independently from their work, and local gates report via standard connections on a remote station. Consider, as illustrated in Fig. 2, a WSN-structured cluster-tree consisting of network coordinator, routers, and devices. Where a nil network depth network coordinator is initiated. The system facilitator and switches can

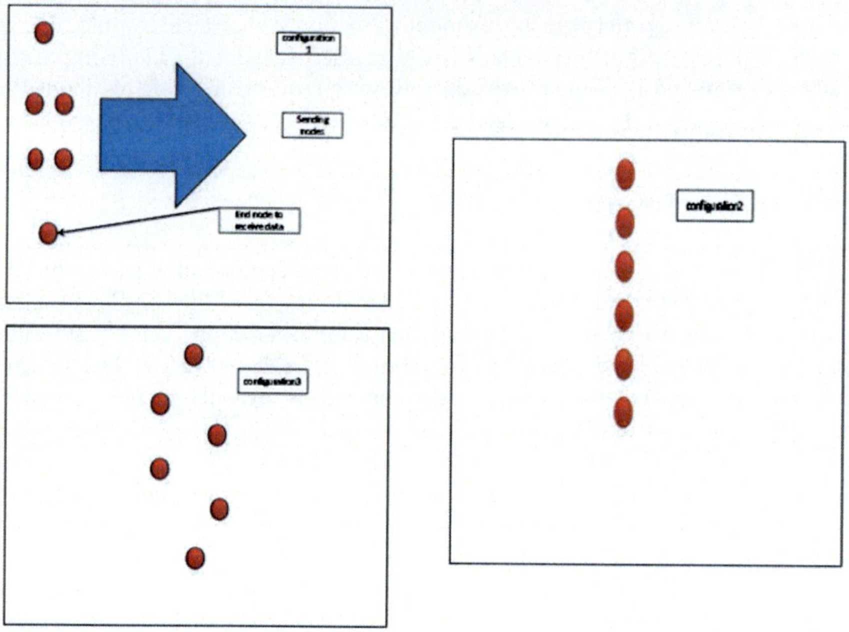

Fig. 1. Network model

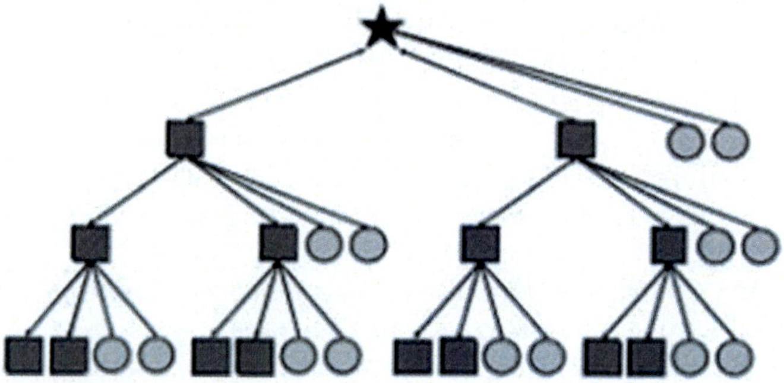

Fig. 2. Example of DAAM-tree structure methodology

have a limit of Cm kid hubs (i.e., Rm hubs as switches, or CmRm hubs as end gadgets) that every ha a system profundity more note worthy than their parent hubs in a steady progression, gave the utilization of a 16-piece DRAM address space. In the event that your system profundity is equivalent to the most extreme system profundity Lm, the switch cannot have a kid hub. The end gadget is a leaf hub that can just speak with its parent hub. The group tree organized WSN can be believed to help N hubs [13]. In this section, you can compare the three different scenarios based on the devices' location on the rocket. Additional scenarios and topology are compared but were not included

Table 1. Topology configuration

Attribute	Value
Transmission band	2.4 GHz
Transmission power	0.05 W
Device type	Coordinator/router
Network topology	Tree network
Application traffic	Bernoulli (0.01)
Packet size (nodes)	48 bits
Simulation time	15 mins

because of the lack of space. For these three scenarios, the network's output and delay were compared. With a less late network, more efficiency is required. The network's performance also allows us to get more data with fewer losses (for example, because of packet collisions). The attributes of each node within the network are described in Table 1. The hubs are put in three unique set ups on the rocket. The rocket is 12 ft tall and contains six hubs altogether. The top hub goes about as an organizer, and the four halfway hubs fill in as switches for sending the information to the base hub at the last part of the rocket, alluded to as the end hub, represents that the initial two situations are almost the equivalent, while the last situation (Configuration 3) is superior to the next two. This is because of the distinction between the hubs inside the system and the utilization of the CSMA/CA convention. This enables the client with less crashes to utilize the channel. In this way, the third topology set up can be utilized to situate the hubs on the rocket. The core of the MWSN is the SN, which must be designed to comply with section requirements.

It is made up of: • Physical sensor, • Semiconductor and radio (hosted by a single chip), • Antenna, • Battery, • Energy efficacy harvesters, indoor/outdoor, • Charging circuit.

We chose silicon carbide (SiC) among all the possible UV sensors since it is a small sensor with high speed and good spectral efficiency. The only disadvantage is that there is very small output power, and if some interference is present, a measurement cannot be done, a pre-amplified sensor can be used to resolve the problem. We have chosen the sensor SGLux TOPCON-ABC1, nano UV with a 2.4 mW power output, a proportionate output voltage to sunlight. For the dust sensor, we have decide to use an optical one which, through a diode and a photo transistor, is able to detect the reflected light of dust in the air. We selected the Sharp GP2Y1010AU0F that detects very fine particles, has 60 mW electricity consumption. The transceiver module ZigBee (Amber Wireless, AMB8420) includes both the radio and the microserver. The requested power, for example, at 868 MHz, is higher (135 mW) than the available power at 245 GHz (75 mW), depending on the frequency. The system on a chip module that includes both the radio and micro-controllers has been chosen for the ad-hoc transmission system.

This module can be utilized with different frequencies, from 433 to 915 MHz (TI, CC 1110F32). Clearly, the power required by the module varies according to the frequency selected, for example, power consumption is 108 mW at an 868 MHz transmitter mode.

The polypropylene offers strength, lightness, and transparency. The use of the polypropylene sheet, with the exception of the sensor, cover all electronic devices ensures overall protection for the whole system which enhances the usability of the device in general and ensures that it can be washed. The process of manufacture is very efficient on the basis of the use of rotary machines for the printing of food packages. These machines are characterized by optimal speed, great accuracy, and cheap assembly costs (up to 200 m per minute). As the SN must be automatically rechargeable, the power supply is an important element to consider. In order to minimize power consumption, all SN components are chosen, but an energy harvester is added to extend the life of the sensor. Various types of energy harvesters were analyzed [9] and two were found suited for the project: one thermo-electric harvester for indoor use and one for the outdoor photo-voltaic harvester. The Seebeck effect is based on a thermo-electric harvester (Micro pelt, MPG D751) which is exposed to the exterior and to the cold side of the device in contact with the human skin. The temperature difference between the body is sufficient for generating electrical energy and charging the battery, normally greater than 10 ccs. The solar harvester is the ideal photo-voltaic harvester (Ixys, SolarMD600H10L), as the situation is usually favorable for maximizing solar modular efficiency.

- Network joining mechanism of ZigBee

ZigBee bolsters two kinds of hubs: The full capacity gadget (FFD) and the RFD. The net as a switch or an end gadget can be joined by a FFD. It attempts first to join the system as a switch and afterward as a gadget when the system is not associated as a switch. A RFD, notwithstanding, just as an end gadget can interface with the system.

3 Existing Scheme

Every switch can choose detached hubs as its kid switch, enabling switches to interface hubs in the entire territory, can mitigate the versatility issue in ZigBee. It might likewise be alluring to join the system as the terminal gadget and not switch for new hubs closer to the system organizer instead of neighboring switches, lessening misfortunes of system profundity. Another hub n prepared to associate with the system initially creates the plausibility of PN guardians another way, organize facilitators and switches decide their child's switches. The system organizer chooses potential youngsters' switches as its kid switches with the most extreme RSSI. It can, in this manner, pick youngster switches with a profundity of 1. The switch chooses potential kids' switches with least guardians hubs as its kids' switches, which make associating hubs simpler. It tends to be seen that numerous hubs join the system as the end gadget as opposed to the switch in the proposed parent determination framework, in this manner making it simple to build organize profundity inclusion.

The existing selection scheme for child routers also distributes routers throughout the entire area, alleviating the Problem of network bias. The scheme is fully distributed, reducing energy consumption while maximizing network scalability. It can be used for ZigBee directly.

4 Result

The networks location and topology have demonstrated an impact on network performance. A way to position the nodes in the network is determined to achieve the desired results. For now, development boards have been used rather than individual components. The preliminary tests nevertheless show that the concept is applicable.

When ZigBee/IEEE 802.15.4 reference points empowered, where the tending to bit is 16-piece, the Beacon request BO 7, the super casing request SOis0 [8, 11], the performance of the proposed plan is checked by PC reenactment. We accept that every switch communicates a reference point message with an interim of 15,362BOms and transmits the sign for an interim of 15,362SOms after guide.

Table 2. Mint-route evaluation statistics

Parameters	Minimum	Maximum	Average
Duty cycle	0.031	0.047	0.044
Delivery (in percentage)	94.1	100	97.4
Re-transmission per packet	0	0.095	0.059
Parent changes	0	5	1.58

Table 2 shows Mint-Route statistics over an 8 h on 29 nodes in a 3-hop network. The results show Mint-Route, as shown by the high success rate, has established reliable delivery and stable routes across a 3-hop network. It can also be concluded that the neighboring table was handled successfully by Mint-Route since very few parent changes and evictions occurred. All the nodes, we assume, are distributed equally in a circular radius of 60 m and have a same 10 m transmission range. the Number for phoned nodes of network representations of average number joining by node number. The conventional ZigBee is not capable of making a correct node connection since the system size additionally increments with the utilization of enough space to address the system. Notwithstanding the one-sided hub association, the system profundity can be squandered. The proposed plan likewise decreases the stranded hubs significantly (Figs. 3, 4 and 5). All the nodes, we assume, are distributed equally in a circular radius of 60 m and have a same 10 m transmission range. The number for phoned nodes of network representations of average number joining by node number. The conventional ZigBee is not capable of making a correct node connection since the system size

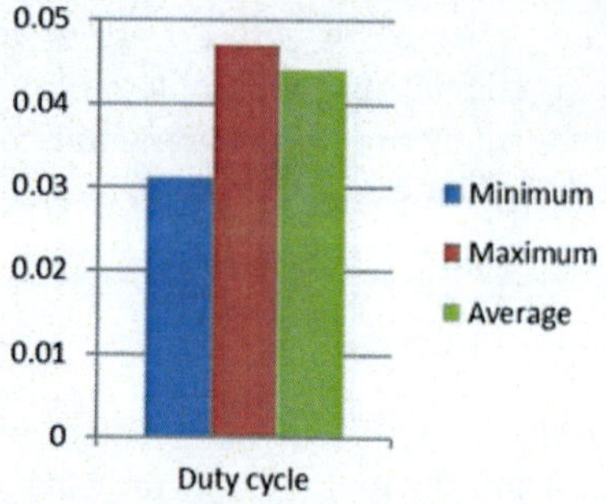

Fig. 3. Duty cycle

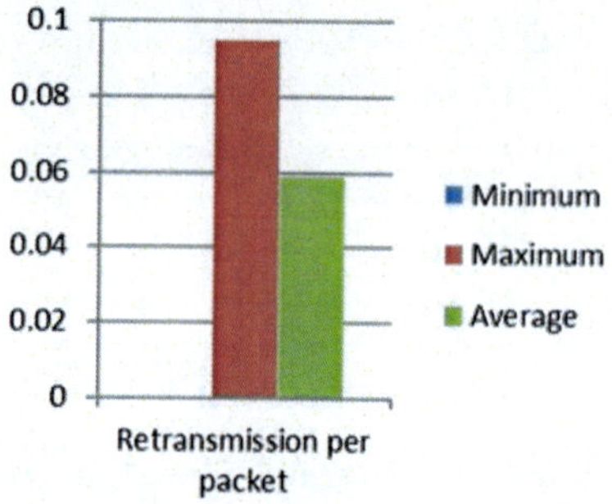

Fig. 4. Re-transmission per packet

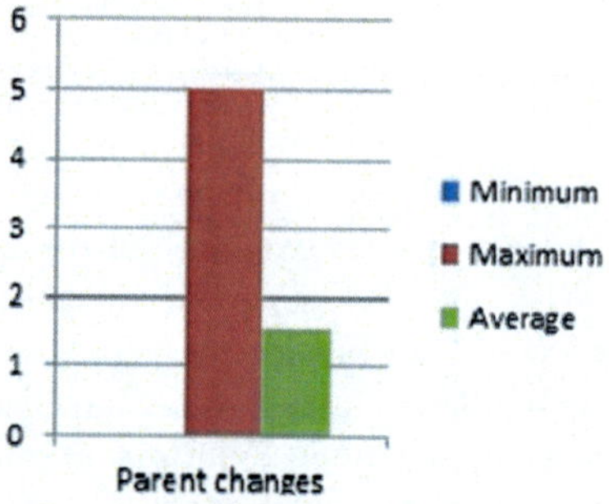

Fig. 5. Parents changes

additionally increments with the utilization of enough space to address the system. Notwithstanding the one-sided hub association, the system profundity can be squandered. The proposed plan likewise decreases the stranded hubs significantly.

5 Applications

There are various applications of WSN. Here, few popular applications can be highlighted. Military applications, health applications, environmental applications, home applications, commercial applications, area monitoring, environmental/earth sensing, air pollution monitoring, land slide detection, and industrial monitoring.

6 Conclusion

It was demonstrated that the area and topology of the system influence the system execution. One technique is resolved to put the hubs inside the system to get the ideal outcomes. Future research includes concentrating different topology of systems. In contrast with wired frameworks, they are firmly prescribed for future frameworks for control frameworks due to the lower weight and cost of WSNs and their comparative presentation. The paper shows the performance analysis of WSN in various setup conditions and limitations. What are the possible future prospect in WSN has been discussed.

References

1. Abedi, A.: Power-efficient-coded architecture for distributed wireless sensing. IET Wirel. Sens. Syst. **1**(3), 129–136 (2011)
2. Abedi, A., da Cunha, M.P.: Hybrid wireless communications with high reliability and limited power constraints in noisy environments. In: NASA/CANEUS Fly-by-Wireless Workshop, March 2007, Grapevine (2007)
3. Akbulut, A., Patlar, F., Zaim, A.H., Yılmaz, G.: Wireless sensor networks for space and solar-system missions. In: Proceedings of 5th International Conference on Recent Advances in Space Technologies-RAST2011, pp. 616–618. IEEE (2011)
4. Akyildiz, I.F., Su, W., Sankarasubramaniam, Y., Cayirci, E.: A survey on sensor networks. IEEE Commun. Mag. **40**(8), 102–114 (2002)
5. Alliance, Z.: Zigbee specification v1. 0. New York, USA (2005)
6. Attia, S.B., Cunha, A., Koubâa, A., Alves, M.: Fault-tolerance mechanisms for zigbee wireless sensor networks. In: Work-in-Progress (WiP) Session of the 19th Euromicro Conference on Real-Time Systems (ECRTS 2007), Pisa, Italy, pp. 37–40 (2007)
7. Droms, R.: Dynamic host configuration protocol, p. 2131. Network Working Group, Request for Comments (1997)
8. Droms, R., Bound, J., Volz, B., Lemon, T., Perkins, C., Carney, M.: Dynamic host configuration protocol for ipv6 (dhcpv6). Tech. rep., RFC 3315 (2003)
9. Harb, A.: Energy harvesting: state-of-the-art. Renew. Energy **36**(10), 2641–2654 (2011)
10. Hur, S., Kim, J., Choi, J., Park, Y.: An efficient addressing scheme and its routing algorithm for a large-scale wireless sensor network. EURASIP J. Wirel. Commun. Netw. **2008**(1), 765–803 (2008)
11. Lee, J.H., Lee, E.S., Kim, D.S.: Network joining algorithm for mobile nodes in ubiquitous sensor networks. In: 5th International Conference on Computer Sciences and Convergence Information Technology, pp. 836–839. IEEE (2010)

12. Lu, J.L., Valois, F., Barthel, D., Dohler, M.: Low-energy address allocation scheme for wireless sensor networks. In: 2007 IEEE 18th International Symposium on Personal, Indoor and Mobile Radio Communications, pp. 1–5. IEEE (2007)
13. Oxman, A.D., Muir, D.C., Shannon, H.S., Stock, S.R., Hnizdo, E., Lange, H.: Occupational dust exposure and chronic obstructive pulmonary disease: a systematic overview of the evidence. Am. Rev. Respir. Dis. **148**(1), 38–48 (1993)
14. Pan, M.S., Tsai, C.H., Tseng, Y.C.: The orphan problem in zigbee wireless networks. IEEE Trans. Mob. Comput. **8**(11), 1573–1584 (2009)
15. Schrader, K., Abedi, A., Caccese, V.: Impact localization and scaling for shm of inflatable habitats. In: CANEUS Fly by Wireless Workshop 2010, pp. 20–21. IEEE (2010)
16. Shen, B., Abedi, A.: Error correction in heterogeneous wireless sensor networks. In: 2008 24th Biennial Symposium on Communications, pp. 111–114. IEEE (2008)
17. Tadokoro, T., Kobayashi, N., Zmudzka, B.Z., Ito, S., Wakamatsu, K., Yamaguchi, Y., Korossy, K.S., Miller, S.A., Beer, J.Z., Hearing, V.J.: Uv-induced dna damage and melanin content in human skin differing in racial/ethnic origin. FASEB J. **17**(9), 1177–1179 (2003)
18. Vaidya, N.H.: Weak duplicate address detection in mobile ad hoc networks. In: Proceedings of the 3rd ACM International Symposium on Mobile Ad Hoc Networking & Computing, pp. 206–216. ACM (2002)
19. Vladimirova, T., Barnhart, D.: Toward space based wireless sensor networks. In: Helvajian, H., Janson , S.W. (eds.) Small satellites: past, Present, and Future (2008)
20. Ye, F., Pan, R.: A survey of addressing algorithms for wireless sensor networks. In: 2009 5th International Conference on Wireless Communications, Networking and Mobile Computing, pp. 1–7. IEEE (2009)

Verification of 32-Bit Memory Using Layered Testbench with Optimum Functional Coverage and Constrained Randomization

Sangeeta Parshionikar, Sheryl Serrao(✉), and Yash Ramesh Kumar

Department of Electronics, Fr. CRCE, Mumbai University, Mumbai, India
{sangeeta05, sherylcorina, laryash99}@gmail.com

Abstract. As intricacy of electronic designs, chips, ASICs increases, an efficient, organized and automated approach is required to create testbench. Verification using System Verilog layered testbench is the systematic way to verify the design without any bugs. Also, it checks whether design meets all the specifications correctly. In this paper, a memory model has been implemented and verified by creating real-world verification environment as per design requirement. At top level, scenarios are created, and testbench is broken down to small pieces using classes. Verification is carried out using System Verilog layered testbench methodology. Firstly, verification plan is prepared considering coverage sequences, constrained random stimulus, application of test cases and verification process for memory module. To improve the coverage which is measure of correctness of verification, randomization methodology is used. All input test cases are randomized, and missed or corner cases are covered by directed test cases. Mentor graphics Questa Sim 10.0b is used to carry out simulation. Both code and functional coverage are analyzed. Our research work achieved 100% code coverage and 85% functional coverage. Also, scoreboard and monitor verification blocks are implemented to check the predicted output with actual one.

Keywords: Verification · Functional coverage · Random stimulus · Layered testbench · Transactor · Monitor · Scoreboard

1 Introduction

1.1 Need of Verification

Currently, the demand for verification engineers is much more than design engineers. The reason is an exponential increase in complexity of verification environment [1].

To verify the functional correctness and validity or to determine if the design is a perfect reflection of the specifications without any bugs, extensive verification is needed. Correct verification avoids surprises at the later stages of design development phases, ensuring that the developed product can reach the market on time promising to be economical and of refined quality.

© Springer Nature Singapore Pte Ltd. 2021
P. Deshpande et al. (eds.), *Next Generation Information Processing System*,
Advances in Intelligent Systems and Computing 1162,
https://doi.org/10.1007/978-981-15-4851-2_25

Verification of chip is equally important as that of the designing of chip. Verification is a process in which a design is tested against a given specification [2]. It starts with the designing right from the time the design architecture build up.

A traditional way to verify a design using directed testbench has few drawbacks like it reports only predicted errors. It is time consuming and needs more number of verification people. This is probably due to hardcoded binary value inputs. The outputs can be easily predicted and observed in the waveforms. In case any input combination is missing and verification engineer is trying to check output for it, it shows errors. Thus, directed testbench reports only predicted errors and not real-time application running errors. For complex large designs, one cannot perform verification using directed testbench as it is not possible to give all hardcoded input combinations. Also, it is very time consuming and needs huge man power.

While using layered or smart testbench, we create the actual environment where the design is used and tested it for all possible cases. This is what we are presenting in our paper by creating a complete environment for memory module using layered testbench. The advantage of smart testbench is that it finds the bugs that are never anticipated [3]. While implementing the smart testbench, verification engineers should ensure that assertion checking is performed on every instance of a DUT module because an engineer can never know which DUT might fail [4].

The steps involved to verify the design are generate stimulus, capture response, determine correctness and measure progress [5]. The testbench creates a real-world around the design, mimicking the entire environment. The key principle behind testbench development is that it simulates everything not in DUT.

1.2 System Verilog Features

Testbench or verification environment is used for generating and passing a predefined input sequence to a design, capturing and comparing the design response to expected output.

System Verilog is basically a hardware verification language (HVL), nowadays also used as hardware description language (HDL). System Verilog provides a complete set of verification blocks needed for layered testbench [2]. SV offers all the necessary functions for the use of constraint random generation, assertion-based verification and coverage-driven verification. These methods significantly refine the verification process. It also provides improved hardware simulation capabilities that increase design efficiency and ease the design process [6].

SV provides useful constructs and new commands like interfaces, new data types (logic, int), enumerated types, arrays, hardware-specific always (always_ff, always_comb). SV has object-oriented programming concept that offers abstraction, encapsulation, inheritance and safe pointer capabilities [7]. One of the robust features of SV is interface that represents a bundle of wires with intelligence used for synchronization and functional code. SV allows multithreading and synchronous communication between them through events, mailbox and semaphores [8].

2 Layered Testbench

Generator is a verification block that falls under the scenario layer of the layered Testbench, which handles both the data and control command stimulus generation using transactions. These are randomized on every stimulus transaction generation. Transactions have constraints to constrain the variables of the transactions. The randomness of constrained random testing can be controlled via the scenario layer. A mailbox is required to pass the randomized packets to the driver [9].

The transaction block falls under the functional layer of the layered Testbench, and it helps with top level executions. Data and control fields required to be randomized are declared in this block. It provides vital data to coverage model regarding the generated stimulus. In our testbench, the data inputs addr [7:0], wdata [31:0] and command inputs wr_en and rd_en have been randomized. A constraint has been added to ensure that read and write operations do not occur simultaneously.

Driver is a verification block that falls under the command layer of the layered Testbench, which receives transaction packets from the generator via a mailbox, and it then forwards the packet level data inside the transaction packet into signal level layer for it to be passed to the DUT. The driver may introduce errors or add delays wherever required. It then bifurcates the commands down into individual signal changes such as bus requests and handshakes [10].

Monitor is a verification block that falls under the command layer of the layered Testbench. It examines the signal level processing via a virtual interface and makes it compatible with functional level to be sent to the scoreboard. The monitor also reports any failures or violations in transactions, alongside; it forwards the output of the DUT for further verification.

Scoreboard is a verification block that falls under the functional layer of the layered Testbench, and it is a self-checking structure. Scoreboard is also referred as tracker. Scoreboard computes the expected DUT output and compares the expected output with actual output from the monitor. It is imperative to note that the DUT and the reference model may follow different data paths. Hence, the data could be received at different instants of time. Owing to this, data storage must be done. The required synchronization needs to be maintained between the expected output and actual output. Post-synchronization, the output from the DUT and the predicted output can be compared to determine if effectives and validity exist. Thereby, the scoreboard block implemented checks if the memory written onto the memory at a specific address location is same as the memory read from that address.

Environment contains the instances of all the verification components along with the required component connectivity. The necessary processes for execution of the components (verification blocks) are initiated here.

Test block is implemented using the program block to prevent race condition between the DUT and the testbench [11]. The test block's primary task involves environment instantiation, setup of testbench configuration, i.e., setting the type and number of transactions to be generated and initiation of stimulus to be driven.

The topmost block is responsible for connecting the DUT and Testbench. It comprises the DUT, the test block and interface instances. The complete structure of layered testbench is shown in Fig. 1 [9].

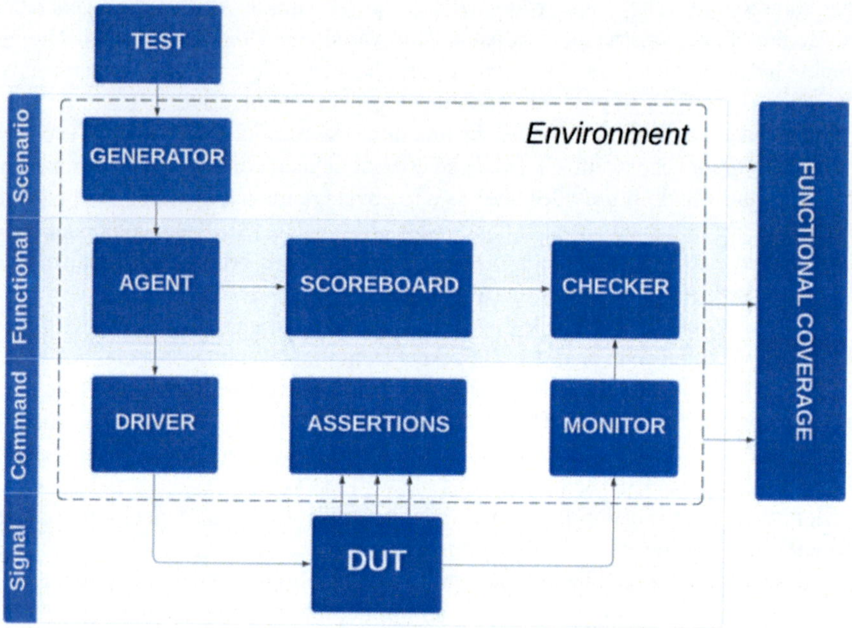

Fig. 1. Block diagram of layered testbench

3 Working of Memory Model

The basic block diagram of 32-bit memory model is shown in Fig. 2. This memory is capable of storing 32-bit data as per address location. It has 256 address locations, write and read enables. Write and read operations are synchronized with the clock.

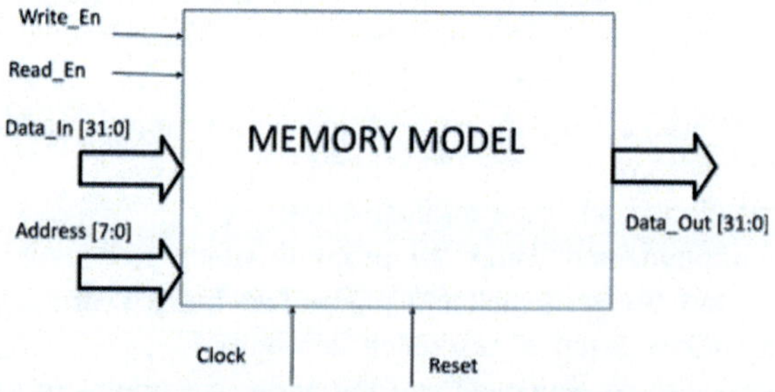

Fig. 2. 32-bit memory model

The working of the memory module is

Write Operation—To write at particular location, wr_en needs to be active, and data should be available on data_in lines.
Read Operation: To read from a particular location, rd_en input signal needs to be active. Address and rd_en are required to be active on same clock.
First step of verification process is to create the verification plan, followed by testbench architecture and then writing the testbench.

As per layered testbench, scenarios are created in verification plan. For our design, we focused on four main scenarios listed below:

- Write to any location, read from the same location, written data should match with read data.
- Write to all locations and read from any location.
- Check for the default value. Default value is set to ones.
- Also, at any instance of time, if RESET is high, that sets the value equal to default.

4 Verification Results and Simulation Analysis

A 32-bit memory is implemented and verified. The read, write, reset functionality of the memory unit was successfully verified via the waveforms on Questa Sim 10.0b. The output was observed for various cases and combinations of the input [7]. The quantitative comparison of both verification methods is shown in Table 1.

Table 1. Quantitative analysis of directed testbench and layered testbench

Verification methods	Functional coverage (%)	Code coverage (%)
Directed testbench	70.7	77.9
Layered testbench	85	100

The timing and synchronization with respect to clock were also observed and verified. The output is shown in Fig. 3.

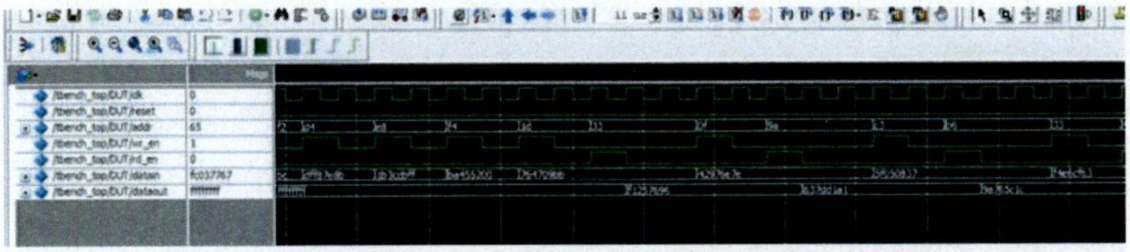

Fig. 3. Output waveforms for write and read operation

The scoreboard, a self-checking structure, was implemented. The scoreboard successfully managed to verify the correctness of the read/write/reset functionality of the memory unit. The same is depicted in Fig. 4. The scoreboard compared the expected output from the data stored with the data read at a specific memory location address. The output was verified via the Transcript Window.

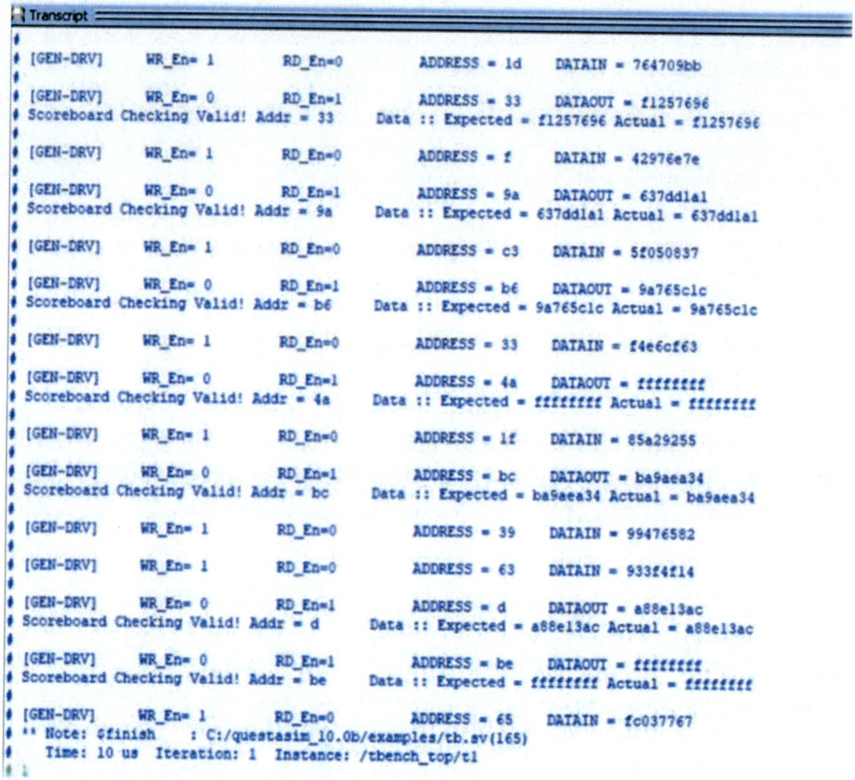

```
Transcript
#
# [GEN-DRV]   WR_En= 1   RD_En=0   ADDRESS = 1d   DATAIN = 764709bb
#
# [GEN-DRV]   WR_En= 0   RD_En=1   ADDRESS = 33   DATAOUT = f1257696
# Scoreboard Checking Valid! Addr = 33   Data :: Expected = f1257696 Actual = f1257696
#
# [GEN-DRV]   WR_En= 1   RD_En=0   ADDRESS = f   DATAIN = 42976e7e
#
# [GEN-DRV]   WR_En= 0   RD_En=1   ADDRESS = 9a   DATAOUT = 637dd1a1
# Scoreboard Checking Valid! Addr = 9a   Data :: Expected = 637dd1a1 Actual = 637dd1a1
#
# [GEN-DRV]   WR_En= 1   RD_En=0   ADDRESS = c3   DATAIN = 5f050837
#
# [GEN-DRV]   WR_En= 0   RD_En=1   ADDRESS = b6   DATAOUT = 9a765c1c
# Scoreboard Checking Valid! Addr = b6   Data :: Expected = 9a765c1c Actual = 9a765c1c
#
# [GEN-DRV]   WR_En= 1   RD_En=0   ADDRESS = 33   DATAIN = f4e6cf63
#
# [GEN-DRV]   WR_En= 0   RD_En=1   ADDRESS = 4a   DATAOUT = ffffffff
# Scoreboard Checking Valid! Addr = 4a   Data :: Expected = ffffffff Actual = ffffffff
#
# [GEN-DRV]   WR_En= 1   RD_En=0   ADDRESS = 1f   DATAIN = 85a29255
#
# [GEN-DRV]   WR_En= 0   RD_En=1   ADDRESS = bc   DATAOUT = ba9aea34
# Scoreboard Checking Valid! Addr = bc   Data :: Expected = ba9aea34 Actual = ba9aea34
#
# [GEN-DRV]   WR_En= 1   RD_En=0   ADDRESS = 39   DATAIN = 99476582
#
# [GEN-DRV]   WR_En= 1   RD_En=0   ADDRESS = 63   DATAIN = 933f4f14
#
# [GEN-DRV]   WR_En= 0   RD_En=1   ADDRESS = d   DATAOUT = a88e13ac
# Scoreboard Checking Valid! Addr = d   Data :: Expected = a88e13ac Actual = a88e13ac
#
# [GEN-DRV]   WR_En= 0   RD_En=1   ADDRESS = be   DATAOUT = ffffffff
# Scoreboard Checking Valid! Addr = be   Data :: Expected = ffffffff Actual = ffffffff
#
# [GEN-DRV]   WR_En= 1   RD_En=0   ADDRESS = 65   DATAIN = fc037767
# ** Note: $finish    : C:/questasim_10.0b/examples/tb.sv(165)
#    Time: 10 us  Iteration: 1  Instance: /tbench_top/t1
# 1
```

Fig. 4. Scoreboard output

There were four assertions that were implemented to validate the behavior of the DUT. Assertions are checks which are embedded in a design to assert the correctness of the design and generate any warnings or errors if they occur [12]. The assertions that were implemented are

- assertion to validate the reset functionality, to ensure data at all addresses reverts to FFh;
- assertion to ensure that read and write operations do not occur simultaneously;
- assertion for write operation and
- assertion for read operation as shown in Fig. 5

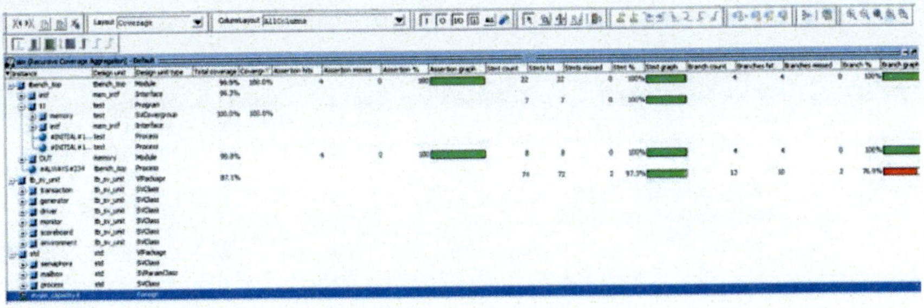

Fig. 5. Assertion output

Functional coverage validates the design specifications. Functional coverage will aid in uncovering hard to find bugs, determine the source of the bugs and provide an effective way to model complex timing checks, along with the design specifications checks [13]. The cover group created in program module encapsulated the cover points for address and data_in. For each cover point, five cover bins are created explicitly. These cover points aided in determining the functional coverage, thereby measuring if the design specifications were met correctly. The functional coverage for our design was found to be 100% as shown in Fig. 6.

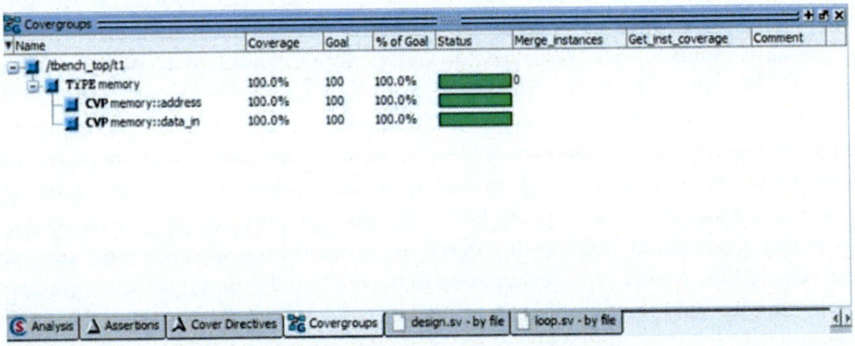

Fig. 6. Coverage output

Filewise detailed design and testbench coverage report are illustrated in Fig. 7

```
Coverage Report Summary Data by file

File: design.sv
    Enabled Coverage          Active      Hits    Misses % Covered
    ----------------          ------      ----    ------ ---------
    Stmts                          8         8         0     100.0
    Branches                       4         4         0     100.0
    FEC Condition Terms            0         0         0     100.0
    FEC Expression Terms           0         0         0     100.0
    Toggle Bins                  152       151         1      99.3

File: tb.sv
    Enabled Coverage          Active      Hits    Misses % Covered
    ----------------          ------      ----    ------ ---------
    Stmts                         74        72         2      97.2
    Branches                      13        10         3      76.9
    FEC Condition Terms            0         0         0     100.0
    FEC Expression Terms           0         0         0     100.0
    Toggle Bins                  156       155         1      99.3


TOTAL COVERGROUP COVERAGE: 100.0%  COVERGROUP TYPES: 1


NEVER FAILED: 100.0%  ASSERTIONS: 4
```

Fig. 7. Detailed design and testbench coverage output

5 Conclusion

A 32-bit memory model is designed and implemented in System Verilog. The design is verified using both directed test cases and layered testbench verification methodology. The design is simulated and analyzed for various input conditions as per verification plan. Default values, reset condition at any instant, proper write and read at any location are checked correctly. All verification blocks in layered testbench are implemented as classes, and interprocess communication is achieved using mailbox. The scoreboard report was generated by analyzing the randomized test cases and matching the predicted result obtained from the DUT. The SV functional coverage methodology successfully managed to verify if the objectives of the design specifications were met effectively. 100% code coverage has been achieved for the design. The constrained random stimulus and the transactions generated yielded 85% functional coverage for the design.

References

1. Chen, W., Ray, S., Bhadra, J., Wang L.-C., Abadir, M., NXP Semiconductors Helic Inc.: Challenges and trends in modern SoC design verification. IEEE CEDA, IEEE CASS, IEEE SSCS, 2168-2356/17 © 2017 IEEE, 13 Sept 2017
2. Akhare, M., Narkhede, N.: Design and verification of generic FIFO using layered test bench anf assertion technique. IJEAT **8**(6), (2019). ISSN: 2249-8958

3. Ahlawat, D., Shukla, N.K.: Performance analysis of verilog directed testbench vs constrained random system verilog testbench. IJCA **118**(22), (2015). (0975–8887)
4. Clifford Cummings, System verilog assertions—Bindfiles and best known practices for simple SVA usage. sunburst design (SNUG)
5. Ahlawat, D., Shukla, N.K.: DUT verification through an efficient and reusable environment with optimum assertion and functional coverage in system verilog. Int. J. Adv. Comput. Sci. Appl. **5**(4) (2014)
6. Sutherland, S., Davidmann, S., Flake, P.: System verilog for design: A guide to using systemverilog for hardware design and modeling. Kluwer Academic Publishers (2003)
7. System Verilog 3.1a, Language Reference Manual
8. Bergeron, J.: Writing testbenches using system verilog. Springer, vol. 1 (2006)
9. Spear, C.: System Verilog for verification. Springer, Berlin, vol. 1 (2005)
10. Kel H., Zhongliang, D, Qiong S.: Verification of AMBA bus model using system verilog. In: The Eighth International Conference on Electronic Measurement and Instruments ICEMI' 2007
11. Sudhish, N., Raghavendra, B.R., Yagain, H.: An efficient method for using transaction level assertions in a class based verification environment. In: International Symposium on Electronic System Design, pp. 72–76, (2011)
12. Questa® SIM User's Manual, Software Version 10.0b
13. Ashok, B.M.: System verilog assertions and functional coverage. (2nd ed.) (Online). Available: http://extras.springer.com

Stochastic Model of a Sensor Node

Rakhee Kallimani[1(✉)], Krishna Pai[1], and Krupa Rasane[2]

[1] KLE Dr. M.S. Sheshgiri College of Engineering and Technology, Belagavi, India
rakhee.kallimani@gmail.com
[2] Jain College of Engineering and Technology, Belagavi, India

Abstract. The emerging area for researchers in the field of power concern embedded system is wireless sensor network (WSN) to develop a platform capable of analyzing and controlling the power behavior of low power embedded system and achieve high performance in terms of battery life. In recent days, the embedded systems use ultra-low powered hardware components at the node level, yet there is a need to analyze the consumption of power in sensor node and performance of the complete network. This motivates us to provide a power analyzer unit on the wireless sensor node based on stochastic process that can be used to monitor, analyze and control the node energy by switching to low power down modes or turn off other peripherals and improve the lifetime of a node. The mathematical model for power consumption and lifetime is developed to observe the effect of proposed system using stochastic approach. This paper presents simulation results that are evident to improve the battery lifetime of a node by operating the node at low power states using dynamic power management schemes.

Keywords: WSN · Dynamic power management · Stochastic model · Lifetime · Power consumption · Sensor node · Analyzer · Event arrival · Change detection probability

1 Introduction

In the recent years, there is a phenomenal growth in the designing of a power concern embedded system as power is considered to be an important design metric. There are many design metrics to be considered while designing the power efficient system. One of the key and challenging tasks is to manage power and design an efficient system. To design a power efficient system, it is needed to reduce power at all the levels of design [1]. In the field of design of the embedded system, wireless sensor network (WSN) is the most considered system which has gained the most attention by the academicians and researchers for the Internet of things has become more needed technology and is estimated that 99.4% of the existing objects are yet to be connected as per [2]. One of the most used power aware embedded systems is WSN that consists of self-monitored and self-organized sensor nodes. This motivates the researchers to think about managing the power of WSN node. A typical WSN node consists of numerous sensor unit, a processing unit and a communication unit as shown in Fig. 1. And these WSN nodes are required to gather/collect information from the uncontrollable area of human.

© Springer Nature Singapore Pte Ltd. 2021
P. Deshpande et al. (eds.), *Next Generation Information Processing System*, Advances in Intelligent Systems and Computing 1162,
https://doi.org/10.1007/978-981-15-4851-2_26

Usually, these nodes are deployed at remote places and are expected to be self-organized and self-reliant. As they need to gather real-time information at regular intervals, it is found that they are power hungry too. So, the challenge to the researcher is to foster the power concern about these nodes in an efficient manner without much change at the hardware level but at system level and make these power-hungry nodes to perform better in terms of battery life. Thus, there is a need for an accurate power analysis module to evaluate the power consumption of WSN node and manage the power dynamically by selecting the appropriate low power down mode of the processor. Today's processors are all ultra-low power processor, and this motivates to dynamically manage power and reduce the consumption of power for continuous monitoring application. To increase the performance of the sensor node in terms of its battery lifetime, it is needed to analyze the power consumption of the node, and this necessitates the need of a platform to perform an accurate analysis of the power and performance of the sensor node at the early stage and dynamically control the power consumption of the node and improve the lifetime of the node. After a node is analyzed, this can be extended to the network and then gradually improve the performance of the network. As shown in Fig. 2, we can modify the existing node by introducing an analyzer unit, which can analyze the power and efficiently manage power by dynamically switching to low power down states or switching the peripherals OFF when not in use.

Fig. 1. Typical wireless sensor node

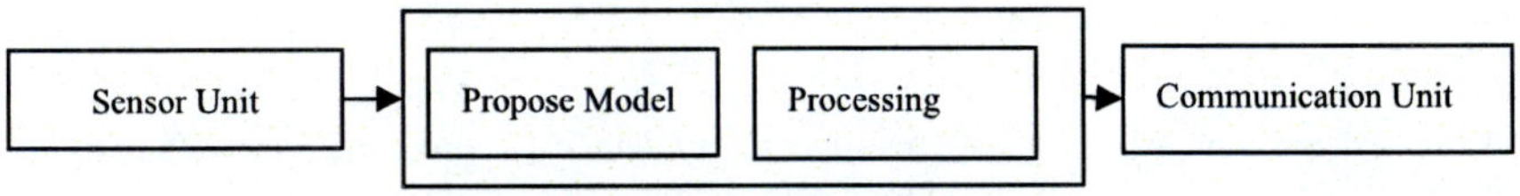

Fig. 2. Modified wireless sensor node

2 Literature Survey

A detailed literature survey is done on the need of analyzing power and methods of analysis. The following Table 1 summarizes the literature been surveyed.

This extensive literature survey motivates to develop a mathematical model and simulate the wireless sensor node for reducing the consumption of power and improve the performance of the WSN node by introducing analyzer unit to detect the desired event. The trade-off between the performance and power consumption of the embedded system is gaining the attention of the researchers using power management techniques.

Table 1. Literature survey

Work done	Method	Future scope
Authors proposed a stochastic model based on discrete time Markov decision processes for power-managed devices. The authors also mentioned the limitation of the related work on discrete as well as on continuous time Markov process [4]	The principle commitment of this paper is to present a limited state, theoretical framework model for control and power-managed systems dependent on Markov choice procedures. Authors have discussed examples of two-state SR models and related definitions	Authors propose the systems consisting of multiple interacting resources that can be modeled and demonstrated by stochastic processes instead of stationary Markov chains. The need of adaptive algorithms to compute optimal policies for nonstationary workloads and SP model
The authors have modeled the power-managed system using a continuous-time Markov decision process [5, 6]	Based on the theory of continuous-time Markov decision process and stochastic networks, the authors have formulated the problem of system-level power management as a controlled optimization	Experimental results encourage the power management method based on Markov decision process and show that it outperforms heuristic approaches in terms of power dissipation savings for a given level of system performance
Authors have developed mathematical and simulation model a WSN node and demonstrated the power saving efficiency compared to the conventional model [3]	A sensor node is modeled using a stochastic approach. The dynamic power management is implemented at the operating system level of the sensor node	The authors have worked on the analyzer unit for single arrival rate and are intending to extend the model and optimize for wireless sensor network
Authors have proposed a node architecture that employs adaptive suspended scheduling approach and demonstrated the battery life extension [7]	Authors focus on simple analytical method to decide the kind of sensor node to be used. They present the simple equations to calculate the battery life of conventional sensor and the proposed node with adaptive sleep/wake-up	Authors encourage to investigate the different desired combinations and develop the sensor node with different components
Authors surveyed the different approaches, i.e., stochastic process, dynamic voltage and frequency, different scheduling methods to wireless sensor network. The authors mentioned the need of accurate model to be	The authors categorize the dynamic power management techniques to minimize the consumption of power after deployment of the sensor node	This paper surveys the stochastic approach and scheduling approach to reduce consumption of power in WSN. Discuss about limitation and advantage of Markov model, the review focuses on the

(*continued*)

Table 1. *(continued)*

Work done	Method	Future scope
developed considering non-geometric transition times of state for optimal power-performance trade off in WSN [8]		need of evaluating average power consumption
Authors surveyed the techniques of power management at system level and the focused at the need of designing optimal power-managed system [9]	The authors classify the techniques of managing the power in a portable embedded system and emphasize the need of power management. Focus of the review was on the power concern function in a portable embedded system	The survey paper covers the necessity of managing the power and improving the performance as the trade off in a portable embedded system. Focus is on the need of modeling and optimization at system level
The author presents power management techniques on the sensing unit that reduce the power consumption of power-hungry sensors. The techniques include hierarchical, adaptive and model based acquisition and data processing [10]	The paper exhibits the state-of-the-art in a part of the PM schemes that concern radio PM, with duty-cycling and a separate wake-up receiver as main approaches. The author also included the hierarchical, adaptive and model-based acquisition and data processing	The paper focuses the interest in managing the energy resources at sensor level, node level and network level. He mentions the flexibility of design to be improved by combining the radio PM and sensing PM methods to be more implemented
Authors investigate the reasons of dissipation of power in a node and investigate the strength and limitation of selective switching and dynamic frequency and voltage scaling factors affecting the DPM technique. Authors mention the difference between addressing the problem of power dissipation at global scope and at local scope [11]	This paper gives a complete comprehensive assessment of dynamic power management techniques in wireless sensor networks. Author discusses the strength and limitation of DPM techniques	The authors mention the demand of DPM in context of embedded systems. And the need of research to describe the resource demand, complication in implementation and processing time of the methods to be surveyed

3 Mathematical Model

To meet the need of modeling a sensor node used for an event-based application, we propose a Markovian model. In semi-Markov model, transition from state to state depends on the present state and the time spent in the present state. The model sensor node is of four state; s0, s1, s2 and s3. The sensor node senses event arrival (λ), and the data is preprocessed by the preprocessor for filtering of data and then passed to the analyzer state the filtered data is analyzed and if the event that occurred is a desired event. The desired event data is further transited from analyzer to the main processor with alpha (α), i.e., probability of change detection else the ($1 - \alpha$) probability is passed back to the preprocessor. Thus, the main processor processes only the desired event, and thus, we can provide a hypothesis function stating that if the event is passed through the analyzer block, then it is desired event, this can reduce the power consumption of the node as it reduces the multiple computation of the main processor. If the event is undesired, then the processor can be in its any of the low power down states and wakes up only on desired event been analyzed by the analyzer. The model also proposes second hypothesis statement that the lifetime of the node improves by minimizing the multiple computation of the processor for the same events occurring.

In the simulated model, we assume

1. As input is random in nature, the event arrival 'λ' is considered to be poison distribution and is varied in the range from 0 to 100.
2. The change probability (α) of the analyzer block is varied from 0 to 1.
3. The complete system models four power states as shown in Table 2.

Table 2. Power state mode table [3]

Power state mode	Ps0	Ps1	Ps2	Ps3
Sensor state	1	0	1	0
Preprocessor state (S0)	1	1	1	1
Analyzer state (S1)	0	1	1	1
Main processor state (S2)	0	0	1	1
Transceiver state (S3)	0	0	0	Tx

4. The cost of transition from analyzer to preprocessor and main processor are assumed to be of smaller value and thus can be neglected.

The equations governing the model:

$$\text{Eave} = \text{TanPs1} + \alpha\text{TprPs2} + \alpha\text{Ttx Ps3} + \text{Cp}\alpha + \text{CR}\alpha \tag{1}$$

$$\text{Tave} = \text{Tan} + \alpha\text{Tpr} + \alpha\text{Ttx} \tag{2}$$

$$\text{Pedn} = \frac{\text{Ps0} + \lambda(\text{Eave})}{1 + \lambda(\text{Tave})} \tag{3}$$

$$\text{Tedn} = \frac{(1 + \lambda\text{Tave})\text{Eresource}}{\text{Ps0} + \lambda(\text{Eave})} \tag{4}$$

where Ps0, Ps1, Ps2, Ps3 are power state modes

Tan is the time taken to analyze the event in analyzer unit.

Tpr	is processing time and Ttx is transmission time.
Eave	is average energy consumption of a node.
Pedn	is power consumption of the node.
Cp and CR	is the energy cost of processing and communication unit.
Tedn	is the lifetime of the node in days.
Tave	is the average time consumption.

4 Performance Evaluation with Results

The proposed model is simulated on Spyder IDE, where the random event arrival and servicing follows Markovian model. Here are few of the simulation parameters of the sensor node as shown in Table 3 taken from [3] for our study. The independent variables are the rate of event arrival (λ) and the change in the detection probability (α). The dependent variables considered are average power consumption (Pedn) and lifetime of the node (Tedn).

Table 3. Simulation parameters [3]

Parameter	Value	Parameter	Value
Ps0	2.3 mW	Ps3	307.9 mW
Ps1	2.3 mW	Cp	2.206 m J
Ps2	237.5 mW	CR	6.9 μJ
Tpr	2 s	Eresource	1.863 kJ
Ttx	0.175 s	Tan	0.5 s

1. The power consumption analysis of the event-driven sensor node model based on a single server queuing by changing the event arrival and change detection probability.
2. The performance analysis of the event-driven sensor node model considering the lifetime and number of events arrived, change detection probability as performance measure parameters.

The dynamic effect of variation of event arrival lambda (λ) on the battery lifetime of the sensor node is as shown in Fig. 3 indicates that, increase in arrival rate would have an impact on the lifetime of the node. The lifetime reduces as the arrival rate increases. Figure 3 also depicts the power block of the node without analyzer, wherein the alpha (α) is assumed to be unity, Table 4 shows the effect of variation of arrival rate with change detection probability to be unity on power consumption and lifetime of node. It is observed that Eave = 0.53121 (J) and Tave = 2.225 (s) are constant.

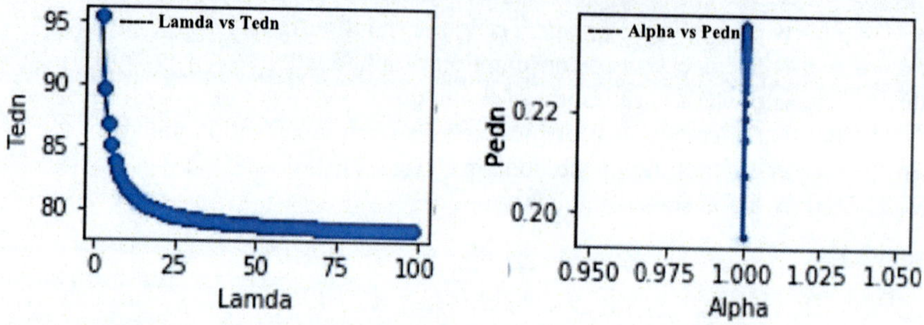

Fig. 3. Effect on Tedn and Pedn on varying independent variables without analyzer unit

Table 4. Effect of varying independent variables without analyzer unit

λ	α	Pedn (mW)	Tedn (days)
λ	α	Pedn (mW)	Tedn (days)
2	1	0.195362	95.36162
3	1	0.207939	89.59362
4	1	0.214863	86.7065
5	1	0.219246	84.97323
6	1	0.222269	83.8173
7	1	0.224481	82.99143
8	1	0.226169	82.37192
9	1	0.2275	81.89001
10	1	0.228577	81.50444
11	1	0.229465	81.18894

The effect of variation of α on the average power consumption and lifetime of the node is shown in Fig. 4. Table 5 shows the variation of independent variables.

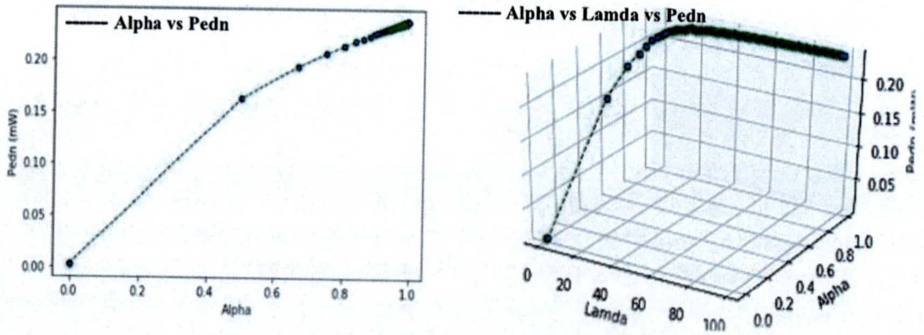

Fig. 4. Effect on Pedn on varying independent variables with analyzer unit

Table 5. Effect of varying independent variables with analyzer unit

λ	α	Pedn (mW)	Eave (J)	Tave (s)	Tedn (days)
0	0	0.0023	0.000115	0.05	8100
1	0	0.0023	0.000115	0.05	8100
2	0.5	0.162939	0.265663	1.1375	114.33
3	0.666	0.193607	0.354179	1.5	96.226
4	0.75	0.206608	0.398437	1.68125	90.170
5	0.8	0.213795	0.424991	1.79	87.139
6	0.833	0.218355	0.442695	1.8625	85.319
7	0.857	0.221505	0.45534	1.91428	84.106
8	0.875	0.223813	0.464823	1.95312	83.239
9	0.888	0.225576	0.4722	1.98333	82.588

5 Conclusion

Dynamic power management techniques are more effective in reducing the power consumption of the sensor node and reduce the power consumption of the network as compared to dynamic voltage/frequency scaling and other power management policies. It can also be practically applied as power is managed at system level without changing any hardware involved as in comparison with other methods. It is evident from the simulation results that the lifetime of the node can be effectively prolonged by detecting the event as desired at analyzer unit and just switching the main processor to different low power down states. This approach can improve the performance of the individual sensor node and eventually can improve the lifetime of the entire network. It is evident that change detection probability has an effect on the power consumption and as well as on the lifetime improvement of individual node. As future plans, authors intend to work towards optimization at network level.

References

1. Arnold, B.: Embedded System Design. CMP Books (2002)
2. Bradley, J., Barbier, J., Handler, D.: Embracing the internet of everything to capture your share of $14.4 trillion, Cisco, San Jose, CA, USA, White Paper, 2013, Accessed on Apr, 2016 (2016)
3. Pughat, A., Sharma V.: Performance analysis of an improved dynamic power management model in wireless sensor node. Digit. Commun. Netw. **3**, 19–29 (2017)
4. Benini, L., Bogliolo, A., Paleologo, G.A., De Micheli, G.: Policy optimization for dynamic power management. IEEE Trans. Comput. Aided Des. **18**,(6), 813–833 (1999)
5. Qiu, Q., Pedram, M.: Dynamic power management based on continuous-time Markov decision processes. In: Proceedings of the 36th Design Automation Conference, pp. 555–561 (1999)

6. Qiu, Q., Wu, Q., Pedram, M.: Stochastic modeling of a power-managed system: construction and optimization. In: International Symposium on Low Power Electronics and Design, pp. 194–199 (1999)
7. Yamawaki, A., Serikawa, S.: Battery Life estimation of sensor node with zero standby power consumption. In: IEEE International Conference on Computational Science and Engineering, IEEE International Conference on Embedded and Ubiquitous Computing, and International Symposium on Distributed Computing and Applications to Business, Engineering and Science, pp. 166–172 (2016)
8. Pughat, A., Sharma, V.: A survey on dynamic power management approach in wireless sensor network. IEEE (2014)
9. Kallimani, R., Rasane, K.: A survey of techniques for power management in embedded systems. IJETCSE **14**(2), 461–464 (2015)
10. Jelicic, V.: Power management in wireless sensor networks with high consuming sensors. JELICIC (2014)
11. Dargie, W.: Dynamic power management in WSN—state of art. IEEE Sens. J. **12**(5) (2012)

Prevention of Replay Attack for Isolated Smart Grid

L. Pavithra(✉) and D. Rekha

VIT University, Chennai, India
pavikotte@gmail.com, rekha.d@vit.ac.in

Abstract. An isolated smart grid with energy flow optimization and high energy quality has developed as the next-generation electricity grid. An isolated smart grid has more number of regular power frameworks and will become a new foundation for power systems and executives. Protecting the isolated smart grids from security threats is an important issue. Replay attack is one of the security threats for isolated smart grids, in which the data transmission is fraudulently delayed. This paper proposes using HMAC-MD5 algorithm a solution to detect replay attack. Further, HMAC-MD5 algorithm is compared to other existing algorithms, and the proposed algorithm is proved better with respect to key generation time and running time.

Keywords: Isolated smart grid · Internet of things (IoT) · Replay attack

1 Introduction

With growing technology, our society is increasingly dependent on power systems. Earlier developed technologies are getting overloaded with data especially during peak hours. This gave rise to network communications and power systems being combined to perform many powerful applications. Power transmission can be managed with the help of communication networks. Failures could be detected as soon as possible. The most vital role is responding to customer and recording power consumption. By having sensors, failures are easily recognized, and issues can be sorted in efficient way [1].

Internet of things is a system of interrelated things that are connected with unique identifier and transfer data without any human or computer interaction [2]. IoT is mainly used to the industrial framework because they are connected to a huge number of devices and sensors which are synchronized over the use of software tools. A smart grid is a power grid that includes many components like smart meters, smart appliances, energy-efficient resources, renewable resources and so on [3]. Smart grid (SG) consists of four components as: generation, transmission, distribution and consumers. SG can be enhanced in reliability, decrease the price and improve the energy resource implementation of the standard power system.

Nowadays, progressively smart grid network has been carried out by power administrations [4–6]. SG is connected using smart devices like smart meters, phaser management units and so on. The smart meter performs a vital role as a part of smart grid because it is used to collect power consumption from various power resources.

© Springer Nature Singapore Pte Ltd. 2021
P. Deshpande et al. (eds.), *Next Generation Information Processing System*,
Advances in Intelligent Systems and Computing 1162,
https://doi.org/10.1007/978-981-15-4851-2_27

The power consumption data is transmitted through network to the data center and stored in server. This information is protected from unauthorized access. SG is of two types (i) network smart grid (ii) isolated smart grid. Network smart grid is a fully automatic, while the isolated smart grid is partially automatic; the reading part alone human can take using handheld devices. SG is a one of the parts of IIoT framework, and it can be used to control the devices and monitor remotely. This paper is concentrating on an attack type called replay attack, wherein an attacker can intercept the smart meters to view the data for certain time duration and replay a similar data when the attack is on. The intruder can also manipulate the user's smart meter by introducing false data to the system, which leads to incorrect energy pricing or inaccurate readings.

Replay attack may be used to manipulate the readings from programmable logic controllers in smart grids. In advanced metering infrastructure (AMI) meter, an authentication scheme is used in between two smart meters, replay attack entails a false host to intervene authentication packets being sent from a smart meter and resending them later in time, expecting to authenticate and attain unauthorized entry inside the network. For such attacks, HMAC-MD5 algorithm can be applied to verify both the authentication of a message and the data integrity by keyed-hash message authentication code.

The overall arrangement of this paper goes as in Sect. 2. The method which is proposed for this paper is available in Sect. 3. The simulation results are evaluated in Sect. 4. End of Sect. 5 serves as a conclusion for the work described in this research.

2 Literature Survey

Sha et al. [7] have proposed enabling secure information from isolated smart grid device using a smart reader and smart grid cloud. This smart grid reader acts in a method that links the SG device and also bridges to cloud in the smart grid. They have considered physical constrains of smart grid devices also. The security analysis is efficient from the various typical attacks. It also evaluates the efficiency and the performance of the smart grid devices.

Tran et al. [8] have proposed that replay attack is one of the significant security attacks in smart grid systems. A smart grid system is controlled by the control system to develop a new detection system for the replay attack and find the solution for smart grid in the estimation part. Finding of the replay attack is based on the signals.

Sechilariu et al. [9] described the system issues, the executives, security and safety and protection are some of the other serious problems in IIoT-based SG systems. The security can be diminished in SG frameworks whenever by a foe utilizing attacks like man-in-the-middle, eavesdropping and denial-of-service. For the occurrence, the vitality cost incurred is too much at pinnacle hours and a little low at a different hour. An enemy would be able to bargain the SG tasks to use vitality at pinnacle hours, which somehow can create an additional heap on the smart meter and the control unit.

Mita et al. [10] have developed an increased communication for new SG applications and also evaluated the QoS, automatic generation control process for peak loads, volt/var optimization and traffic load steady-state operation. Implementation of this concept is done with SDN-based dynamic system. The disadvantage of this work is it is only applicable for dynamic SG applications.

Yuanion et al. [11] proposed the data security for SG in the lightweight quantum encryption process. They also combined the process of quantum cryptography, quantum number generator. They have developed the new cipher encryption algorithm for the SG. Finally, they proved the security for power data transmission using lightweight stream cipher encryption algorithm.

Chunqiang et al. [12] introduced a concept to protect the data communication between smart meters, service provider and companies in a secure way. In this, the author protected the data communication by using this method of Linear Secret Sharing Scheme matrix. They have analyzed the performance and efficiency in terms of computational cost.

3 System Design

3.1 Replay Attack

Replay attacks could be propelled when an aggressor picks up entry benefit into smart meters and could infuse control sign inside the set framework. The assailant immediately records information sent from the client's hardware to the smart meters and dissects them so as to accomplish client's qualities of intensity use. After in-depth examining, the aggressor may manipulate the information and introduce relating control sign into the framework. The replay attacks are completed for two purposes. One objective can be to take vitality; replay attacks could distort the inert supplies' status to occupied state so as to misroute the ability to somewhere else. The other is to make physical harm to the framework. Stuxnet is one of the most famous instances of that sort of attacks [13]. To understand and infer the effect of replay attacks to the brilliant lattice framework, think about a virtual framework with adjusted information and control sign produced by the intruder. For the instance of replay attacks, the adjusted sign will essentially be time moved renditions of the first flag. Figure 1 shows how the hash code is intercepted and the same is replayed later.

To understand the impact of replay attack, we have compared actual data with modified data.

3.2 HMAC-MD5 Algorithm

For preventing attacks, encryption and decryption process are followed [14]. This makes the data more secured. SHA1 and MD5 algorithm are commonly used hash function for checking data integrity. HMAC works on keyed-hash operations using hash functions and key inputs. HMAC automatically generates a backup word and repeats the key for the following HMAC jobs that need the same key. HMAC-MD5 is a combination of algorithm. It has been widely used in cryptographic application and composition of efficient message authentication code (MAC). A secret key is used to calculate HMAC which is only known to the sender and receiver. It makes it impossible for the attacker to get into HMAC as attacker will not be able to create the

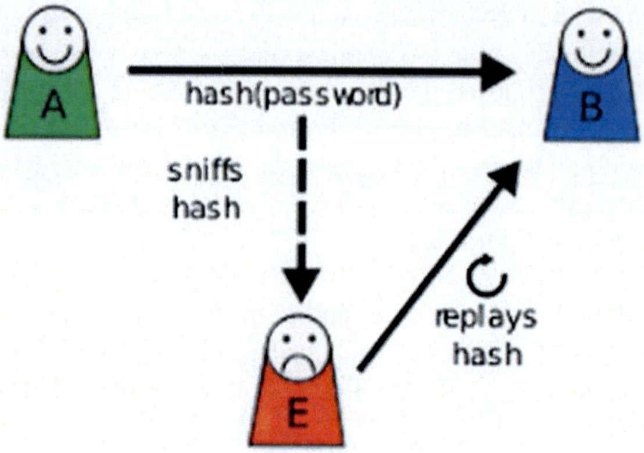

Fig. 1. Replay attack

message sets offline without being aware of the key. During an attack, until the attacker predicts the key, HMAC cannot be broken into as offline generated message pairs are required based on the key.

Currently, among the best technique, HMAC is used for certification that message not be altered or distorted for forgery over unprotected networks. MD5 algorithm provides a receiver confidence of unaltered message by resulting in a 128 bit hash value. The sender then calculates the hash value of a message then concatenates it and sends in the network. Receiver has to recalculate the hash value from message and cross checks the hash value present along with the said message to ensure the data integrity.

3.3 Proposed Scheme for Preventing Replay Attack

A detection scheme has been developed against replay attack to secure the customers device and the data of core system. The conventional scheme causes huge load on the system in the SG. Therefore, we have proposed using HMAC-MD5 algorithm for efficient prevention of replay attack. The advantage of the proposed scheme is that the customers meter will work normally when the system is not under attack. This highly brings down the system management issues and waste of power.

A SG system manages working state of equipment, to predict system failure and to control customer usage. Prevention scheme for replay attack has been created for the core system, wherein excess burden is not an issue. However, in SG, one also needs to secure the customer device. HMAC-MD5 algorithm is proposed to efficiently prevent replay attack in SG system. A random signal is periodically added for short time within the period time. Thecustomer's equipments remain normally operational. The benefits of the proposed scheme are that the customer data is prevented from further replay attack.

After analyzing all the methods, to prevent the replay attack, it is proposed to use authentication protocol called the HMAC-MD5 algorithm. This works with random key generation function over IP protocol authentication header. The authentication header (AH) offers IP datagrams with integrity and authentication. The system utilizes the MD5 hash feature (keyless) that generates a digest message. Authentication data is generated when coupled with an AH key. This value is put in the AH's field of authentication data. This value happens to be the basis for the data integrity service of the AH protocol. A replay prevention field is included as an alternative for transforming, to provide security and protection against replay attacks. To assist avoid attacks where a message is subsequently stored and refused to replace or repeat the initial message, this field is used in Fig. 2.

HMAC-MD5 Algorithm

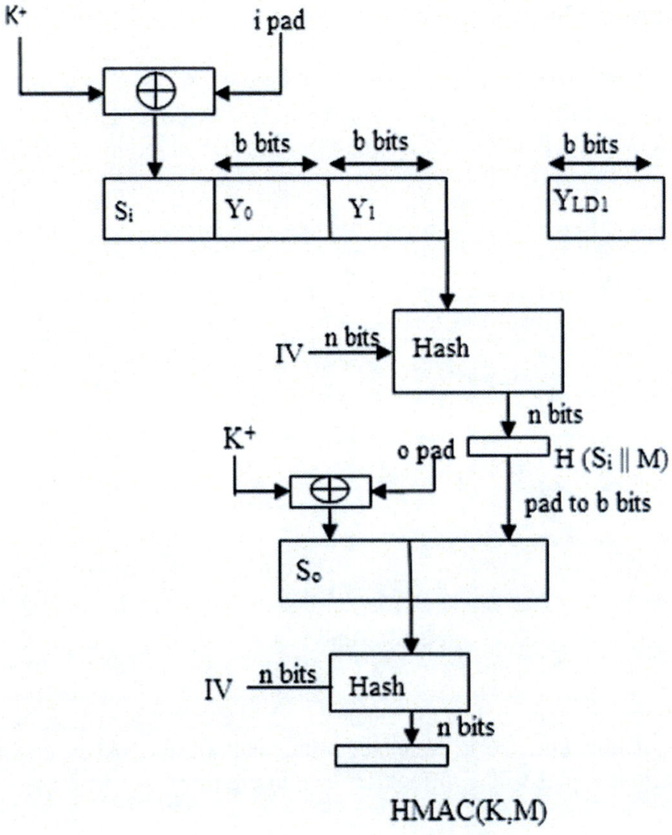

Fig. 2. HMAC structure

Algorithm:

Input: 0X36 byte frequent 64 times (ipad)
Output: 0X5C byte frequent 64 times (opad)
Step 1: Append the value zeros to the end of k values to generate 64 bytes.
Step 2: XOR operation computed the 64 byte string value with step (1) ipad.
Step 3: Include the information flow text to the 64 byte strings and ensuing from the step (2).
Step 4: Incorporate to the MD5 and generated from the step (3).
Step 5: XOR string was computed from the step (1) with output (opad).
Step 6: The MD5 solutions from the step (4) to the 64 byte string ensuing from step (5).
Step 7: The MD5 algorithm to the stream generated from step (6) and output results.

4 Simulation Results

This Fig. 3 compares the running time of proposed and existing algorithm. MD4, MD5, SHA1 and SHA2 are existing algorithm. HMAC-MD5 is the proposed algorithm. It is found the running time for the HMAC-MD5 algorithm is the least.

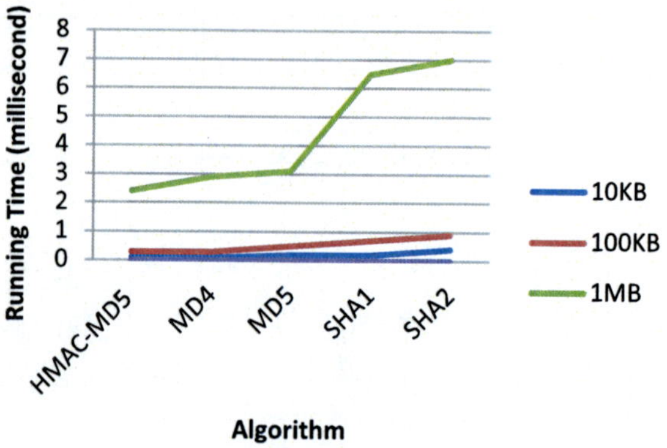

Fig. 3. Replay attack

Figure 4 implies the simulation result of HMAC-MD5 algorithm using replay attack. Data of different size is compared with running time. When compared with existing algorithms, HMAC MD5 took lesser running time.

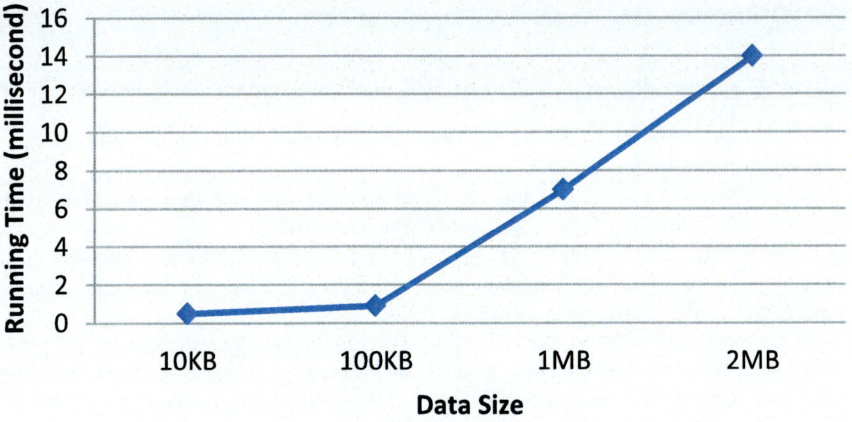

Fig. 4. Simulation result for HMAC-MD5 algorithm

Table 1 shows that the performance evaluation for encryption and decryption process on existing and proposed algorithm.

Table 1. Performance comparison of encryption and decryption

Algorithms	Encryption (ms)	Decryption (ms)
MD4	0.641201	0.630312
MD5	0.501210	0.500912
SHA1	0.541224	0.532134
SHA2	0.510123	0.522104
HMAC-MD5	0.4013131	0.396612

5 Conclusion

With huge amount of energy collected in SG, preventing the vulnerable data from inaccurate readings has been an overwhelming issue. This research work focuses on efficient prevention of replay attacks in SG data. In the current setup, the reader undergoes authentication process for each smart device. Therefore, this elaborate framework that can reliably authenticate the reader to an area of many smart devices is created. The paper has compared efficient prevention of replay attacks in smart grids by comparing existing algorithm with proposed algorithm. The earlier algorithm needed a longer performance time. The proposed algorithm greatly reduced this time, thereby preventing the replay attack at the earliest. HMAC-MD5 algorithm has been proved to substantially save the data and simultaneously provide the earliest prevention.

References

1. Wang, W., Lu, Z.: Survey cyber security in the smart grid: Survey and challenges. Comput. Netw. **57**(5), 1344–1371 (2013)
2. Whitmore, A., Agarwal, A., Da Xu, L.: The internet of things—a survey of topics and trends. Inf. Syst. Front. 1–14 (2015)
3. Fang, X., Misra, S., Xue, G., Yang, D.: Smart grid—the new and improved power grid: a survey. IEEE Commun. Surv. Tutor. **14**(4), 944–980 (2012)
4. Quilumba, F.L., Lee, W., Huang, H., Wang, D.Y., Szabados, R.: Using smart meter data to improve the accuracy of intraday load forecasting considering customer behavior similarities. IEEE Trans. Smart Grid **6**(2), 911–918 (2015)
5. Rouf, I., Mustafa, H., Xu, M., Xu, W., Miller, R.: Neighborhood watch: security and privacy analysis of automatic meter reading systems. In: Proceedings of the 2012 ACM Conference on Computer and Communications Security, Raleigh, NC, USA, pp. 462–473 (2012)
6. Taysi, Z.C., Guvensan, M.A., Melodia, T.: Using circuit-level power measurements in household energy management systems. In: Proceedings of the First ACM Workshop on Embedded Sensing Systems for Energy-Efficiency in Buildings, pp. 7–12 (2010)
7. Sha, K., Alatrash, N., Wang, Z.: A secure and efficient framework to read isolated smart grid devices. IEEE Trans. Smart Grid **8**(6), 2519–2531 (2017)
8. Tran, T.T. Shin, O.S., Lee, J.H.: Detection of replay attacks in smart grid systems. In: Proceedings of the International Conference on Computing Management and Telecommunications, pp. 298–302 (2013)
9. Sechilariu, M., Wang, B., Locment, F.: Building-integrated microgrid: advanced local energy management for forthcoming smart power grid communication. Energy Build. **59**, 236–243 (2013)
10. Cokic, M., Seskar, I.: Analysis of TCP traffic in smart grid using SDN based QoS. In: IEEE 26th Telecommunications Forum (2018). https://doi.org/10.1109/telfor.2018.8611800
11. Ma, M., He, D., Kumar, N., Choo, K.-K.R., Chen, J.: Certificate less searchable public key encryption scheme for industrial internet of things. IEEE Trans. Ind. Inf. (2017). https://doi.org/10.1109/TII.2017.2703922
12. Fan, C.I., Huang, S.Y., Lai, Y.L.: Privacy-enhanced data aggregation scheme against internal attackers in smart grid. IEEE Trans. Ind. Inf. **10**(1), 666–675 (2014)
13. Li, H., Lu, R., Zhou, L., Yang, B., Shen, X.: An efficient merkle tree based authentication scheme for smart grid. IEEE Syst. J. (2013)
14. Joseph, D.P., Krishna, M., Arun, K.: Cognitive analytics and comparison of symmetric and asymmetric cryptography algorithms. Int. J. Res. Stud. Comput. Sci. Eng. **2**(3), 63–68 (2015)

Dangers of Bias in Data-Intensive Information Systems

Baekkwan Park[1], Dhana L. Rao[2], and Venkat N. Gudivada[3](✉)

[1] Center for Survey Research, East Carolina University, Greenville, NC 27855, USA
[2] Department of Biology, East Carolina University, Greenville, NC 27855, USA
[3] Department of Computer Science, East Carolina University, Greenville, NC 27855, USA
gudivadav15@ecu.edu

Abstract. Data-intensive information systems (DIS) are pervasive and virtually affect people in all walks of life. Artificial intelligence and machine learning technologies are the backbone of DIS systems. Various types of biases embedded into DIS systems have serious significance and implications for individuals as well as the society at large. In this paper, we discuss various types of bias—both human and machine—and suggest ways to eliminate or minimize it. We also make a case for digital ethics education and outline ways to incorporate such education into computing curricula.

Keywords: Human bias · Algorithmic bias · Information systems · Digital ethics

1 Introduction

Data-intensive information systems (DIS) represent a broad range of software systems. Though their functionality varies across domains, the systems share some common characteristics: large datasets, complex data transformations and integration, data heterogeneity, and compute-intensive. For strategic decision making, a DIS gathers and integrates heterogeneous data from disparate data sources, transforms and integrates data, and recommends solutions. In some cases, the final decision is made by a human. For example, DIS for admission decision on college applications falls in this category. On the other hand, there are DIS which operate in standalone mode and make decisions in real-time. A DIS for a self-driving car fall in this category.

The DIS are pervasive and affect people in all walks of life including law enforcement [12], granting parole [17], political election [10], a loan decision [41], employment [25], child protection welfare systems [20], health care, and insurance [17], among others. For example, Google, Facebook, Instagram, and

© Springer Nature Singapore Pte Ltd. 2021
P. Deshpande et al. (eds.), *Next Generation Information Processing System*,
Advances in Intelligent Systems and Computing 1162,
https://doi.org/10.1007/978-981-15-4851-2_28

Twitter affect millions of lives everyday. In some cases, people are not even aware of the fact that DIS make decisions for them either directly or indirectly. With the recent resurgence of artificial intelligence (AI) and machine learning (ML), some components of DIS are based on these technologies. This entails an increased concern for the social and ethical issues engendered by DIS [25,29,34].

Though DIS are essential to our societal functioning, they are viewed as mere blackboxes by the end-users of these systems. This view entails many dangers. For example, police officers may use data-driven predictive policing systems to make their jobs more efficient. However, they do not necessarily understand how DIS systems actually work [5,14,44]. Loan officers in financial institutions may use the credit score systems to determine who is eligible for loans, but how credit scoring and loan decision algorithms work remains opaque to them [9]. Therefore, the problem needs to be addressed both at the source as well as at end-user level. Addressing the problem at the source entails educating the computing professionals about the dangers of bias in DIS and ways to eliminate or minimize the bias. The DIS system end-users need to be aware of the biases that are intrinsic to the system and how to interpret the system outputs. Given the above backdrop, the goal of this paper is to identify and discuss various types of bias—both human and algorithmic—that have implications for DIS and ways to minimize the bias.

The remainder of the paper is organized as follows. In Sect. 2, we briefly describe types of human bias and indicate two high-profile cases of machine bias. The goal of this discussion is to motivate the significance of human bias embedded into DIS and its serious implications. Data science is an emerging interdisciplinary academic discipline and is the foundation for many components of DIS systems. In Sect. 3, we discuss data science as the primary DIS infrastructure. Types and sources of bias—both human and algorithmic—are presented in Sect. 4. The need for digital ethics education and how to incorporate it into computing education are sketched in Sect. 5. Lastly, we provide a summary and conclude the paper in Sect. 6.

2 Human and Machine Bias

There are two major types of bias: human and machine. Human bias is also referred to as *social bias*. Machine bias, also called *algorithmic bias*, refers to systematic and repeatable errors of an algorithm that generates unfair outcomes. The outcomes may create an advantage for one social group over other groups.

We first list various types of human bias that are relevant to the DIS context. This is important as human biases systematically get embedded into DIS either implicitly or explicitly. Twenty human cognitive biases are discussed in [23]. The following are a subset of the twenty biases that are of interest to DIS:

(1) **Anchoring bias** is the use of the first piece of data received as a criterion to determine the validity of subsequent data received, (2) **availability bias** refers to assigning illogical information values to more accessible and convenient data. In the domain of probability and statistics, convenience data sampling

techniques best exemplify the availability bias, (3) **bandwagon effect** is the human tendency to believe in something simply because several people believe in that belief. The veracity of the belief does not arise, and results in lead to illogical reasoning and false conclusions, (4) **confirmation bias** is valuing only that data which confirms to one's own preconceptions, (5) **Conservatism bias** is putting more emphasis on already available evidence more than newly available evidence, (6) **information bias** entails seeking that information which does not necessarily contribute to action. What is needed is the right data, but not large size irrelevant data, (7) **blind-spot bias** is not being able to recognize one's own bias, which is a bias in itself, (8) **recency bias** is valuing newly available data than preexisting data, (9) **choice-supportive bias** refers to rationalizing the choice one have made, though the data may be faulty or not optimal, (10) **clustering illusion** is hypothesizing non-existing patterns in random events, (11) **ostrich effect** is ignoring data that is not pleasant or palatable, (12) **outcome bias** refers to determining the value of a decision based solely on the outcome, disregarding the process of generation, (13) **pro-innovation bias** refers to over-emphasizing the value and usefulness, ignoring the inherent limitations, (14) **salience bias** is focusing on statistically improbable events over probable events, (15) **selective perception bias** refers to seeing and taking information from the world based on one's own subjective expectations, (16) **stereotyping bias** is over-generalization though considerable amount of differences are in the data, (17) **survivorship bias** refers to making a illogical and inaccurate decision based on corrupted data without recognizing the presence of it, and (18) **zero-risk bias** is not quantifying uncertainty and reasoning with uncertain data.

All data is inherently biased. The machine learning algorithms trained with biased data naturally become biased. In the DIS context, *bias in, bias out* is the counterpart to the proverbial saying: *garbage in, garbage out.* Knowing about and minimizing human biases in the data collection process lead to minimizing the machine bias. Most of the bias originates in the data collection process and makes it way into DIS as the machine learning models are developed using the biased data. Care should be exercised in collecting the data. It is often tempting to go with the available data or data that is easy to collect (e.g., convenient sampling rather than stratified sampling). When a third-party data is used, we need to be cognizant of the original purpose for which the data is collected. The data needs to be assessed for quality—fit for intended use. This brings up a range of other issues including missing data, inconsistent and conflicting data, and they need to be effectively addressed [22].

Next, we indicate two high-profile cases related machine bias. Facial recognition technology provides a means to recognizing human faces. It has a large number of commercial applications from surveillance to marketing. However, this technology is mature and reliable only for certain groups of people. Recently, gender and racial bias was found in Amazon's facial recognition technology. The Amazon's system was unable to correctly identify female faces as well as gender in darker-skinned people. On a related note, in May 2019, San Francisco banned facial recognition technology given its potential for abuse.

The second case is about cost-effectively patching potholes in the city of Boston. The city released StreetBump smartphone app with the goal of detecting potholes through crowd sourcing. The app draws on accelerometer and GPS data to detect potholes and instantly reports them to the city. But the data collected by the StreetBump was biased in the sense that the data was not representative of all the neighborhoods. As of 2013, people in lower-income groups in the USA are less likely to have smartphones, and therefore, the potholes in their neighborhoods did not get reported. Hence, the collected data missed inputs from significant parts of the economically disadvantaged population.

3 Data Science as DIS Infrastructure

Data science is a new academic discipline. As shown in Fig. 1, it is a confluence of three other academic disciplines: Computer Science, Mathematics/probability/statistics, and application domain(s). Since data science provides the digital infrastructure for DIS components, addressing bias in data science will have far reaching positive implications in eliminating or minimizing bias in DIS. The danger posed by data science is that it enables building systems using high-level software libraries and application frameworks. Such tools include Weka, scikit-learn, TensorFlow, and Keras. The engineers who build DIS systems are often oblivious to how the machine learning models built with this infrastructure make actual decisions and the associated caveats. Worse yet are the self-proclaimed machine learning experts who have no understanding of the limitations of various algorithms packaged into the libraries and application frameworks.

Notice that in Fig. 1, data science is at the intersection computing, mathematics/probability/statistics, and application domain expertise. Often models are built and deployed with improper or inadequate representation of the domain experts. Domain experts are the ones who can help assess the quality of data as well as identify any biases embedded into the data.

Given the nascency of data science, it can avoid repeating the mistakes that were made in disciplines which came before data science. For example, consider the pharmaceutical industry. It has evolved over hundreds of years and along the way established several best practices. Consider the information that is provided to drug users, which includes capabilities, limitations, and side effects of the drug; active ingredients present in the drug; and dosage—how and when to use the drug. The pharmaceutical industry social responsibility also include child-resistant packaging through safety caps, and drug recall when it is learned that the drug does not work as claimed or causes harm than cure. Though this analogy may seem far fetched, it is inevitable for machine learning libraries and application frameworks to move in this direction.

Markham et al. [35] defined ethics as methods and emphasize “ethically correct actions in technology design and research contexts.” That is, digital ethics is about “how we act and what positions we take toward our technology” [35]. In this context, the issues of digital ethics are a little different from traditional

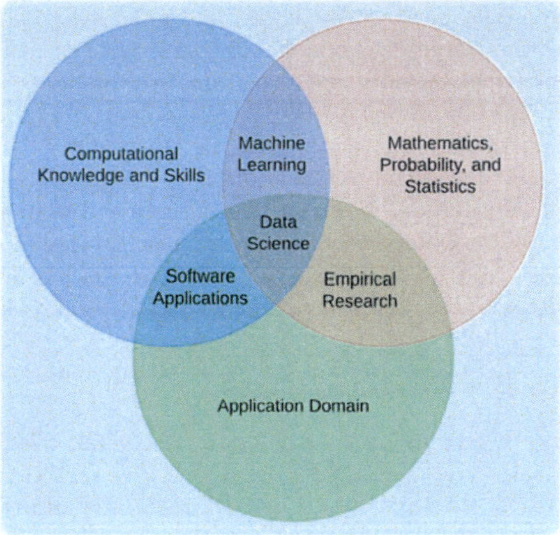

Fig. 1. The data science academic discipline

ethical concerns because these issues often arise in the process in which humans make an effort to avoid human errors in decision making.

There is a widespread belief that information systems based on algorithms can overcome human biases and help people to make impartial and unbiased decisions [2]. While humans tend to be biased and prejudiced in their decision making, it is fair to assume that DIS systems are fair, unbiased, impartial and thus lead to more desirable and fair results. But, there has been growing body of research that examines the ethical issues that these systems can bring to society [13]. Since the DIS systems are vital to many aspects of society from business to medicine, health care, and law, these ethical concerns from the algorithm-based analytics are relevant to all individuals and society.

4 Sources and Types of Bias in Data-Intensive Information Systems

In this section, we illustrate how certain types of bias manifests in DIS systems. As indicated earlier, it is often believed that algorithms can be designed to avoid the harmful effects of cognitive biases that human decision makers often fall to. Relying on DIS, we expect to avoid common human cognitive biases. However, even if leverage the massive amount data and the unbiased algorithms, we still have to deal with the following biases: availability bias, zero-risk bias, outcome bias, and overconfidence bias along with relevant example cases [23].

Unlike human decision makers, there is a optimistic belief that algorithms do not discriminate, thus enhance transparency. This is because algorithms are based on computation, which is neutral and unbiased [28]. Similarly, "the widespread belief that large datasets offer a higher form of intelligence" often leads people to infer that big data and algorithms are "objective" and untainted by human biases [7,31]. However, it is as dangerous as it is inaccurate to believe that, because algorithm-based decision making is computational, it is always neutral and unbiased. A central criticism of DIS systems lies in its framing as a neutral form of information. This assumption ignores the inherent cognitive human biases at all stages of the processes, from initial design of the systems through to analyses and reporting. In particular, we discuss availability bias, zero-risk bias, outcome bias, and overconfidence bias.

4.1 Availability Bias: Sampling Bias

An algorithm can only be as good as the data it inputs. When an algorithm collects dataset that is not necessarily a representative of the population, any output generated from the algorithm is non-representative in nature.

Many studies using big data primarily rely on social network media content such as Twitter, Facebook, Instagram, among others [4,6,38,45]. A few studies have paid attention to the potential problems of those data collection strategies [19,26], it is remarkable that sampling bias is still commonly present in many data collection process. Another line of research has illustrated that sampling bias can exist from one platform to another. According to [48], data from Facebook users are demographically different from Google. User data from LinkedIn are socieconomically different from those from Twitter [26].

Big data is not just about collecting large amounts of data and using it for training algorithms. For more meaningful and accurate analysis, data must be properly acquired and processed to be used for finding solutions to social problems [18,30]. When there is sampling bias, we cannot expect to find just and equitable solutions.

Similarly, the availability of enormous amounts of surveillance data from online sources and the advancement of information retrieval algorithms have generated unprecedented amounts of personal data. Acquisti et al. [1] indicated that big data has made it possible that seemingly unrelated multiple datasets be used for cross-referencing to generate completely new data. That is, even when there is no identifiable personal information or indication in a given dataset, by combining multiple datasets, it generates "some" identifiable information. For example, data from facial recognition applications can be used to search and match anonymous profiles on online dating contents with their public Facebook pictures, although they are necessarily related [46]. In particular, when combined with sampling bias, it may lead to racial profiling and preemptive criminalization of people of color living in marginalized communities [21].

4.1.1 Sampling Bias in Predictive Policing Systems Predictive policing systems are the application of quantitative and computational techniques to predict where criminal activities are likely to occur [5,11]. The predictions are based on previous historical criminal activity data and analytical algorithms and help police forces to use their finite resources more efficiently to prevent crimes. Although it is not going to completely replace traditional policing, it has been more increasingly used across the world [36].

A few empirical evidences indicate that the use of forecasting and mapping software may increase the probability of detecting potential criminal cases compared to traditional police practices, resulting in a net decrease in overall crime rates. [3,40]. It has been found that the system predicted "hotspot monitoring," which identifies high-risk locations and concentrates police resources in these areas, leads to the suppression of criminal offenses not only in the field but also in the surrounding areas [24] in UK. Using historical crime activities, the Richmond Police Department in the USA attempted to predict where guns would be fired on New Year's Eve 2003 and modified their patrol routes to these predictions. According to their data, random shots fell 47% that night, the number of confiscated weapons increased 246%, while the police was able to save more than $15,000 [43]. Similarly, analyzing the New York Police Department (NYPD) and their usage of the predictive policing software, [32] find that the total crime index of New York has fallen 6% since the predictive policing systems were introduced.

Despite the fact that lots of law enforcement agencies and scholars are optimistic about a future of predictive policing systems, some people are concerned about the use of machine learning for predicting criminal behaviors and indicate evidence that proactive policing could have important problems. First of all, the prediction systems are primarily driven by existing and historical data, which is generally targeting a certain racial or ethnic groups. The data-driven approaches could lead to over-emphasizing correlations rather than causality [3].

Relatedly, because predictive policing algorithms are based on historical (preexisting) data, it is highly probable that any biases in the previous data influence on the prediction outputs. This may lead to stigmatization of individuals or groups [44]. Instead of producing "neutral" or unbiased predictions, predictive algorithms could simply "amplify" the existing biases [5]. Biases in preexisting data can generate biased predictions. Of particular concern is the role algorithms might play in "amplify[ing] feedback problems" because of the use of prior outputs to shape future inputs. The complexity of the models also makes it more difficult for police officers to identify and address these issues, since correcting the problem necessarily requires some understanding of the algorithms.

4.2 Zero-Risk: Interpretation

Even if there is no bias in data collection, because prediction algorithms are basically about statistical probabilistic predictions, they inevitably produce errors and uncertainty. However, when it comes to social issues, the inevitability of predictive errors is not simple. The stakes of those errors are not just unpalatable or unpleasant outcome. Zero-risk bias refers to the tendency that people are likely to avoid qualifying uncertainty and reasoning with it [23]. Although algorithm-based decision making can reduce the level of errors, it would not be able to remove all the errors. Thus, it is important not only to recognize the presence of those errors but also the social implication of them. Because when it comes to social issues, it is much more than just numeric errors.

4.2.1 Example: Child Welfare Systems Predictive algorithms have been incorporated into child welfare decision-making systems. Many agencies utilize these algorithms to avoid human errors and produce more consistent and better decisions [20]. Though misclassification errors are common in classification algorithms, the risks of false negatives and false positives in child protection systems are not trivial. For example, when a child protection service worker investigates a family, "If they err in interrupting parental custody, they may be accused of infringing the parents' constitutional rights. If they err in not removing child, they risk injury to the child and may be accused of infringing the child's rights" [20]. Even if it is common and inevitable for predictive algorithms to make errors, when it comes to sensitive issues like child custody and intervention, it would be very difficult to justify the takes of any small errors.

4.3 Outcome Bias: Operationalization

It is often indicated that algorithms' biases are based on the information that human feed them. Thus, the source of algorithms' bias is the input. Unlike humans, algorithms simply update their previous decisions when they have the new incoming data. In other words, algorithms outputs are simply dependent on human inputs. Any decision-making algorithms need humans input to make any predictions which is basically under a conditional probability [33]. Understanding the exact process is crucial to how human identify the decision-making tasks and conceptualize them to build algorithms. Without clearly defined concepts and processes, algorithms can only generate incorrect outputs.

4.3.1 Example: Measuring Concept Processes Since it is humans that determine what outcomes an algorithm is supposed to look for, humans must clearly define the problem and make a decision about prediction results before processing. Without the well-defined prediction outcomes, it would be difficult to expect algorithms to perform well. However, operationalizing social concepts into some measurable variables is not always straightforward. Finding a proxy is much more complicated than it is usually expected. For example, in child welfare systems, it has been always difficult to define and properly measure "child maltreatment" [20]. This can be anything between something small such as a child missing school and something very serious such as severe deprivation of basic life necessities [20]. Even if there were agreed-upon concept of "child maltreatment," it would not be as simple as it may sound. Identifying a data point that "accurately reflect when and where maltreatment takes place" in real conditions would be extremely difficult [20]. Similarly, machine algorithms have been widely used to help in hiring decisions and to predict "good," "employable," and "personable" candidates, but the meaning of "good" or "employable," or "personable" is vague [42].

4.4 Pro-innovation Bias and Stereotyping: Statistical Discrimination

The primary assumption of statistical discrimination is that applying rational Bayesian updating to decision making based on algorithms is straightforward. Since big data provides the massive amounts of information, some scholars believe that simply having more data would reduce the problems of statistical discrimination. More to the point [49]. In this vein, privacy protection in data collection would simply worsen the problems of statistical discrimination. Gudivada et al. [23] called this pro-innovation bias which overestimates the usefulness of the data or algorithms, thus often ignores the limitations of them.

4.4.1 Example: Employment and Disparate Impact An algorithmic process can produce a disparate impact even when trained with representative data [8]. For example, an algorithm-based hiring system can make its decision after observing candidates group identity and each candidate's "productivity" which is imprecise and noisy. In this case, we see how statistical discrimination can lead to a disparate impact for two workers with different ethnic background and the same level of productivity. Stereotyping occurs when each worker sends equally informative signals of productivity, but one worker belongs to a specific ethnic background that is generally associated with low skilled labor. Since the hiring systems will take both the group identity data and the productivity level data, it is more likely to lead to end up with hiring one worker over the other based on the group identity [16].

Some argue that once we have more data and provide more information for the algorithms, the issue of statistical discrimination would disappear [27]. In other words, according to rational Bayesian update, algorithms can produce more unbiased and unprejudiced outputs with more data. However, as we discussed above, the problems of sampling bias exist in data, because the biases already exist in the data, simply feeding more data would not lead to unbiased prediction.

5 Digital Ethics Education: Incorporating Digital Ethics into Computing Curricula

As we have discussed, there exist the risk of biases in algorithmic decision-making systems. Although the systems help to reduce the human cognitive errors, they do not remove them completely. It is primarily because the biases in DIS systems precede the system which rely on human input data. And, even with the help of the systems, it is the humans who need to interpret the outputs.

Thus, it is important to be aware that these biases get in the way of DIS systems. Instead of making these potential sources of bias as opaque or a blackbox, we need to make them more explicit and incorporate into computer science and software engineering curricula. Students should be able to check and acknowledge the potential presence of sampling bias in the data that they are processing. This has to be the first step in any DIS system development. Second, it is important to be more sensitive about the social implications of errors in prediction. As we briefly discussed in the previous section, having prediction errors is inevitable in conditional probability-based analyses. Students should be more informed about the types of errors and their meaning in the related social issues. Social issues cannot be simply dismissed as some numeric errors. Third, computer science and software engineering curricula, especially when it is going to be applied to deal with social issues, should understand how to conceptualize or operationalize social problems.

Unlike the *hard sciences*, social issues such as defining and measuring "inequality" or "injustice in criminal justice system" is often vague and not easily converted to numerical values. Social problem-based, and specifically targeted domain knowledge education is required. Instead of applying general computational education to all issues, substantively more specifically targeted subject education is needed. Lastly, with the ever-growing massive data and optimism, it is essential for students to understand the limitations of the system.

It is timely and relevant that educational institutions develop and offer a full-fledged course in *digital ethics*. It should be noted that *digital ethics* is not same as *philosophy-based ethics* courses. Resources to support digital ethics education are available and will grow with time [15,37,39]. Other tools include The Tarot Cards of Tech,[1] Google People + AI Research (PAIR),[2] LIME—Local Interpretable Model-Agnostic Explanations,[3] and AI Fairness 360.[4]

6 Summary and Conclusions

The data-intensive information systems have become very important in our daily lives. They form and shape our decision-making processes and outcomes. It is critical to understand how and where cognitive human biases play a role in DIS systems. Many people are simply very much optimistic about the prominent future of these technological developments. They expect that "neutral" and "unbiased" information systems would remove all human errors and biases. However, errors and biases still exist. Some of them are deeply embedded in the data and simply amplified when it is processed by the algorithms. Some of them are not foreseen previously and have become more apparent with the intensive usage of massive data.

As Silva and Kenney describe [47], these new tools and systems provide new dangers and risks for "reinforcing old biases with new tools." However, these new tools also enable making these potential sources of risks more explicit so that we could deal with them better than before. It is also imperative that universities impart the knowledge and skills needed to eliminate or mitigate bias in DIS systems to future computing professionals, software engineers, and data scientists.

[1] http://tarotcardsoftech.artefactgroup.com/.
[2] https://ai.google/research/teams/brain/pair.
[3] https://homes.cs.washington.edu/~marcotcr/blog/lime/.
[4] https://developer.ibm.com/open/projects/ai-fairness-360/.

References

1. Acquisti, A., Gross, R., Stutzman, F.: Faces of facebook: Privacy in the age of augmented reality. BlackHat USA **2**, 1–20 (2011)
2. Alarie, B.: The path of the law: towards legal singularity. Univ. Toronto Law J. **66**(4), 443–455 (2016)
3. Andrejevic, M.: Digital citizenship and surveillance— to pre-empt a thief. Int. J. Commun. **11**, 18 (2017)
4. Asur, S., Huberman, B.A.: Predicting the future with social media. In: Proceedings of the 2010 IEEE/WIC/ACM International Conference on Web Intelligence and Intelligent Agent Technology-Volume 01, pp. 492–499. IEEE Computer Society (2010)
5. Bakke, E.: Predictive policing: the argument for public transparency. NYU Ann. Surv. Am. L. **74**, 131 (2018)
6. Bakshy, E., Messing, S., Adamic, L.A.: Exposure to ideologically diverse news and opinion on facebook. Science **348**(6239), 1130–1132 (2015)
7. Barocas, S., Boyd, D.: Engaging the ethics of data science in practice. Commun. ACM **60**(11), 23–25 (2017)
8. Barocas, S., Selbst, A.D.: Big data's disparate impact. Calif. L. Rev. **104**, 671 (2016)
9. Burrell, J.: How the machine 'thinks': understanding opacity in machine learning algorithms. Big Data Soc. **3**(1), 2053951715622512 (2016)
10. Cadwalladr, C., Graham-Harrison, E.: The Cambridge analytics files. The Guardian (2018)
11. Camacho-Collados, M., Liberatore, F.: A decision support system for predictive police patrolling. Decis. Support Syst. **75**, 25–37 (2015)
12. Chandler, S.: The AI chatbot will hire you now. Wired.com (2017)
13. Chen, W., Quan-Haase, A.: Big data ethics and politics: Toward new understandings. Soc. Sci. Comput. Rev., p. 0894439318810734 (2018)
14. Datta, A., Sen, S., Tschantz, M.C.: Correspondences between privacy and nondiscrimination: why they should be studied together. arXiv preprint arXiv:1808.01735 (2018)
15. Eubanks, V.: Automating Inequality: How High-Tech Tools Profile, Police, and Punish the Poor. St. Martin's Press, New York, NY (2018)
16. Fang, H., Moro, A.: Theories of statistical discrimination and affirmative action: a survey. In: Benhabib, J., Jackson, M.O., Bisin, A. (eds.) Handbook of Social Economics, vol. 1a (2011)
17. Fry, H.: Hello World: Being Human in the Age of Algorithms. WW Norton & Company (2018)
18. Gandomi, A., Haider, M.: Beyond the hype: Big data concepts, methods, and analytics. Int. J. Inf. Manage. **35**(2), 137–144 (2015)
19. Gayo-Avello, D.: A meta-analysis of state-of-the-art electoral prediction from twitter data. Soc. Sci. Comput. Rev. **31**(6), 649–679 (2013)
20. Glaberson, S.K.: Coding over the cracks: predictive analytics and child protection. Fordham Urb. LJ **46**, 307 (2019)
21. Goffman, A.: On the run: fugitive life in an American city. Picador (2015)
22. Gudivada, V., Apon, A., Ding, J.: Data quality considerations for big data and machine learning: going beyond data cleaning and transformations. Int. J. Adv. Softw. **10**(1), 1–20 (2017)

23. Gudivada, V.N., Ramaswamy, S., Srinivasan, S.: Data management issues in cyber-physical systems. In: Transportation Cyber-Physical Systems, pp. 173–200. Elsevier (2018)
24. Guerette, R.T., Bowers, K.J.: Assessing the extent of crime displacement and diffusion of benefits: a review of situational crime prevention evaluations. Criminology **47**(4), 1331–1368 (2009)
25. Hamilton, M.: The biased algorithm: evidence of disparate impact on hispanics. Am. Crim. L. Rev. **56**, 1553 (2019)
26. Hargittai, E.: Is bigger always better? potential biases of big data derived from social network sites. Ann. Am. Acad. Polit. Soc. Sci. **659**(1), 63–76 (2015)
27. Hersch, J., Shinall, J.B.: Something to talk about: Information exchange under employment law. U. Pa. L. Rev. **165**, 49 (2016)
28. Kleinberg, J.: Inherent trade-offs in algorithmic fairness. In: ACM SIGMETRICS Performance Evaluation Review, vol. 46, pp. 40–40. ACM (2018)
29. Kroll, J.A., Barocas, S., Felten, E.W., Reidenberg, J.R., Robinson, D.G., Yu, H.: Accountable algorithms. U. Pa. L. Rev. **165**, 633 (2016)
30. Labrinidis, A., Jagadish, H.V.: Challenges and opportunities with big data. Proc. VLDB Endowment **5**(12), 2032–2033 (2012)
31. Lazer, D., Pentland, A., Adamic, l., Aral, S., Barabasi, A.L., Brewer, D., Christakis, N., Contractor, N., Fowler, J., Gutmann, M., Jebara, T., King, G., Macy, M., Roy, D., Van Alstyne, M.: Computational social science. Science **323**(5915), 721–723 (2009)
32. Levine, E., Tisch, J., Tasso, A., Joy, M.: The New York city police department's domain awareness system. Interfaces **47**(1), 70–84 (2017)
33. Lipton, Z.C.: The mythos of model interpretability. arXiv preprint arXiv:1606.03490 (2016)
34. Madden, M., Gilman, M., Levy, K., Marwick, A.: Privacy, poverty, and big data: a matrix of vulnerabilities for poor americans. Wash. UL Rev. **95**, 53 (2017)
35. Markham, A.N., Tiidenberg, K., Herman, A.: Ethics as methods: doing ethics in the era of big data research-introduction. Social Media Soc. **4**(3), 2056305118784502 (2018). https://doi.org/10.1177/2056305118784502
36. Meijer, A., Wessels, M.: Predictive policing: Review of benefits and drawbacks. Int. J. Pub. Adm., pp. 1–9 (2019)
37. Noble, S.: Algorithms of Oppression: How Search Engines Reinforce Racism. NYU Press, New York, NY (2018)
38. O'Connor, B., Balasubramanyan, R., Routledge, B.R., Smith, N.A.: From tweets to polls: linking text sentiment to public opinion time series. In: Fourth International AAAI Conference on Weblogs and Social Media (2010)
39. O'Neil, C.: Weapons of Math Destruction: How Big Data Increases Inequality and Threatens Democracy. Crown Publishing Group, New York, NY (2016)
40. Oswald, M., Babuta, A.: Data Analytics and Algorithmic Bias in Policing (2019)
41. Pasquale, F.: The Black Box Society. Harvard University Press, Cambridge (2015)
42. Passi, S., Barocas, S.: Problem formulation and fairness. In: Proceedings of the Conference on Fairness, Accountability, and Transparency, pp. 39–48. ACM (2019)
43. Pearsall, B.: Predictive policing: the future of law enforcement. Nat. Inst. Justice J. **266**(1), 16–19 (2010)
44. Schlehahn, E., Wenning, R.: Gdpr transparency requirements and data privacy vocabularies. In: IFIP International Summer School on Privacy and Identity Management, pp. 95–113. Springer (2018)

45. Schwartz, H.A., Eichstaedt, J.C., Kern, M.L., Dziurzynski, L., Ramones, S.M., Agrawal, M., Shah, A., Kosinski, M., Stillwell, D., Seligman, M.E., et al.: Personality, gender, and age in the language of social media: the open-vocabulary approach. PloS one **8**(9), e73791 (2013)
46. Shahin, S., Zheng, P.: Big data and the illusion of choice: Comparing the evolution of India's aadhaar and China's social credit system as technosocial discourses. Soc. Sci. Comput. Rev., p. 0894439318789343 (2018)
47. Silva, S., Kenney, M.: Algorithms, platforms, and ethnic bias: an integrative essay. Phylon (1960-) **55**(1 & 2), 9–37 (2018)
48. Stern, M.J., Bilgen, I., McClain, C., Hunscher, B.: Effective sampling from social media sites and search engines for web surveys: demographic and data quality differences in surveys of google and facebook users. Soc. Sci. Comput. Rev. **35**(6), 713–732 (2017)
49. Strahilevitz, L.J.: Reputation nation: law in an era of ubiquitous personal information. Nw. UL Rev. **102**, 1667 (2008)

Performance of Routing Protocols Using Mobility Models in VANET

Bhushan Yelure[1](✉) and Shefali Sonavane[2]

[1] Department of Computer Science & Engineering, Walchand College of Engineering, Sangli, MH 416415, India
bhushan.yelure@walchandsangli.ac.in
[2] Department of Information Technology, Walchand College of Engineering, Sangli, MH 416415, India
shefali.sonavane@walchandsangli.ac.in

Abstract. The vehicular ad hoc network (VANET) is a wireless technology adopted by the research community to implement smart transportation applications. Due to the variation in the vehicular speed, topology in the VANET continuously changes that make a routing complex task. It is better to use the position-based routing protocol that is adaptive and mostly suitable for dynamically changing road environments. The objective of the paper is to analyze an impact of real mobility traces on position-based routing protocols using simulation. It is difficult to perform a simulation in the VANET due to the restricted movement of vehicles. Vehicular mobility plays a major role in the VANET. Realistic mobility traces such as intelligent driver model-intersection management (IDM-IM), IDM–lane changing (LC) and ManhattanGrid are generated using VANETMobiSim and Bonnmotion tool. These traces are applied to the intersection-based routing (IBR), greedy perimeter stateless routing (GPSR) and ant colony optimization-intersection-based routing (ACO-IBR). The impact of the mobility traces has been evaluated with parameters such as delay, throughput and successful delivery ratio (SDR). The result shows that the selection of proper mobility pattern is a key to achieve the realistic performance of the simulation. ACO-IBR minimizes delay and produces better successful delivery ratio and throughput for all real mobility patterns.

Keywords: Vehicular ad hoc network · Vehicular routing · Mobility model · Position-based routing

1 Introduction

Smart transportation is a core and emerging task of the smart city. The smart city provides new and economical amenities to its users [1, 2]. The latest information about accidents, weather, traffic flow situation and road obstacles can help drivers. Thus, the driver is able to make decisions and attain a smooth driving experience by avoiding road mishaps. VANET considers the timely acquisition of accurate information as a key component of ITS [3, 4]. It is a network of vehicles that interacts with each other without fixed infrastructure. WAVE and DSRC [5, 6] are the contemporary techniques used for ITS that lets vehicles to participate in wireless communication and establish

© Springer Nature Singapore Pte Ltd. 2021
P. Deshpande et al. (eds.), *Next Generation Information Processing System*,
Advances in Intelligent Systems and Computing 1162,
https://doi.org/10.1007/978-981-15-4851-2_29

VANETs. The vehicle is a key component of VANET, which acts as a routing node. It shares information about the direction of the vehicle, traffic pattern, speed and density of vehicles. Communication in the VANET is inter- and intra-vehicular. Inter-vehicle communication implies communication between vehicles, and intra-vehicle indicates communication within a vehicle. An on-board unit is mounted inside the vehicle and helps to achieve V_2V communication. Less deployment cost, support for the short message delivery and minimum latency in the communicating links are some features of V_2V communication. V_2I communication is feasible by using road-side unit (RSU) [7, 8]. Wi-Fi [8] and Wi-MAX [9] are popular technologies used in the V_2I.

The mobility model imitates the real movement of vehicular traffic. There are several aspects that affect mobility such as the type of traffic, speed limits, layout of the road, random distribution of vehicles, assistance at the intersection and communication interruption due to the obstacle interference. In addition to that speed-up and slow-down models are considered since vehicles abruptly break and move. The urban environment has roads comprising traffic lights and intersections. The highway environment has roads having multiple lanes, rare intersections and traffic lights. It is necessary to use the real traces in the simulation so that simulation replicates the real-time performance of the VANET. Thus, the selection of the mobility model to attain a realistic performance is another challenge [7]. The contribution of the paper is to study various mobility models and to generate various real-time traces for the city environment. Apply generated traces on the GPSR, IBR and ACO-IBR to evaluate their comparative QoS performance by using simulation.

The structure of the paper is organized into four sections. Section 2 is related work. It consists of a discussion about the various position-based routing protocols. Section 3 is about the experimental details and results with discussion. It includes a description of the mobility traces used with their specifications. The conclusion is the last section.

2 Related Work

Routing performs a significant role in the deployment of the VANET. It establishes an optimized route from one vehicle to another. Mostly, it takes less communication time and fewer resources. Mobility of vehicles and dynamic network topology changes affect the QoS in the VANET routing [7]. Information about topology and the channel of communication is the major concern in topological routing protocols. Cluster-oriented protocol is a group of vehicles with similar characteristics [10]. It has issues about the cluster head selection and the cluster formation.

Position-based routing (PBR) protocol uses the location of the vehicle in order to ease communication. It is typically appropriate for the urban and highway environment. The source vehicle collects position information through the periodical beacon message. The packet header comprises position information of the source and destination vehicle to achieve routing. The heart of the PBR protocol is greedy forwarding. The PBR protocol suffers from local maxima situations, network partitioning and frequent interruption in sparse networks because of obstacles and high velocity in the city environment [11, 12].

2.1 Greedy Perimeter Stateless Routing (GPSR)

It is a classic example of a PBR protocol. The packet forwarding decision depends upon the position of a source, adjacent and destination vehicle. Greedy forwarding and perimeter forwarding are two prime phases in GPSR [13]. Greedy forwarding uses information about the closest neighbor of the source in the network structure. It has identified a failure of the neighborhood or out of transmission range when a beacon message is not received for a longer duration than timeout interval. In case of multi-hop routing, the node located inside the range is substantially closer than other available nodes in the network. The greedy forwarding approach fails when a local maxima problem occurs. The local maxima problem suggests that the source or current node is the node having the shortest distance to the destination. This condition is also referred to as a communication void. If the number of nodes is more, then there are fewer chances of communication void. But if there is a shortage of fixed density with time, the occurrence of obstacles and less reliable nodes, then such packets are easily removed. Such a situation is handled with the perimeter of the region. The working of the greedy phase is depicted in Fig. 1a.

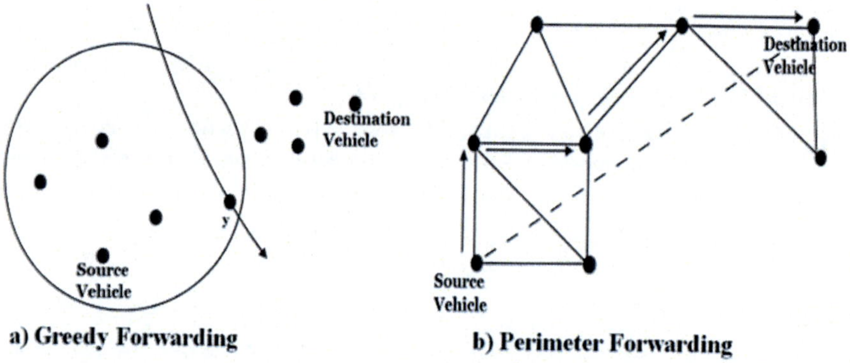

Fig. 1. Working of GPSR: **a** greedy forwarding and **b** perimeter forwarding

GPSR uses perimeter mode as a recovery mechanism [13] as represented in Fig. 1b. In perimeter forwarding, a packet traverses continuously nearby faces of the planner subgraph of the full radio network connectivity. When the packet is destined from source to the destination, the next edge traversed is selected in the counter-clockwise direction about the source from edge source to the destination. Previously, the perimeter mapping is done by making packet tours using the right-hand rule. The said approach works by assuming that two edges in the graph do not cross each other. If such a situation arises, then it blindly removes the second edge in the crossing edge pairs. After the removal of one of the crossing edge, still, there is a partition observed in the topology, the algorithm fails to search routes across the partition.

2.2 Intersection-Based Routing (IBR)

Intersection routing is a subcategory of PBR protocols. These are mainly appropriate for city areas. It involves intersections on the road in the city. It consists of a traffic light at intersections; road accident scenarios; and three-dimensional roads and streets [14, 16].

A vehicle located at an intersection picks a neighboring intersection that is nearby to the destination. It forwards information through an intersection. For that purpose, an explicit criterion is decided and used to forward the information. It improves route stability, adaptive to changes in the topology and minimizes routing overhead [15, 16].

Path decision in IBR [15] uses certain key factors, such as vehicle direction, vehicle location and vehicle density. If the traveling route of the next communicating vehicle varies from the transmission direction of packet routing, then opposite direction forwarding incurs an extensive delay and high packet loss. This exclusively takes place in scant traffic road segments. To evade the aforesaid condition, it is essential to use vehicle direction and route the messages to the desired vehicles. Thus, the IBR protocol uses vehicle travel direction and the projected packet delay. Delay is the summation of the carry and the forwarding delay. It evaluates forwarding delay for high-density roads. The road with fewer vehicles requires both forwarding and carry delay. Additionally, vehicles use a new route for all packets when they are at the intersection according to the packet delay. Then, the use of travel direction of the next relay vehicles is done for routing of messages. Thus, the determined route has less packet delay toward destination.

2.3 ACO-IBR

It is a hybrid approach. It uses ant colony optimization technique due to its foraging behavior of ants that is applicable to determine the shortest route. The shortest route is determined according to the pheromone value and intersection rating. The method adopted to forward the packet is greedy and intersection-oriented. An intersection rating is helpful in making the decision on the next intersection to whom the packet is forwarded. It uses the direction of a vehicle, distance and delay towards neighbor intersection closer to the destination. The source uses forward ant to send RREQ toward intermediate intersection to attain the global pheromone at the destination. If the global pheromone is recent and the intersection is selected according to their ratings, then the path is established. The path is sent as a RREP by using backward ant toward the source. Then, the source uses that path to achieve actual communication between vehicles. The working of ACO-IBR is represented in Fig. 2.

3 Performance Analysis

3.1 Experimental Results

The traces used in the simulation are ManhattanGrid, IDM-IM and IDM-LC. These traces are generated using widely adopted trace generator tools such as Bonnmotion and VANETMobiSim [17–19]. The trace files description is mentioned in Table 1.

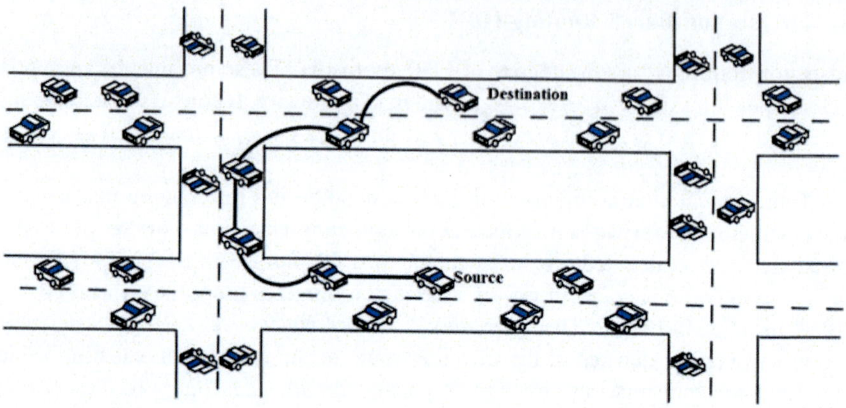

Fig. 2. Working of ACO-IBR (shortest path is shown between source and destination vehicle)

Simulation has been conducted to attain the effect of real-time traces on the QoS of the routing protocol. It uses the state-of-the-art simulation parameters that are applied to conduct the simulation [20]. The simulation parameters are mentioned in Table 2.

Table 1. Mobility models with their specifications

Mobility pattern	Area (m^2)	Time (s)	Nodes	Intersections	Edges	Traffic lights	Lanes	Tool used
IDM-IM [19]	2000m^2	600	50	27	42	06	One	VANET MobiSim
IDM-LC [19]	2600m^2			37	60	06	Two	
ManhattanGrid [18]	2000m^2			25	42	–	One	Bonn motion

Table 2. Simulation environment

Parameter	Simulation value
Transmission range	250 meters
Antenna model	Omni-directional antenna
Radio propagation model	TwoRayGround
MAC protocol	IEEE 802.11p
Interface queue type	Droptail, priority queue
Evaluated routing protocols	GPSR, IBR, ACO-IBR
Number of vehicles (nodes)	50
The speed of the vehicle	Up to 20 m/s
CBR connections	15
Packet size	512 kb

SDR defines the ratio of the successful exchange of messages with sent messages. Delay defines the time required to reach the message from the source to the destination vehicle. Throughput is the ratio of total bytes transmitted successfully through the link in the entire simulation. Table 3 shows the performance of routing protocols for various real-time traces.

Table 3. Performance of routing protocols for various real-time traces (600 s)

Protocol	ACO-IBR	GPSR	IBR	ACO-IBR	GPSR	IBR	ACO-IBR	GPSR	IBR
	IDM-IM			IDM-LC			ManhattanGrid		
SDR	91.70	88.92	91.19	92.92	89.94	90.15	93.85	89.96	92.34
Delay	0.073	0.102	0.082	0.071	0.101	0.083	0.273	0.101	0.081
Throughput	49.84	47.69	49.30	53.83	52.10	53.47	61.51	58.96	60.52

For IDM-IM city scenarios as compared to the GPSR and IBR, ACO-IBR produces 2.78% and 0.51% better SDR. It takes 0.029 and 0.009 s less delay to reach the message toward the destination. It produces 2.15 and 0.54 kbps extra throughput (Fig. 3).

For IDM-LC city scenarios as compared to the above-mentioned protocols, ACO-IBR produces 2.98 and 2.77% better SDR. It takes 0.03 and 0.012 s minimum delay. It produces 1.73 and 0.36 kbps better throughput. For the ManhattanGrid scenarios, ACO-IBR produces 3.89 and 1.51% better SDR as compared to the above-mentioned protocols. IBR produces 0.19, 0.02 s less delay to reach the message to the destination. ACO-IBR produces 2.55 kbps, 0.99 kbps extra throughput.

3.2 Justification on Analysis

ACO-IBR achieves superior connectivity between vehicles, repair partitions, reduction in disconnections due to its hybrid and intrinsic behavior of ACO. Because of improved connectivity, more opportunities are there to transmit a message to the destination successfully. It forwards messages to the intersection optimally. It handles the routing through greedy as well as intersection. It helps to distribute traffic load over the network that reduces congestion and keeps a greater SDR. It can make the dynamic decision based on the pheromone and intersection rating at the intersection. Also, it can effectively reduce holes in the road segments used for the packet transfer. Thus, ACO-IBR is dominant in the SDR, delay and throughput with protocols in the study.

GPSR undergoes more packet transmission by establishing longer routes toward the destination in perimeter mode. Thus, it causes packet drop and routing loops. Even in the connected graph, an interruption occurs in the perimeter mode with a planner graph. Thus, it requires more delay to transmit the message to the destination. For more vehicles, a reduction in the delay is possible because the greedy mode becomes more active.

IBR uses a carry and forward mechanism that provides a better chance to determine proper vehicles to forward the messages. High-speed vehicles can cut down the carrying delay as carry vehicles can travel quicker in the carrying distance. Sometimes in

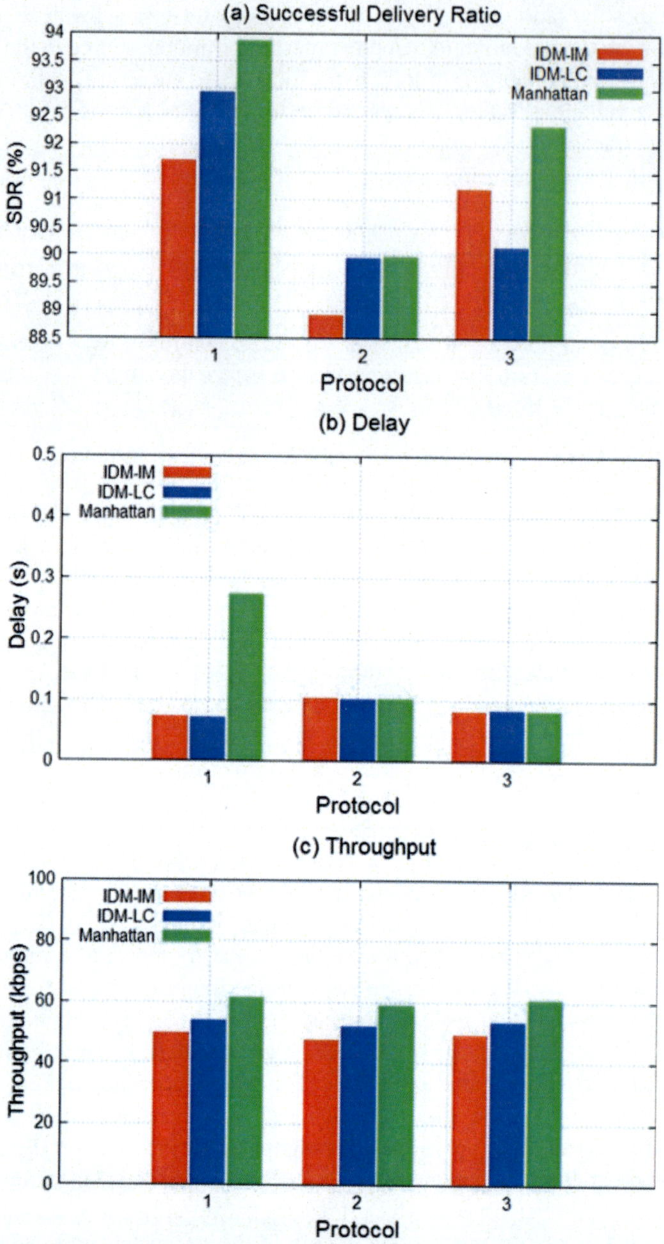

Fig. 3. Simulation results: **a** performance of SDR, **b** performance of delay and **c** performance of throughput note: 1-ACO-IBR, 2-GPSR and 3-IBR

low vehicle density, carrying delay is more that causes more contention, queue waiting and packet loss. Occasionally, if a collision takes place, then the SDR falls, that causes packets to be retransmitted and thus delay increases.

4 Conclusion

The position-based routing protocols with their issues are discussed. GPSR, IBR and ACO-IBR are extensively simulated and compared their performance using real traces such as IDM-IM, IDM-LC and ManhattanGrid. Simulation results show that ACO-IBR outperforms in terms of throughput and SDR as compared to GPSR and IBR for all traces. ACO-IBR witnessed minimum delay for nearly all mobility patterns except ManhattanGrid scenarios. Each protocol under study shows relatively better QoS performance for ManhattanGrid scenarios due to the restricted movement pattern of the vehicles on the road. It is also observed that the ManhattanGrid mobility pattern is a promising solution for the simulation of protocols in the city environment for sparse density. Furthermore, the mobility pattern improves the connectivity in a vehicular network. In future, there is a scope to determine the connectivity duration probability for road segments.

References

1. Rawal, T.: Intelligent transportation system in India—A review. J. Dev Manage Commun. **2** (299) (2015)
2. Roor, T.: Intelligent transportation system in India—A review. (2016)
3. Suman Malik, Prasant Kumar Sahu.: A comparative study on routing protocols for VANETs. Heliyon, 5(8), (2019)
4. A. Thakur and R. Malekian.: Internet of Vehicles Communication Technologies for Traffic Management and Road Safety Applications. In: Wireless Personal Communications, (2019)
5. J. B. Kenney.: Dedicated Short-Range Communications (DSRC) Standards in the United States. In: the Proceedings of the IEEE, pp. 1162–1182, (2011)
6. Ramakrishnan, B., Nishanth, R.B., Joe, M.M., Shaji, R.S.: Comprehensive analysis of Highway, Manhattan and Freeway mobility models for vehicular ad hoc network. Int. J. Wireless Mobile Comput. **9**(1), 78–89 (2015)
7. F. Li and Y. Wang.: Routing in vehicular ad hoc networks: A survey. IEEE Vehicular Technology Magazine, 2(2), pp. 12–22, (2007)
8. G. D. Singh, R. Tomar, H. G. Sastry, and M. Prateek.: A review on VANET routing protocols and wireless standards. Smart Innov. Syst. Technol., 78(1), pp. 329–340, (2018)
9. C. Cooper, D. Franklin, M. Ros, F. Safaei and M. Abolhasan.: A Comparative Survey of VANET Clustering Techniques. In: IEEE Communications Surveys & Tutorials, 19(1), pp. 657–681, (2017)
10. N. Goel, G. Sharma and I. Dhyani.: A study of position based VANET routing protocols. 2016 International Conference on Computing, Communication and Automation (ICCCA), pp. 655–660, Noida, (2016)
11. Kumar, S. & Verma, A.K.: Position Based Routing Protocols in VANET: A Survey. Wireless Pers Commun (2015)
12. B. N. Karp and H. T. Kung.: GPSR: Greedy perimeter stateless routing for wireless networks. In: Proceedings ACM MOBICOM, pp. 243–254, (2000)
13. H. Saleet, R. Langar, K. Naik, R. Boutaba, A. Nayak and N. Goel.: Intersection-Based Geographical Routing Protocol for VANETs: A Proposal and Analysis. In: IEEE Transactions on Vehicular Technology, 60(9), pp. 4560–4574, (2011)

14. L. Der Chou, J. Y. Yang, Y. C. Hsieh, D. C. Chang, and C. F. Tung.: Intersection-based routing protocol for VANETs. Wireless Personal Communications, 60(1), pp. 105–124, (2011)
15. G. Li, L. Boukhatem and J. Wu.: Adaptive Quality-of-Service-Based Routing for Vehicular Ad Hoc Networks With Ant Colony Optimization. IEEE Transactions on Vehicular Technology, 66(4), pp. 3249–3264, (2017)
16. J. Harri, F. Filali, and C. Bonnet.: Mobility models for vehicular ad hoc networks: a survey and taxonomy. In: IEEE Communications Surveys & Tutorials, 11(4), pp. 19–41, (2009)
17. S. A. Ben Mussa, M. Manaf, K. Z. Ghafoor and Z. Doukha.: Simulation tools for vehicular ad hoc networks: A comparison study and future perspectives. International Conference on Wireless Networks and Mobile Communications (WINCOM), pp. 1–8, Marrakech, (2015)
18. Martinez, F. J., Toh, C. K., Cano, J.-C., Calafate, C. T. and Manzoni, P.: A survey and comparative study of simulators for vehicular ad hoc networks (VANETs). Wireless Communication Mobile Comput, pp. 813–828, (2011)
19. Mobile Ad-hoc Networking group- https://www.ietf.org/rfc/rfc3561.txt last accessed Oct-2019
20. http://www.isi.edu/nsnam/ns/index.html last accessed Oct-2019

Graph Partitioning Using Heuristic Kernighan-Lin Algorithm for Parallel Computing

Siddheshwar V. Patil[1(✉)] and Dinesh B. Kulkarni[2]

[1] Department of Computer Science and Engineering, Walchand College of Engineering, Sangli, MH, India
siddheshwar.patil@walchandsangli.ac.in
[2] Department of Information Technology, Walchand College of Engineering, Sangli, MH, India
dinesh.kulkarni@walchandsangli.ac.in

Abstract. The goal of parallel computing is to distribute the load on available processors such that all processors should be utilized in a fair manner. This minimizes the overall execution time required to execute a complex task. So, the load balancing issue becomes an important aspect of parallel computing. It is abstracted as a graph partitioning problem in which the nodes represent computation cost, edges represent communication cost, and number of partitions should be equal to number of available processing units. So, the objective is to cut the graph into k-partitions such that—(i) total node weight should be equal for each partition and—(ii) minimize total edge weight across the partitions. A heuristic Kernighan-Lin graph partitioning algorithm for two-way partitioning is evaluated in this paper. It starts with an initial random graph partition and consecutively exchanges the nodes between partitions, determines cut size at each stage and saves the minimum cut found so far. After the desired number of swaps has been performed, the saved minimum cut will give optimal partitions. The graph data for experimental work is obtained from DIMACS tenth implementation challenge Web site. The experimental results show minimum cut-cost with respect to balance constraint. The results are compared with ground truth results for validation.

Keywords: Graph partitioning · Parallel computing

1 Introduction

In many applications, problem modeling is abstracted by using the graphs. One of the fundamental operations is to cut the graph into smaller subgraphs. In applications such as social networks, scientific simulations, very large scale inte-

© Springer Nature Singapore Pte Ltd. 2021
P. Deshpande et al. (eds.), *Next Generation Information Processing System*,
Advances in Intelligent Systems and Computing 1162,
https://doi.org/10.1007/978-981-15-4851-2_30

gration (VLSI) circuit placement, sparse matrix factorization and parallel computing, graph partitioning is important and challenging. In parallel computing, load balancing is an important issue to be addressed [1–3]. The aim is to distribute the computational load on various available processors so that it will minimize total execution time. Graph partitioning models map load balancing problem in parallel computing. Graph partition problem is NP-complete, and the solutions for this problem are derived from approximation/heuristic algorithms. The graph partitioning problem has been researched for many years to achieve near-optimal (balanced) partitions. In this paper, we tried to obtain graph two-way partitioning (bipartitioning) which satisfies the constraint of a minimum number of an edge-cut and balanced sets at the same time.

1.1 Graph Partitioning

In mathematics, $G = (V, E)$ is a graph, in which V is set of vertices, E is set of edges. Dividing the graph into smaller components is called graph partitioning. For example, in k-way partitioning, vertex set is divided to k small sets. The sets are called partitions, and the set of edges across the partitions is called as a cut. The quality of partition is good if there are minimum edge-cut (cut-cost) across the partitions.

The graph partitioning problem is becoming important due to its application for scientific computing, design of electrical circuits, parallel computing, detection of cliques and clustering in biological and social networks, etc. [4].

An example of a simple graph and a possible balanced partition is shown in Fig. 1. It shows cut-cost of three assuming that each edge and node has weight 1.

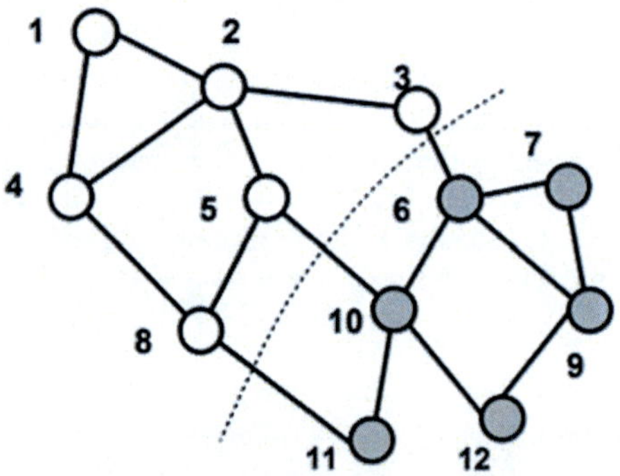

Fig. 1. Example of a partition

1.2 Load Balancing

In multi-CPU or multi-core environments like data centers, there is a heavy workload running and executing [5]. For getting maximum efficiency or throughput, there is a desire to balance the workloads and lowering interprocess communication. Interprocess communication takes up more time and overhead which slows down the performance of systems. Secondly, we need to use all the computing power uniformly as the running costs of all these machines are the same. Basically, we want to get the best use of resources with respect to both time and money—hence the load is partitioned to different processors. In this work, assume the nodes as the different workloads and edges as the use of shared data. Let k be the number of cores or CPUs the work needs to be put on. Therefore, k-way graph partition will give a solution such that those activities which have most use of shared data are running on one CPU such that the data is made locally available to them. Likewise, it has extensive use in data clustering, telephone network design and related similar avenues [6,7].

Remaining part of the paper is organized as follows: Section 2 discusses related work. Graph partitioning algorithm and results are presented in Sects. 3 and 4. Section 5 concludes.

2 Related Work

There are some prominent minimum-cut graph partitioning techniques found in the literature. They are discussed in this section.

Sheblaev and Sheblaeva [8] proposed a new method for the Fiduccia–Mattheyses (FM) algorithm for two-way partitioning as well as for multi-way partitioning problems. They have presented a method to find initial partitioning for Fiduccia–Mattheyses algorithm. They have obtained solutions for a balanced hypergraph partition problem. Their methods use geometric properties and dimensionality reduction methods for large data.

Bui and Moon [9] proposed a genetic algorithm for graph partitions. Local improvement heuristic methods are used in this work. The space searching capability is improved by the schema preprocessing phase, further improves algorithm performance. The results show that this technique performs better as compared to the simulated annealing algorithm.

Schloegel and Karypis [10] proposed multilevel partition algorithm. The problem is extended by incorporating balancing constraints static and dynamic load balancing on high performance computing platforms. Every vertex is weighted vertex. Goal is to achieve k-way partitioning so that weights across each partition should be balanced while minimizing the edge-cut. This multilevel paradigm consists of graph coarsening, uncoarsening, initial partitioning, refinement, etc. The experimental work evaluates the effectiveness of parallel multi-constraint partitioners on the number of synthetic graph problems.

Leng et al. [11] presented a multilevel paradigm-based graph two-way partitioning algorithm. The algorithm uses a tabu search technique and a boundary refinement approach. This algorithm gives excellent results as compared with the partitioning results obtained from the metis tool.

Andreev and Racke [12] solved a balanced graph partitioning problem. Their work shows the bicriteria polynomial-time approximation algorithm for partitioning. This work extends for the graph with weighted nodes, and there should be weight balance among the partitions. It is a challenging part to improve the running time of these algorithms.

Andersen and Chung [13] presented local graph partitioning technique using PageRank variation for given initial distribution. They have derived results for PageRank vectors similar to random walks. Small conductance cut is obtained by vertices ordering generated using PageRank vectors. The partition can be recursively applied to generate a new algorithm for multi-way partitioning, solving diagonally dominant linear systems and graph sparsification.

Peng et al. [14] proposed a spectral clustering algorithm to partition a graph into k-clusters. It first calculates the graph laplacian of the input graph. Then, it finds eigenvalues and eigenvectors and partitions the embedded points via k-means algorithm. It shows that spectral clustering algorithm gives a good approximation of the optimal clustering.

3 Kernighan-Lin Algorithm

The Kernighan-Lin algorithm is an iterative algorithm to partition the graph which produces near-optimal results [15]. Here, the basic version of the algorithm which partitions the input graph into two equal-sized blocks is presented. It is an iterative improvement algorithm which can have initial balanced partition (chosen randomly) A and B. We need new partition $A*$ and $B*$ such that number of edge-cut across partitions is minimum.

After initial partition, if vertex is moved from one partition to other partition, then gain for each vertex is calculated to achieve reduction of edge-cut. At every iteration, it moves unlocked vertex having maximum gain, from partition in surplus (partition which has more vertices) to the partition in deficit. Then, vertex is locked, and it updates gains.

This procedure is repeated till the highest gain becomes negative and all vertices are locked. Then, last few moves which have negative gain are undone, and bisection is reverted to one which has small edge-cut found so far. So, it completes outer iteration, and it will restart iterative procedure. If outer iteration will fail in minimization of edge-cut or if it makes load imbalance, then the algorithm will terminate.

3.1 Rationale Behind the Kernighan-Lin Algorithm

It is an order of complexity—$O(n^2 \log n)$ and can take $O(V^2)$ in terms of memory space. It requires—Equal disjoint subsets (let A & B) at initial step. Swapping two vertices ($a \in A, b \in B$) will maintain both the sets to contain same number of elements.

Let E_a be the sum of crossing or external edges from A into B and I_a be the sum of internal edges from A into A.

Similarly, let E_b is the sum of crossing or external edges from B into A and I_b be the sum of internal edges from B into B.

Let $D_a = E_a - I_a$ & $D_b = E_b - I_b$

Therefore, gain is $g = D_a + D_b - 2 \times c(a, b)$.
For the best partition, it is needed to maximize $D_a + D_b - 2 \times c(a, b)$.
The pseudocode for Kernighan-Lin two-way graph partitioning is shown below.

Algorithm 1 Kernighan-Lin two-way graph partitioning pseudocode

1: *Initial partition nodes into the two equal sets A and B*
2: i=*1*
3: *Calculate D for all nodes in A and B*
4: *Choose* a_i *from A and* b_i *from B such that* $g_i = D(a_i) + D(b_i) - 2 * c(a_i b_i)$ *is maximized*
5: *Swap and fix* a_i *and* b_i *to make equal set*
6: *If all nodes are fixed, go to Step 10. Else*
7: *Computation of D values and updation of it for all nodes which are connected to* a_i *and* b_i *(Not fixed nodes)*
8: $i{+}{+}$
9: *Go to* 4th *step*
10: *Find move sequence* $1 \dots m$ *and* $(1 \le m \le i)$, *so* $G_m = \sum_1^m g_i$ *is maximized*
11: *If* $G_m > 0$, *go to step 12. Else, STOP*
12: m *swaps are executed and remaining nodes reset*
13: *Go to Step 2*

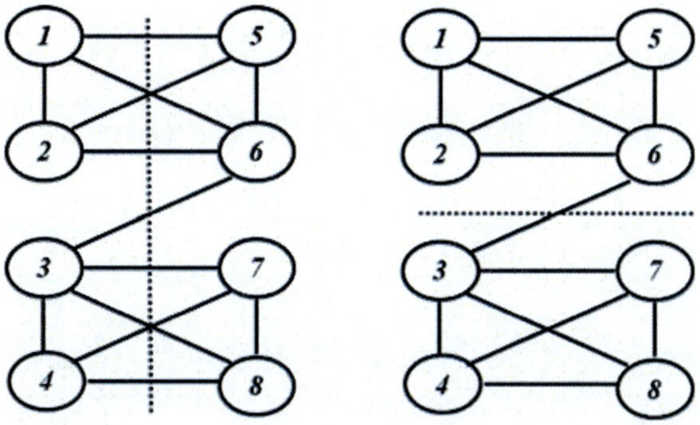

Fig. 2. **a** Initial graph partition (cut-cost = 9). **b** Final graph partition (cut-cost = 1)

Figure 2 shows an example of graph partitioning using a Kernighan-Lin algorithm. The dotted line shows the number of edges cut. The initial two-way partitioned graph (random partitioning) with a cut-cost = 9 is shown. In the end, the algorithm gives the final cut-cost = 1 which satisfies both balanced and minimum edge-cut constraint.

4 Results

The performance evaluations were run on Intel Xeon E5520@2.27GHz (8 cores) with 16GB RAM. The algorithm was applied on several graphs obtained from DIMACS tenth implementation challenge Web site [16]. These are undirected and unweighted graphs. Following parameters are considered to obtain the results:
G—Input graph in metis format
N—Number of nodes
E—Number of edges
T—CPU execution time required for partitioning (in s).

In metis graph file format, total node and edge count are shown at first line. If any additional extra value is present at the first line, then it is considered as weight. The second line to the last line lists "neighbors" of a node.

Let us consider the graph with 2851 nodes and 15,093 edges. The initial cut-cost is 609. For this graph, it took total four iterations to achieve minimum cut-cost. After the first iteration, it has a cut-cost of 355. After the second, third and fourth iteration, it shows a cut-cost of 279, 274 and 270, respectively. For all the above iterations, there is iterative incremental gain. So, the final result shows cut-cost as 270 to obtain balanced two-way partitions.

The results for final cut-cost and execution time required for partitioning are shown in Table 1. The results are compared with ground truth results (Metis benchmark) for validation. It shows approximate results as compared to results from Metis. It is also seen that to partition a graph with 10,498 nodes and 45,878 edges, the experiment hardly takes 1 minute to obtain optimal partitions.

Table 1. Two-way graph partitioning results

G	N	E	Initial cut-cost	Final cut-cost	T (s)
jazz.graph	198	2742	2740	868	0.05
1e3.in	1000	1000	1237	641	1.83
cq9.rig	9278	106,156	92,544	43,162	21.66
vsp.graph	10,498	53,868	31,384	15,068	47.11
4elt.graph	15,606	45,878	5525	4359	51.30
vspl.graph	32,212	101,805	101,758	47,922	1315.10

The two-way Kernighan-Lin partitioning algorithm shows a time complexity of $O(n^3)$. The k-way partitioning is obtained by recursively applying the same algorithm on partitioned graphs. This extension is not worth as it leads to added complexity with each cycle of an iterative call to the algorithm. So, it is essential to explore further algorithms and come up with a novel approach to solve the partitioning problem.

5 Conclusion

The Kernighan-Lin algorithm for graph partitioning produces two-way graph partitions with minimum edge-cut. The experimental work is evaluated, and it shows promising results for two-way graph partitioning in terms of a balance constraint and a minimum edge-cut constraint.

This work can be extended in the future to k-way graph partitioning. Partitioning with different node and edge weights can be challenging future work. For large graphs, the parallel implementation can be focused to speed up the execution time.

References

1. Doe, J.: Load balancing strategies in parallel computing: short survey (2015)
2. Kushwaha, M., Gupta, S.: Various schemes of load balancing in distributed systems—a review. Int. J. Sci. Res. Eng. Technol. **4**(7), 741–748 (2015)
3. Prasad, V.: Load balancing and scheduling of tasks in parallel processing environment. Int. J. Inf. Comput. Technol. **4**(16), 1727–1732 (2014)
4. Sakouhi, C., Khaldi, A., Ghezal, H.B. : An overview of recent graph partitioning algorithms. In: Proceedings of the International Conference on Parallel and Distributed Processing Techniques and Applications, pp. 408–414 (2018)
5. Kumar, S., Das, S. K., Biswas, R. : Graph partitioning for parallel applications in heterogeneous grid environments. In: Proceedings of International Parallel and Distributed Processing Symposium, pp. 7-pp. IEEE (2002)
6. Patil, S.V., Kulkarni, D.B.: A review of dimensionality reduction in high-dimensional data using multi-core and many-core architecture. In: Workshop on Software Challenges to Exascale Computing, pp. 54–63. Springer (2018)
7. Patil, S.V., Kulkarni, D.B.: Parallel computing approaches for dimensionality reduction in the high-dimensional data. Int. J. Comput. Sci. Eng. **7**(5), 1750–1755 (2019)
8. Sheblaev, M.V., Sheblaeva, A.S.: A method of improving initial partition of Fiduccia-Mattheyses algorithm. Lobachevskii J. Math. **39**(9), 1270–1276 (2018). Springer
9. Bui, T.N., Moon, B.R.: Genetic algorithm and graph partitioning. IEEE Trans. Comput. **45**(7), 841–855 (1996)
10. Schloegel, K., Karypis, G., Kumar, V.: Parallel static and dynamic multi-constraint graph partitioning. Wiley Concurrency Comput. Pract. Experience **14**(3), 219–240 (2002)
11. Leng, M., Yu, S., Chen, Y.: An effective refinement algorithm based on multilevel paradigm for graph bipartitioning. In: International Conference on Programming Languages for Manufacturing, pp. 294–303. Springer (2006)
12. Andreev, K., Racke, H.: Balanced graph partitioning. Theor. Comput. Syst. **39**(6), 929–939 (2006). ACM
13. Andersen, R., Chung, F., Lang, K.: Local graph partitioning using pagerank vectors. In: 47th Annual Symposium on Foundations of Computer Science, pp. 475–486. IEEE (2006)
14. Peng, R., Sun, H., Zanetti, L.: Partitioning well-clustered graphs: spectral clustering works! In: Conference on Learning Theory, pp. 1423–1455 (2015)

15. Kernighan, B.W., Lin, S.: An efficient heuristic procedure for partitioning graphs. Bell Syst. Techn. J. **49**(2), 291–307 (1970)
16. Bader, D., Kappes, A., Meyerhenke, H., Sanders, P., Schulz, C., Wagner, D.: Benchmarking for graph clustering and partitioning. In: Encyclopedia of Social Network Analysis and Mining, pp. 73–82. Springer (2014)

Experimenting with Reordering Model of Phrase-Based Machine Translation System for English to *Hindi*

Arun R. Babhulgaonkar[1(✉)] and Shefali P. Sonavane[2]

[1] Department of Computer Science & Engineering, Walchand College of Engineering, Sangli, India
arun.babhulgaonkar@walchandsangli.ac.in

[2] Department of Information Technology, Walchand College of Engineering, Sangli 416415, MH, India
shefali.sonavane@walchandsangli.ac.in

Abstract. *Hindi* is the national language of India. However, most of the government records, resolutions, news, etc. are documented in English which native urban users may not understand. This fact motivates to develop an automatic language translation system from English to *Hindi*. Grammatical structure of the *Hindi* language is very much complex than the English language. This structural difference makes it difficult to achieve good quality translation results. Translation, reordering and language model are the main working components of a translation system. The translation quality depends on how these individual components of the system are configured. Many times the values of these components are language-dependent. Hence, proper settings of these components are very much essential. This paper discusses various settings of the reordering model and through experimentation demonstrates the proper values of the parameters for getting a quality translation from English to *Hindi*. The freely available n-gram-based BLEU metric and TER metric is used for evaluating the results.

Keywords: Machine translation (MT) · Language modeling (LM) · Word alignment · Reordering model

1 Introduction to Phrase-Based Machine Translation (PBMT)

Machine translation is the process of translating a text in one natural language into another language using the computer system. PBMT is a data-driven approach in which statistical information of corpora is used as the main knowledge source. This model is evolved from word-based models. This model uses the phrase as the atomic unit of the translation instead of words. Any contiguous sequences of words are termed as a phrase. Figure 1 illustrates word-based model, and Fig. 2 illustrates the segmentation of a sentence in phrase-based model. In this paper, various reordering model

© Springer Nature Singapore Pte Ltd. 2021
P. Deshpande et al. (eds.), *Next Generation Information Processing System*, Advances in Intelligent Systems and Computing 1162,
https://doi.org/10.1007/978-981-15-4851-2_31

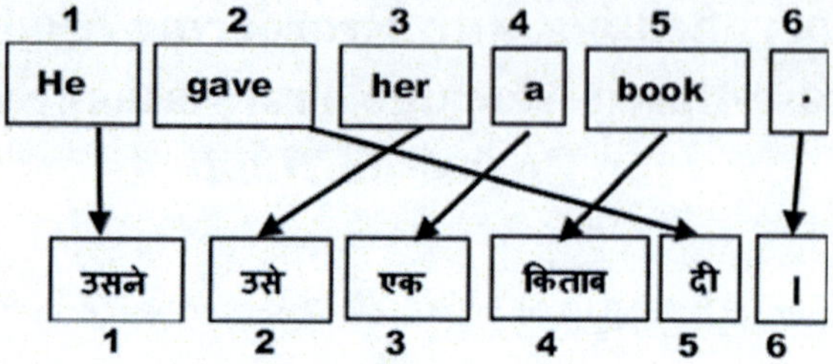

Fig. 1. Word-based model

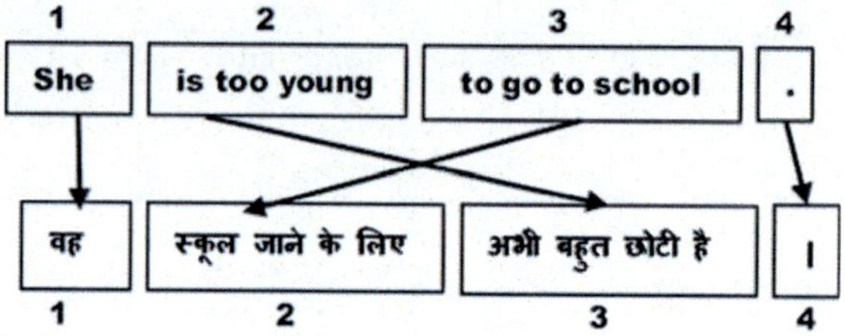

Fig. 2. Phrase-based model

configurations of the PBMT system are tested to check their effects on the translation quality of a sentence from the English language to the *Hindi* language. If an English sentence '*e*' is translated into *Hindi* language sentence '*h*,' mathematically this problem is defined as:

$$h_{\text{best}} = p(h/e) = \arg\max_{e} p(e/h)\, p_{LM}(h) \tag{1}$$

This is the application of Bayes theorem for PBMT which uses the source channel approach of information theory [1]. Bayes theorem divides the translation task in two independent parts, viz. $p(e|h)$ and $p_{LM}(h)$.

While translating a sentence from one natural language into another language the position of the phrases changes. To accommodate this reordering task of phrases, the translation component $p(e|h)$ in Eq. 1 is expressed further into two parts as:

$$h_{\text{best}} = \arg\max_{e} \prod_{i=1}^{I} \phi(\bar{e}_i|\bar{h}_i)\, d(\text{start}_i - \text{end}_{i-1} - 1)\, p_{LM}(h) \tag{2}$$

Equation 2 consists of three components:

(a) The first component is the phrase translation model $\phi(\bar{e}_i|\bar{h}_i)$ which takes care of the actual translation of English phrases into *Hindi* phrases.

(b) The second component $d(\text{start}_i - \text{end}_{i-1} - 1)$ takes care of the reordering of phrases. Reordering is considered relative to the previously translated phrase.

Here, $start_i$ and end_i are the positions of first and last word of the English phrase that translates to the *i*th *Hindi* phrase. The grammatical and structural difference between the source and target language requires a reordering of translated words. Reordering of words during translation is language-dependent. English uses subject-verb-object ordering, while *Hindi* uses subject-object-verb ordering. Hence, reordering of words is very important to get the correct translation of an English sentence to the *Hindi* sentence. The distance-based reordering model calculates the cost based on the number of words dropped when phrases are selected out of order. This is basic reordering model which is illustrated by Fig. 3.

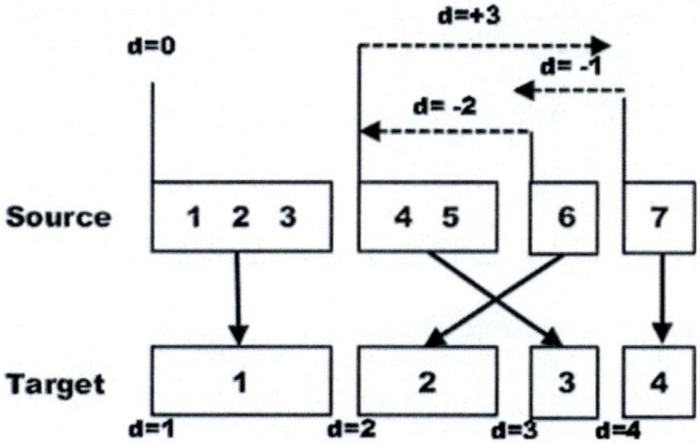

Fig. 3. Distance-based reordering

(c) The language model $p_{LM}(h)$ takes care of well-orderedness and fluency of translated *Hindi* sentence. The leading method of language modeling is n-gram, where n is the degree of the language model. n is the number of previous words considered for predicting the next word in a sentence. Thus, the probability of the next word w_n depends on the history 'h' of previous $n-1$ words $(w_1 w_2 ... w_{n-1})$ in the sentence.

Lexicalized Reordering Models:

The basic reordering model is conditioned only on movement distance. Lexicalized reordering model considers three reordering types based on alignment point in the alignment matrix. Figure 4 illustrates the orientation types in parallel sentence pair.

1. Monotone (m): Word alignment point exists to the top left.
2. Swap (s): Word alignment point exists to the top right.
3. Discontinuous (d): Word alignment point neither exists to top left nor top right.

When the phrase pair is extracted from the word alignment matrix, its orientation pattern is also extracted along with it. To put it more formally, if {m, s, d} denotes a set

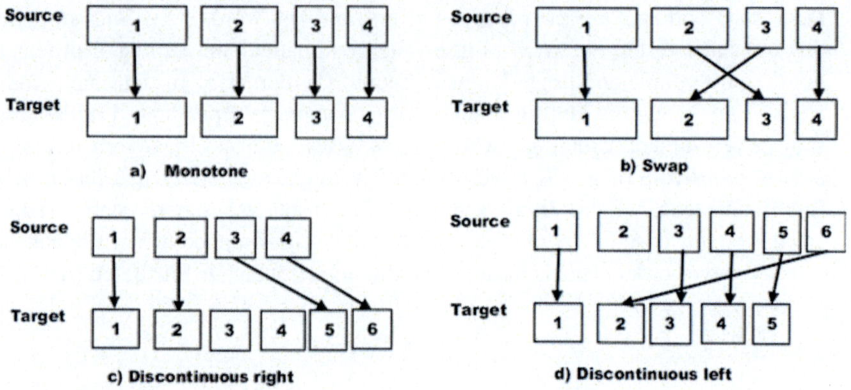

Fig. 4. Illustration of monotone, swap and discontinuous orientations

of orientations and (*e*, *h*) denotes the source and target phrase pair considered currently for translation, then reordering model to predict an orientation will be:

$$p_o(\text{orientation}/e, h) = \text{count}(\text{orientation}, h, e)/\Sigma_{\text{orientation}}\text{count}(\text{orientation}, h, e) \quad (3)$$

2 Related Work

Brown et al. [1] proposed lexicon-based translation models and introduced the notion of word alignment in SMT. Expectation maximization-based implementation of IBM models was published by Johns Hopkins University in 1999 with the name GIZA. Similar approach is described by Tom´as and Casacuberta [2] with the monotonic constraint on segmentation for finding phrases. This monotonic constraint is not appropriate for language pairs that are not close to grammatical structure. In 2006, JHU released open-source decoder for PBMT with the name Moses [3]. Vogel developed a PBMT system with the reordering constraint window of up to three words [4]. Tillmann first proposed the lexicalized reordering models [5]. He called source and target phrase pair as a block and proposed the reordering models based on these blocks. Each block is also associated with an orientation with respect to its previous block. Discriminative lexicalized reordering model was presented by Zens and Ney [6]. Kumar and Byrne implemented the block orientation using weighted finite-state transducers but the model failed to capture all possible phrasal movements [7]. A global reordering model for long-distance reordering was presented by Nagata et al. [8]. Al-Onaizan and Papineni proposed distortion models for reordering in SMT. Xiong et al. used the maximum entropy classifier for the reordering problem [9].

3 Experiments and Results

For experimentation, two English to *Hindi* parallel datasets are combined for training, validation and testing purposes. First dataset is obtained from IIT, Mumbai. Second dataset is HindEnCorp 0.5, LINDAT/CLARIN digital library (ÚFAL) which was also provided in WMT14 translation task. Table 1 gives details of the test and development dataset after all preprocessing. Sentence count is the number of sentences in each of the *Hindi* and English dataset. Unique word count is the vocabulary of each dataset. Number of tokens is total words in each dataset.

Table 1. Dataset used for experimentation

Dataset type	Sentence count		Unique words count		Number of tokens	
	English	Hindi	English	Hindi	English	Hindi
Training set	60,003	60,003	52,290	60,997	685,325	751,013
Development set	5205	5205	12,392	13,700	61,039	66,949
Test set	250	250	1191	1342	2555	2772

Evaluation Metrics used:
Quality of the translation refers to the accuracy and fluency of the MT system-generated output. It is measured by finding the similarity between the human-generated reference translations of the source language sentences with machine-generated output.

1. **Bilingual Evaluation Understudy (BLEU)**: The BLEU metric finds correlation by identifying exact word matches. After finding n-gram matches, n-gram precision is calculated. BLEU score is calculated as [10]:

$$\text{BLEU} = \text{brevity penalty} * \exp\sum_{i=1}^{n} \lambda_i \log \text{precision} \tag{4}$$

   Brevity penalty handles the dropping of words problem. The weights λ_i for various precisions are typically set to 1. Greater BLEU score of system generated translation represents a high correlation with reference translation and is considered as a better translation.
2. **Translation Edit Rate (TER)**: TER metric calculates the minimum number of changes that are required in the system generated translation to make it similar to the reference translation. The typical operations used for editing are delete, insert, substitute, shifts. TER score is calculated by using Eq. 5 as:

$$\text{TER} = \frac{\text{Number of edits(delete, insert, substitute, shifts)}}{\text{Average number of reference words}} \tag{5}$$

Greater TER score indicates that more edit operations are required to make the system translation correct and similar to reference translation. Hence, system translation with low TER score is considered as a better translation.

Experimental setup:
The language model component of the PBMT system is developed using freely available language modeling toolkit SRILM. Starting from 3-gram up to 5-gram, totally three n-gram language models are developed using available *Hindi* language corpus. All the language models are utilized for experimentation with various settings of reordering models to find the effect of various n-grams on translation quality. The parallel sentences are word-aligned using freely available word alignment toolkit GIZA ++. This tool applies an EM algorithm to find word alignment between parallel sentences. Two types of word alignment heuristics are tested with all reordering and language models, viz. grow-diag and grow-diag-final. For extracting phrases and developing PBMT system, freely available MOSES toolkit is used. The parameters of the PBMT system are tuned using minimum error rate training technique. For experimenting with reordering of words during translation, the following settings of lexicalized reordering models that are provided in MOSES are used.

1. **Modeltype**—specify the type of model.
   phrase: phrase-based model.
2. **Orientation**—specify which orientations to be used.
   mslr: monotone, swap, discontinuous-left, discontinuous-right.
   msd: monotone, swap, discontinuous.
   monotonicity: monotone or non-monotone.
   leftright: left or right.
3. **Directionality**—determines direction of orientation.
   backward: orientation with respect to preceding phrase.
   forward: orientation with respect to next phrase.
   bidirectional: use both forward and backward directions.
4. **Language**—specify the language to use for model construction.
   fe: both the source and target languages are used.
5. **Collapsing**—determines how to use the scores.
   allff: use the scores as individual feature functions.
   collapseff: collapse all scores in one direction into one feature function.

One distance based and 24 lexicalized reordering models are used to develop 25 separate PBMT systems. All reordering models are tested with three language models, and translation quality is assessed. The BLEU and TER scores for the PSMT systems developed using various reordering and language model configurations are shown in Table 2 for grow-diag word alignment and in Table 3 for grow-diag-final word alignment heuristic. Figure 5 compares BLEU scores, and Fig. 6 compares TER scores of systems. Aim of the experimentation is to find effective settings of the lexicalized reordering model for English to *Hindi* translation. For testing of the PBMT system performance, a completely different domain corpus is used hence poor BLEU scores are produced. Figure 5 shows that phrase-monotonicity–bidirectional-fe-allff setting of

Table 2. BLEU and TER score for PBMT system with grow-diag alignment

S. No.	Reordering model	3-gram language model		4-gram language model		5-gram language model	
		BLEU	TIE	BLEU	TER	BLEU	TER
1	Phrase-mslr–backward-fe-allff	18.95	71.09	19.85	70.74	21.08	69.87
2	Phrase-mslr–backward-fe-collapseff	18.95	71.09	19.85	70.74	21.08	69.87
3	Phrase-mslr–bidirectional-fe-allff	18.69	71.67	20.05	70.48	21.16	69.62
4	Phrase-mslr–bidirectional-fe-collapseff	18.22	71.96	19.49	71.17	20.99	70.05
5	Phrase-mslr–forward-fe-allff	19.10	70.92	20.45	70.45	21.25	69.73
6	Phrase-mslr–forward-fe-collapseff	19.10	70.92	20.45	70.45	21.25	69.73
7	Phrase-msd–backward-fe-allff	19.04	71.02	20.29	70.48	21.14	69.87
8	Phrase-msd–backward-fe-collapseff	19.04	71.02	20.29	70.48	21.14	69.87
9	Phrase-msd–bidirectional-fe-allff	18.60	71.49	20.04	70.59	21.39	69.69
10	Phrase-msd–bidirectional-fe-collapseff	18.52	71.92	19.60	70.92	21.01	69.87
11	Phrase-msd–forward-fe-allff	19.19	71.13	20.55	70.45	21.39	69.76
12	Phrase-msd–forward-fe-collapseff	19.19	71.13	20.55	70.45	21.39	69.76
13	Phrase-monotonicity–backward-fe-allff	19.15	70.95	20.50	70.30	21.34	69.69
14	Phrase-monotonicity–backward-fe-collapseff	19.15	70.95	20.50	70.30	21.34	69.69
15	Pbrase-monotonicity–bidirectioua1-fe-a11ff	19.00	71.09	20.78	70.30	22.32	69.19
16	Phrase-monotonicity–bidirectional-fe-collapseff	18.70	71.45	20.68	70.02	21.15	69.76
17	Phrase-monotonicity–forward-fe-allff	19.37	71.13	21.10	69.69	21.99	69.12
18	Phrase-monotonicity–forward-fe-collapseff	19.37	71.13	21 10	69.69	21.99	69 12
19	Phrase-1eftright–backward-fe-allff	18.95	70.81	20.68	70.27	21.10	69.84
20	Phrase-1eftright–backward-fe-collapseff	18.95	70.81	20.68	70.27	21.10	69.84
21	Phrase-1eftright–bidirectional-fe-allff	18.86	71.27	20.45	70.63	20.97	70.34
22	Phrase-1eftright–bidirectional-fe-collapseff	19.10	71.17	20.73	69.87	21.03	69.66
23	Phrase-1eftright–forward-fe-allff	18.86	70.88	20.62	70.30	21.33	69.94
24	Phrase-1eftright–forward-fe-collapseff	18.86	70.88	20.62	70.30	21.33	69.94
25	Distance	18.90	70.63	21.08	69.98	21.43	69.62

lexicalized reordering model with 5-gram language model and grow-diag word alignment produces the highest BLEU score of 22.32. Out of all developed systems, the PBMT system with these settings produces more accurate translation output. 5-gram language model provides more contextual information due to which more accurate translations are produced.

Figure 6 shows that phrase-monotonicity–forward-fe-allff setting of the lexicalized reordering model with 5-gram language model and grow-diag word alignment produces the lowest TER score of 69.12. From the editing point of view, the PBMT system

Table 3. BLEU and TER score for PBMT system with grow-diag-final alignment

S. No.	Reordering model	3-gram language model		4-gram language model		3-gram language model	
		BLEU	TER	BLEU	TER	BLEU	TER
1	Phrase-mslr–backward-fe-allff	17 13	74.15	18.26	73.18	19.08	72.50
2	Phrase-mslr–backward-fe-collapseff	17.13	74.15	18.26	73.18	19.08	72.50
3	Phrase-mslr–bidirectional-fe-allff	16.68	74.04	17.74	74.51	18.79	73.50
4	Phrase-mslr–bidirectional-fe-collapseff	16.61	74.26	17.47	74.37	17.95	73.97
5	Phrase-mslr–forward-fe-allff	17.26	73.29	18.75	72.82	19.75	71.74
6	Phrase-mslr–forward-fe-collapseff	17.26	73.29	18.75	72.82	19.75	71.74
7	Phrase-msd–backward-fe-allff	17.39	73.68	18.30	73.14	19.25	72.17
8	Phrase-msd–backward-fe-collapseff	17.39	73.68	18.30	73.14	19.25	72.17
9	Phrase-msd–bidirectional-fe-allff	17.29	73.32	18.32	73.29	19.27	72.57
10	Phrase-msd–bidirectional-fe-collapseff	17.25	73.83	17.68	73.75	18.54	73.29
11	Phrase-msd–forward-fe-allff	17.83	72.86	18.96	72.71	20.06	71.67
12	Phrase-msd–forward-fe-collapseff	17.83	72.86	18.96	72.71	20.06	71.67
13	Phrase-monotonicity–backward-fe-allff	17.39	73.40	18.53	72.64	19.71	71.99
14	Phrase-monotonicity–backward-fe-collapseff	17.39	73.40	18.53	72.64	19.71	71.99
15	Phrase-monotonicity–bidirectioua1-fe-a1 1ff	17.76	72.78	IS.79	72.68	20.08	71.60
16	Phrase-monotonicity–bidirectional-fe-collapseff	16.75	74.08	17.96	73.61	18.94	72.89
17	Phrase-monotonicity–forward-fe-allff	18.05	72.46	19.18	72.32	20.42	71.20
18	Phrase-monotonicity–forward-fe-collapseff	18.05	72.46	19.18	72.32	20.42	71.20
19	Phrase-1eftright–backward-fe-alIff	17.33	73.75	18.62	72.96	19.26	72.35
20	Phrase-1eftright–backward-fe-collapseff	17.33	73.75	18.62	72.96	19.26	72.35
21	Phrase-1eftright–bidirectional-fe-allff	16.77	74.58	17.77	73.75	18.67	73.04
22	Phrase-1eftright–bidirectional-fe-collapseff	17.39	73.72	18.29	73.11	19.12	72.46
23	Phrase-1eftright–forward-fe-allff	17.02	73.90	18.70	72.75	19.04	72.25
24	Phrase-1eftright–forward-fe-collapseff	17.02	73.90	18.70	72.75	19.04	72.25
25	Distance	17.75	73.11	18.69	72.78	19.58	71.71

with these settings produces translation output that is more similar to reference translation. But, this is approximately equal to 69.19 TER score of PBMT system with phrase-monotonicity–bidirectional-fe-allff reordering setting discussed above. Hence, to summarize it can be said that collectively phrase-monotonicity–bidirectional-fe-allff setting of lexicalized reordering model with 5-gram language model and grow-diag

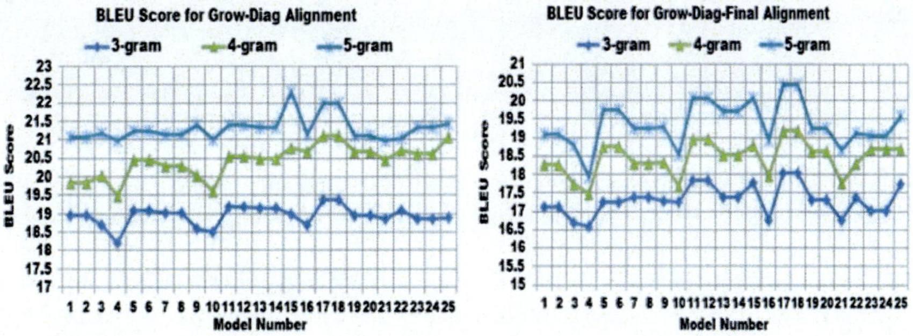

Fig. 5. BLEU scores for PBMT systems

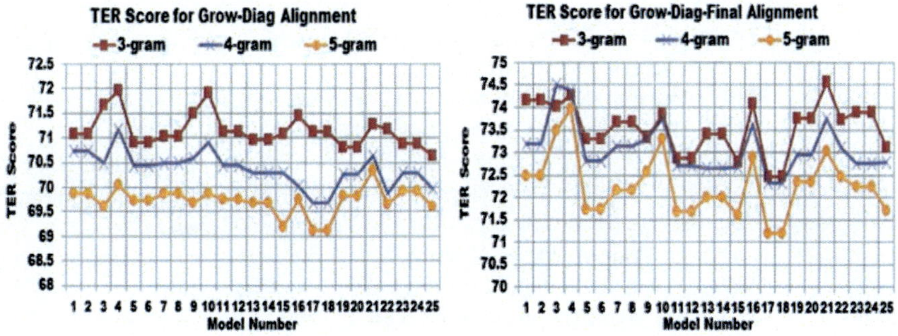

Fig. 6. TER scores for PBMT systems

word alignment symmetrization heuristic is a better configuration for getting a high-quality translation from English to *Hindi* language. When the test data is translated using a neural network-based Google translator it produced the output with 23.41 BLUE score and 67.21 TER score.

4 Conclusion and Future Scope

Translation quality depends on the performance of individual working components involved in phrase-based system. While translating the English language text into structurally and grammatically very different language like *Hindi*, reordering performs very crucial role for getting a quality translation. For BLEU score improvement, phrase-monotonicity–bidirectional-fe-allff setting is good, and for TER score improvement, phrase-monotonicity–forward-fe-allff setting is good. For both metrics, grow-diag word alignment is giving better results. Future scope of this work is to check the effect of other alignment heuristics on translation quality of English to *Hindi* language.

References

1. Brown, P.F., Della Pietra, S.A., Della Pietra, V.J., Mercer, R.L.: The mathematics of statistical machine translation: Parameter estimation. Comput. Linguist. **19**(2), 263–311 (1993)
2. Tom´as, J., Casacuberta, F.: Monotone statistical translation using word groups. In: Proceedings of Machine Translation Summit VIII, pp. 357–361, Santiago deCompostela, (2001)
3. Koehn, P., Hoang, H., Birch, A., Callison-Burch, C., Federico, M., Bertoldi, N., Cowan, B., Shen, W., Moran, C., Zens, R., Dyer, C., Bojar, O., Constantine, A., Herbst, E.: Moses: Open source toolkit for statistical machine translation. In Proceedings of 45th Annual Meeting of the Association for Computational Linguistics (ACL): Poster Session, pp. 177–180, Prague, Czech Republic, (2007)
4. Vogel, S.: SMT decoder dissected: Word reordering. In Proceedings of International Confences on Natural Language. Processing and Knowledge Engineering, pp. 561–566, Beijing, China, (2003)
5. Tillmann, C.: A unigram orientation model for statistical machine translation. In Proceedings of HLT-ACL, Short Papers, pp. 101–104, (2004)
6. Zens, R., Ney, H.: Discriminative reordering models for statistical machine translation. In: Proceedings of HLT-NAACL Workshop on SMT, New York (2006)
7. Kumar, S., Byrne, W.: Local phrase reordering models for statistical machine translation. In: Proceedings of Human Language Technology Conferences on Empirical Methods in Natural Language Processing (HLT/EMNLP), pp. 161–168, Vancouver, Canada, (2005)
8. Nagata, M., Saito, K., Yamamoto, K., Ohashi, K.: A clustered global phrase reordering model for statistical machine translation. In: Proceedings of ACL, Sydney, Australia, (2006)
9. Xiong, D., Liu, Q., Lin, S.: Maximum entropy based phrase reordering model for statistical machine translation. In Proceedings of ACL, Sydney, Australia, (2006)
10. Papineni, K., Roukos, S., Ward, T., Zhu, W.: Bleu: A method for automatic evaluation of machine translation. In: Proceedings of 40th Annual Meeting Association for Computational Linguistics, Philadelphia, PA, pp. 311–318, (2002)

Investigation of Imbalanced Big Data Set Classification: Clustering Minority Samples Over Sampling Technique

Sachin Patil[1,2(✉)] and Shefali Sonavane[2]

[1] Rajarambapu Institute of Technology, Rajaramnagar, Urun Islampur 415409, MH, India
sachin.patil@ritindia.edu
[2] Walchand College of Engineering, Sangli 416415, MH, India
shefali.sonavane@walchandsangli.ac.in

Abstract. Most of the real-world data sets exhibit a skewed scenario of data distribution in contrast to the well-established data sets. The total number of instances of a particular class extremely surpasses the count of other classes. This uneven dispersal of classes leads to a state of imbalance data sets posing an extreme difficulty for learning procedures. Additionally, due to its intrinsic complex data features, analyzing such imbalanced data sets has setup an avenue for focused researchers. Imbalanced class distribution is effectively handled with over sampling of minority class data which is usually independent of the classifiers. A over sampling technique: Clustering minority samples over sampling technique (CMSOT) is proposed to enhance the classification of imbalanced data sets. The projected technique is implemented on Apache Hadoop under mapreduce environment. The data sets are mainly encompassed from the UCI repository. The effect of True Positive rates justifying the imbalance ratio including the examination of improved classification from the generated pool is studied. The achieved experimental results along with its corresponding statistical analysis of over sampled data sets clearly mark the supremacy of the planned technique to the selected benchmarking techniques.

Keywords: Imbalanced big data sets · Relative difference · Imbalance ratio · Over sampling · Safe-level

1 Significance of Handling Imbalanced Big Data Sets

Handling of imbalanced Big Data sets is principally difficult in the skewed data set which asserts a low learning performance of the understated class. Uncommon patterns, abnormal conducts like fraud transactions [1], medical diagnosis [2, 3], network attacks [4], software faults, and natural tragedies carry the sources of rare events. They are of high demand for timely tracking but are merely difficult to detect. They occur very seldom and are produced as a trivial part (minority class) of a Big Data set, setting up a

© Springer Nature Singapore Pte Ltd. 2021
P. Deshpande et al. (eds.), *Next Generation Information Processing System*, Advances in Intelligent Systems and Computing 1162,
https://doi.org/10.1007/978-981-15-4851-2_32

focal point for advanced mining. Learning from this untraced knowledge gains a useful insight for concentric learning beyond a hidden spectrum. This terminology of constituting intelligence from imbalanced Big Data sets is termed as learning from it. Classifying correctly the insignificant examples of such marginal classes has attempted as the focal point of study [5].

The work in this paper comprises an enhanced technique viz. Clustering minority samples over sampling technique (CMSOT) is planned which competently deals with both classes (binary/multi-class) data sets issue. Clustering of the minority classes is conceded as an underlying over sampling process. The planned technique helps to discourse different data characteristics as like data disjuncts, cluster cohesiveness, and borderline impacts. Likewise, it also handles the density dearth and borderline cases along with maintaining the proportion of imbalance by inspecting better instances. Two classifiers (random forest/multilayer perceptron) are selected for classification task. The experimental recitation is conducted on the mapreduce architecture [6, 7]. The precision of classification outcomes is assessed using area under the curve (AUC) measure [8].

The flow of the paper work is further described in brief as follows. Section 2 provides an overview of literature of concerns related to imbalanced data sets. The comprehensive methodology encompassing an experimental design with a detailed outline of CMSOT in stated in Sect. 3. Section 4 states the environment of experimental bed. The last section deals with the conclusion along with the foreseen insight for extending the work.

2 Literature Review

An unequal distribution of class samples within huge data sets has coined a terminology of imbalanced data sets [9–11]. Identifying unusual and rare activities within huge sets is very hard and unfeasible [12, 13]. Furthermore, to detect and address them in a particular time frame is often a priority of concern. The handling of imbalanced data classification is treated with numerous techniques mainly functioning at three levels (data/algorithmic/cost-subtle level) [5, 6].

The technique discussed (CMSOT) deals fundamentally at the data level. Data level techniques are further classified into three types viz. under sampling, over sampling, and hybrid techniques based on the pre-processing method [5, 6]. Each category of approach specifies their respective advantages and disadvantages. The survey of performance outcomes of the advanced over sampling techniques across last two decades has shown a great upliftment compared to under sampling and hybrid approaches.

Synthetic Minority Oversampling Technique (SMOTE) [14] is considered to be one of the early stage over sampling techniques to efficiently handle the issue of imbalanced data set. It devices to synthesize the minority samples while achieving the balanced data set. A number of SMOTE-based flavors were instigated and evaluated, namely Borderline SMOTE/Safe-level SMOTE/ADASYN/ROSE [15–18]. Borderline SMOTE helps to oversample only the borderline minority examples. Safe-level SMOTE sensibly chooses minority instances nearby samples having higher safe-level improving classification accuracy. ADASYN adaptively evolve learning near the hard examples by analyzing the data distribution. ROSE executes better with respect to kernel density assessment implicitly conserving the features of each class [18]. It provides an inflated decision region along with the higher flexibility to over sample. A boundary-based over sampling approach to discourse the issues in imbalance data set is articulated in [19]. Some ensemble methods [20–22] adjunct to SMOTE assist in proficient handling of imbalanced data sets disputes. RHSBoost is mostly an active ensemble classification technique which uses unplanned under sampling and ROSE as a boosting scheme for forming classification rules. In [23], MWMOTE utilizes the weighted instructive minority instances for performing over sampling underlying a clustering approach. It exclusively performs to direct all the over-sampled instances implicitly under minority class clusters.

3 Methodology: CMSOT

The design of anticipated technique deliberates with samples from both types (minority/majority) and categorizes beneath a clustered-based approach. It reflects simultaneously on both class imbalances (between-class and within class) which exist in almost of the data sets. Technique (CMSOT) put forward to cluster classes in advance to perform over sampling. The basic idea behind the suggested technique is first to cluster all the minority classes. Thereafter, it normally perceives the non-duplicate means of clusters as the synthetic set samples. This set is further considered for the subsequent iteration of over sampling until it attains the required I.R. The clustering may be performed using any type of clustering algorithms. Basic explanation of technique (CMSOT) is given as:

I_s—Initial data set consisting of ‘S’ instances

D_{mjr}—Set of majority class instances

D_{mnr}—Set of minority class instances

Input: Over sampled instance set (($I_o \leftarrow \emptyset$))
Output: I_o

1. CMSOT (I_s) *//helps to address within-class imbalances including lack of density*
2. **if** I_s != binary-class data set
3. $D_{mjr} \leftarrow$ Highest $D_{mjr.mi}$ *//using binarization method*
4. $D_{mnr} \leftarrow$ Lowest $D_{mnr.nj}$
5. **end if**
6. **for** each $D_{mnr.nj}$
7. cluster (D_{mnr}[v])
8. O.S.$_{fi} \leftarrow$ Calculated medians of D_{mnr}[v]
9. **if** O.S.$_{fi}$ = Duplicate
10. O.S.$_{fi} \leftarrow$ delete (O.S.$_{fi}$)*//delete the duplicate medians*
11. D_{mnr}[v] $\leftarrow I_{io} \leftarrow$ O.S.$_{fi}$*//append the remaining O.S.$_{fi}$ to D_{mnr} and I_{io}*
12. recluster (D_{mnr}[v]) *//till achieving required I.R. per cluster*
13. goto step 7
14. **end if**
15. **end for**
16. **return** I_{io}
17. $I_o \leftarrow$ sorting.SL (I_{io}) *// Based on Safe-level (SL). Select high SL instances while retaining I.R. = 1.5*
18. Repeat: residual ($D_{mnr.nj}$) *// as per step 6.*

The scheduled technique (CMSOT) reinforces with better over samples to improve classification accuracy by delivering comprehensive over-sampled minority instances. It additionally avoids replication and uplifts issues related to centroids grounded over sampling process (DBSMOTE [24]). The technique uniquely promotes to select the better learning over sampled instances from the projected excess I.R. Fig. 1 illustrates the executional flow of proposed over sampling technique (CMSOT).

4 Experimental Environment

The objective of the work is to authenticate the efficacy of planned technique for handling the imbalanced issue. The examination of performance is presented in comparison to different benchmarking techniques.

The results are analyzed for assessing the impact of proposed technique (CMSOT) over some selected benchmarking techniques (Borderline SMOTE/ADASYN/MWMOTE). They are quantified in Tables 2, 3, 4, 5, 6 and 7 which indicate improved outcomes of CMSOT over AUC values.

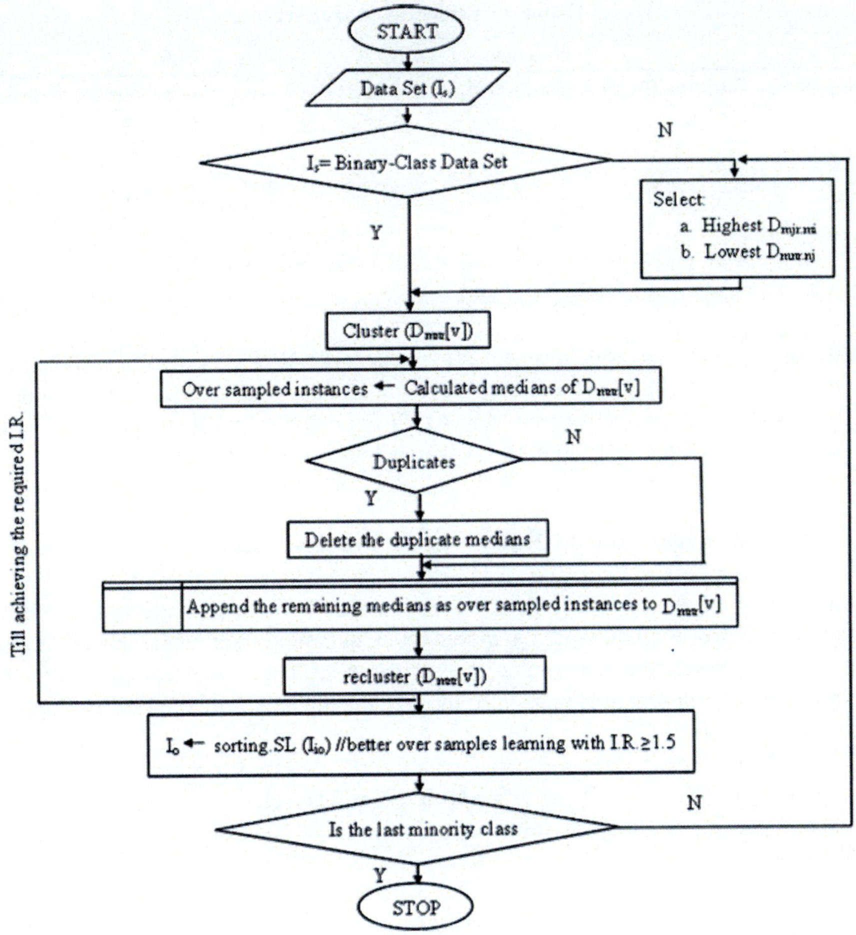

Fig. 1. Executional flow of CMSOT

4.1 Data Set Details

Five data sets from UCI repository [25] are preferred for experimental analysis. The data sets consist of two categories (Binary-class (B_{ss})/Multi-class (M_{ss})). They describe comprehensive samples along with prevailing attributes and wide-ranging I.R. The data set details stated in Table 1 are as follows:

4.2 Experimental Analysis

The analysis of the projected technique is carried out using a Hadoop (v. 2.7.4.) framework underlying a mapreduce cluster on Ubuntu (v. 14.04) with 12 nodes. Each node comprises a hardware set up of i7-x@3.4x GHz and a software set up of

Table 1. Particulars of data set

Category	Data set	$\#IN_c$	$\#ATR_c$	I.R.	$\#NC_c$
Binary-class data sets (B_{ss})	Credit card	284,808	31	577.87	2
	MiniBoone	130,064	51	2.56	
	SIDO	12,678	4932	27.04	
Multi-class data sets (M_{ss})	PAMAP2	385,0505	54	14.35	19
	KEGG-D	53,413	23	13,156.5	13

Note $\#IN_c$—Sample count, $\#ATR_c$—Attribute count, $\#NC_c$—Count of classes

NetBeans IDE (v. 8.2), and Java (v. 1.8) assisted with Weka 3.6 packages. The projected technique (CMSOT) is responsive to handle issues related to the both categories of data sets. The assessment parameter (AUC) are experimentally attained using two classifiers ('K' = 5 and tenfold cross-validation).

5 Justifying Imbalance Ratio: True Positive Rates

The imbalance of classes within data sets is stated as the ratio of majority-to-minority samples. It helps to predict the required over sampling rate while lessening the imbalance form. Justifying the value of over sampling rate which complies with the essential I.R. is experimentally analyzed through breakpoints for True Positive rates of majority-to-minority classes.

Table 2 provides the detail of breakpoints for True Positive rates on MiniBoone data set using CMSOT over random forest classifier. The values for average ratio of the number of cutoff point majority class instances to respective minority class instances are close to 1.5.

Table 2. Breakpoints of true positive rates for MiniBoone data set

Minority class: Instance count (MC_I)	Majority class instances cutoff points (MC_{Icp})	Ratio of MCIcp to MC_I
5000	7432	1.49
15,000	22,639	1.51
25,000	37,112	1.48
35,000	52,890	1.51
45,000	67,298	1.50
52,000	77,799	1.50
Average		1.51

The detail stage-wise study of breakpoints while fixing the number of minority class instances to 35,000 are depicted in table. The cutoff of True Positive rates from majority-to-minority classes is represented at 52,890 (between the range of 50,000–

55,000) majority instances over minority class across a range of full set of majority class instances (5000–75,000 instances). The values in Table 3 aid in visualizing the required I.R. value close to 1.5 justifying a balanced scenario.

Table 3. Breakpoints of true positive rates for MiniBoone data set (minority class-35,000 samples)

TP rate	MiniBoone data set (minority-35,000 samples)							
	Majority samples							
	35,000	40,000	45,000	50,000	55,000	60,000	65,000	70,000
TP rate (majority)	0.84	0.85	0.85	0.86	0.87	0.88	0.9	0.91
TP rate (minority)	0.87	0.87	0.86	0.86	0.85	0.83	0.77	0.69

Figure 2 indicates the breakpoint of True Positive rates, i.e., majority class to minority class instances, while keeping the count of minority instance constant to 35,000. X-axis indicates the majority samples count while Y-axis denotes the TP rate. The graph in Fig. 2. represents the changeover between 50,000 and 55,000 instance counts indicating the ratio close to 1.5.

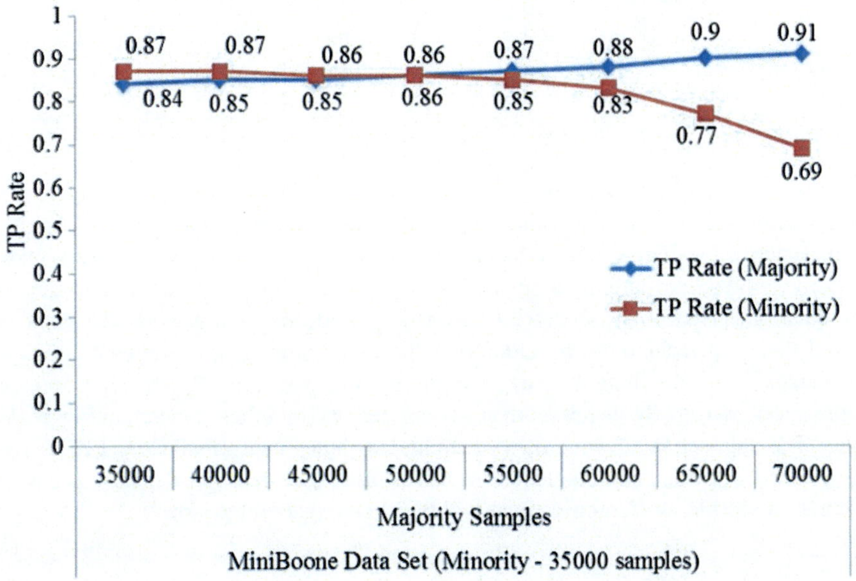

Fig. 2. Breakpoints of true positive rates for MiniBoone data set (Minority class-35,000 samples)

6 CMSOT Versus Benchmarking Techniques: Based on Area Under the Curve (AUC)

A group of five data sets exemplifying varied number of I.R., samples, and feature sets are selected for experimental evaluation. Proposed technique is equipped with the clustering method for over sampling purpose. The results based on AUC values are represented in Table 4 on the two classifiers (random forest and multilayer perceptron).

Table 4. AUC values (cross validation = 10 and 'K' = 5)

Classifier	Data set	Over sampling techniques			
		Borderline SMOTE	ADASYN	MWMOTE	CMSOT
Random forest	Credit card	0.81	0.83	0.86	0.89
	MiniBoone	0.94	0.95	0.97	0.98
	SIDO	0.88	0.91	0.92	0.94
	PAMAP2	0.73	0.74	0.77	0.80
	KEGG-D	0.89	0.90	0.94	0.96
Multilayer perceptron	Credit card	0.79	0.81	0.84	0.87
	MiniBoone	0.87	0.89	0.90	0.92
	SIDO	0.92	0.93	0.95	0.97
	PAMAP2	0.71	0.73	0.75	0.78
	KEGG-D	0.88	0.89	0.93	0.95
Average		0.84	0.85	0.88	0.90

The results in Table 5 specify the significance of proposed technique compared to benchmarking techniques. The values in Table 5 intuit a 6% average rise in classification performance. The CMSOT technique effectually discourses the between-class imbalance issue while attaining a greater impact compared to all others. Its tendency to accomplish the needful I.R. assists in performing well, that is a peripheral lacuna in some of the other techniques. In addition to the I.R., the technique addresses various data characteristics enriching the classification performance. The MWMOTE technique demonstrates marginally better corresponding results to other benchmarking techniques. The data set MiniBoone offers extreme attributes with satisfactory I.R. which aids in added flexibility for augmented AUC results. Other data sets depart with fewer attributes to sample ratio intimidating rigidity for over sampling process.

Table 5. Investigation of % rise in performance (AUC Values)

Avg. AUC value (Benchmarking techniques)	Avg. AUC value (CMSOT)	$\%RD_{GM}$
0.85	0.90	6

$\%RD_{GM}$: % relative difference in GM values

Figure 3 illustrates the comparison graph of AUC values (Table 4) between proposed and benchmarking techniques. The over sampling techniques are signified over X-axis while Y-axis states the average AUC values. The performance of Borderline SMOTE is hampered in view of concentrating boundary samples while over sampling. The MWMOTE technique precincts the performance edges of CMSOT. Overall, the proposed technique (CMSOT) exhibits a better performance in terms of AUC values.

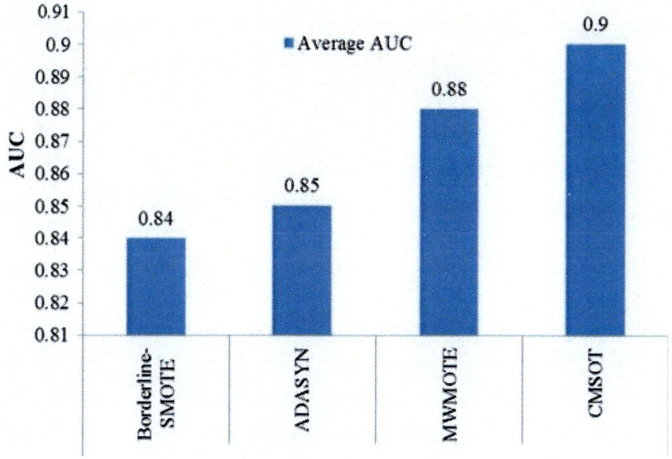

Fig. 3. Average AUC values

7 Consequences of the Varied Number of Mappers

The effect of the different number of mappers is planned and estimated in Table 5. The prescribed subset of data sets (PAMAP2 and KEGG-D) from the M_{ss} category is chosen for experimental assessment on the CMSOT. The outcomes are measured transversely by AUC values with the multilayer perceptron classification model (Cross-validation count = 10 and ‘K’ = 5).

The AUC values in Table 6 help to position the impression of increasing mappers count. The increase in count of mappers from 8 to 64 evidences a negligible drop (6%) in classification performance as stated in Table 7. The identical behavior is seen for almost all data sets.

Table 6. AUC values (varied number of mappers)

Data set	Number of mappers			
	8	16	32	64
PAMAP2	0.79	0.78	0.76	0.74
KEGG-D	0.96	0.95	0.92	0.91

Table 7. Investigation of % drop in performance

Data set	Number of mappers		$\%RD_{AUC}$
	8	64	
PAMAP2	0.79	0.74	6.53
KEGG-D	0.96	0.91	5.34
Average			5.93

$\%RD_{AUC}$: % Relative difference in AUC values

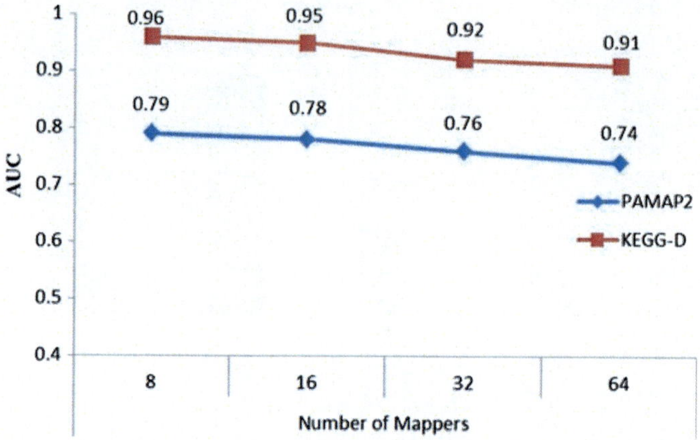

Fig. 4. Varied number of mappers (CMSOT)

Line chart in Fig. 4 depicts the analysis of the prediction model based on the availability of sample count per mapper (Table 6). The lesser the existence of the minority over samples per mapper tends to decline in classification outcomes and AUC values.

8 Conclusion

The work in the paper (CMSOT) aids to visualize the enhanced handling of binary/multi-class imbalanced data sets. The outcomes in Tables 4 and 5 indicate an average rise of 6% in classification performance by CMSOT. The experimentation is recognized considering data sets from UCI source. The selected data sets have provided a wide-spread angel for experimentation representing huge sample size with I.R. and varied class structures. Two classifiers are preferred for modeling prediction based on several dimensions such as boosting, ensemble, and probabilistic impression. The investigation (Table 2) aids to suffice the I.R. across the breakpoints of TP rates. The random forest classifier achieves greater results (Table 4) with a difference of 2% in parallel to other residual. The increase in count of mappers affects negligible performance drop in classification. The intended assessment terminology viz. AUC values for the obtained results help to perceive the superiority benchmarks. The underlying bed of experimental work is upheld on scalable and distributed mapreduce architecture enduring the basic need of Big Data sets.

The insights on investigational results for handling imbalanced Big Data sets help to authorize the efficiency of the projected technique (CMSOT). The attained outcomes emanate to further scrutinize a profound incremental learning scheme addressing the dynamic disorders of data size leading to an imbalance in real-world applications.

References

1. Wei, W., Li, J., Cao, L., Ou, Y., Chen, J.: Effective detection of sophisticated online banking fraud on extremely imbalanced data. World Wide Web. **4**, 449–475 (2013)
2. Tomczak, J., ZięBa, M.: Probabilistic combination of classification rules and its application to medical diagnosis. Mach. Learn. **1–3**, 105–135 (2015)
3. Chen, Y.: An empirical study of a hybrid imbalanced-class DT-RST classification procedure to elucidate therapeutic effects in uremia patients. Med. Biol. Eng. Compu. **6**, 983–1001 (2016)
4. Elhag, S., Fernández, A., Bawakid, A., Alshomrani, S., Herrera, F.: On the combination of genetic fuzzy systems and pairwise learning for improving detection rates on intrusion detection systems. Expert Syst. Appl. **1**, 193–202 (2015)
5. López, V., Fernández, A., García, S., Palade, V., Herrera, F.: An insight into classification with imbalanced data: empirical results and current trends on using data intrinsic characteristics. Inf. Sci. **250**, 113–141 (2013)
6. Del Río, S., López, V., Benítez, J., Herrera, F.: On the use of MapReduce for imbalanced big data using random forest. Inf. Sci. **285**, 112–137 (2014)
7. Jiang, H., Chen, Y., Qiao, Z., Weng, T., Li, K.: Scaling up MapReduce-based big data processing on multi-GPU systems. Cluster Comput. **1**, 369–383 (2015)
8. Huang, J., Ling, C.: Using AUC and accuracy in evaluating learning algorithms. IEEE Trans. Knowl. Data Eng. **3**, 299–310 (2005)
9. Japkowicz, N., Stephen, S.: The class imbalance problem: A systematic study. Intell. Data Anal. **5**, 429–449 (2002)
10. He, H., Garcia, E.: Learning from imbalanced data. IEEE Trans. Knowl. Data Eng. **9**, 1263–1284 (2008)
11. Sun, Y., Wong, A., Kamel, M.: Classification of imbalanced data: A review. Int. J. Pattern Recognit Artif Intell. **04**, 687–719 (2009)
12. Maalouf, M., Trafalis, T.: Robust weighted kernel logistic regression in imbalanced and rare events data. Comput. Stat. Data Anal. **55**, 168–183 (2011)
13. Japkowicz, N., Myers, C., Gluck, M.: A novelty detection approach to classification. InIJCAI **1**, 518–523 (1995)
14. Chawla, N., Bowyer, K., Hall, L., Kegelmeyer, W.: SMOTE: synthetic minority over-sampling technique. J. Artif. Intell. Res. **16**, 321–357 (2002)
15. Han, H., Wang, W., Mao, B.: Borderline-SMOTE: a new over-sampling method in imbalanced data sets learning. In: International Conference on Intelligent Computing, Springer, Berlin, pp. 878–887 (2005)
16. Bunkhumpornpat, C., Sinapiromsaran, K., Lursinsap, C.: Safe-level-smote: Safe-level-synthetic minority over-sampling technique for handling the class imbalanced problem. In: Pacific-Asia Conference on Knowledge Discovery and Data Mining, Springer, Berlin. pp. 475–482 (2009)

17. He, H., Bai, Y., Garcia, E., Li, S.: ADASYN: Adaptive synthetic sampling approach for imbalanced learning. In: IEEE International Joint Conference on Neural Networks, pp. 322–1328 (2008)
18. Menardi, G., Torelli, N.: Training and assessing classification rules with imbalanced data. Data Min. Knowl. Disc. **1**, 92–122 (2014)
19. Hu, F., Li, H.: A novel boundary oversampling algorithm based on neighborhood rough set model: NRSBoundary-SMOTE. Math. Problems Eng. (20130
20. Chawla, N., Lazarevic, A., Hall, L., Bowyer, K.: SMOTEBoost: Improving prediction of the minority class in boosting. In: European Conference on Principles of Data Mining and Knowledge Discovery, Springer, Berlin. pp. 107–119 (2003)
21. Xiang, H., Yang, Y., Zhao, S.: Local clustering ensemble learning method based on improved AdaBoost for rare class analysis. J. Comput. Inf. Syst. **4**, 1783–1790 (2012)
22. Gong, J., Kim, H.: RHSBoost: Improving classification performance in imbalance data. Comput. Stat. Data Anal. **111**, 1–3 (2017)
23. Barua, S., Islam, M., Yao, X., Murase, K.: MWMOTE—majority weighted minority oversampling technique for imbalanced data set learning. IEEE Trans. Knowl. Data Eng. **2**, 405–425 (2012)
24. Bunkhumpornpat, C., Sinapiromsaran, K., Lursinsap, C.: DBSMOTE: density-based synthetic minority over-sampling technique. Appl. Intell. **3**, 664–684 (2012)
25. UCI machine learning repository. https://archive.ics.uci.edu/ml/datasets.html Accessed 13 Nov 2019

Blended Learning and Analysis of Factors Affecting the Use of ICT in Education

Sushama Deshpande and Amit Shesh(✉)

Department of English, Dr. Babasaheb, Ambedkar Technological University, Lonere 402103, India
sushamabindu@gmail.com, apshesh@dbatu.ac.in

Abstract. The increasing availability of Information and Communication Technologies (ICT) presents teachers with exciting opportunities to transform pedagogical practices. In many parts of southern Asia, specially, in India, students are from rural background where the teaching aids limited. This will affect the quality of English language understanding in higher learning. The demand for teachers to integrate ICT into their teaching and learning programs is high and places additional pressures on teachers in an already challenging profession. The pervasive nature of ICT in daily life has made society reliant on systems and tools that provide challenges for the way people think and work. This paper presents the detailed analysis of factors affecting the use of ICT in education. The results provided in this study will be helpful for the teachers, academicians and policy makers to frame a policy to improve the understanding of English language and its effective delivery.

Keywords: Computers · Education · Engineering · ICT · Pedagogy · Teacher

1 Introduction

India is still a developing country. Though India has achieved tremendously in technological, financial and few other sectors, a lot of work is still required in education sector. India once used its traditional Gurukul method of teaching which was based on knowledge and respect for the teacher. With the changing times, India also accepted the new methodology of teaching. Today, education sector is one of the biggest sectors with around 789 universities in the 29 states of India with 20 million students graduating every year. With the advent of ICT around 25 years ago, there came a remarkable change in the education sector. Today, use of ICT is very common right from schools to higher education. ICT uses the use of hardware and software for effective management of information like storage, retrieval, processing, communicating and sorting information to plan and execute a particular task. This wonderful tool is a diverse mixture of technology tools and resources to create, differentiate, store and manage information or communication. It is very important for the students of twenty-first century to be able to use ICT effectively for a better career. As teachers are at the centre of any change in the educational sector, it is a prime requisite for teachers that they get awareness about importance of ICT and should be able to blend this technology in the classroom.

© Springer Nature Singapore Pte Ltd. 2021
P. Deshpande et al. (eds.), *Next Generation Information Processing System*,
Advances in Intelligent Systems and Computing 1162,
https://doi.org/10.1007/978-981-15-4851-2_33

2 Blended Learning

2.1 What is Blended Learning?

Students are used to the traditional brick and mortar classroom which is led and controlled by teachers. But due to the technological inventions, the use of technology is increased at the higher speed. It has become the integral part of almost all the sectors including educational sector. Everybody has easy access to technology and all the students use it and teachers are facing a major problem of seeking attention of students for a long time. Teachers have been using technology extensively in order to create interest among the students.

Due to the innovative teachers and learners, the concept of blending various methodologies is not new. Teachers have been using innovative and creative ideas in their lesson planning and have tried to blend lecture and activities during their teaching. Blended learning would be effective when a traditional classroom teaching is blended with technology to the fullest extent.

Bonk and Graham [1] Blended learning is the teaching and learning experience that includes face-to-face as well as mobile or online learning. It is a blend or mix of various methods of teaching and learning one of which is, necessarily, technology based. Though technology is used, human intervention cannot be avoided. The widespread adoption and availability of digital learning technologies have enhanced the standard of teaching and learning; it has also supported face-to-face learning experience.

2.2 Why Blended Learning?

The use of technology has also brought in enhancement in the teaching–learning process in the traditional classrooms as it has taken place in the other sectors. Technology-based environment provides a better atmosphere for the learners and better scope for the teacher to showcase his/her creativity in the classroom sessions with the help of technology. There are some other reasons for using blended learning.

1. Increased Access: Nowadays, the access to technology has become very common. Almost everyone has a smart phone and availability of Internet access on a mobile which has brought people from all the countries together. They can share their thoughts, opinions, knowledge and talent with the people across the world and explore the knowledge or information from all walks of life. It has become very common in the traditional classrooms in the form of e-learning software and other technology-based learning modes.
2. Learner Flexibility and Convenience: In this era of information technology, many learners are taking some courses while working somewhere. Some are learning to enhance their professional knowledge. Such people find flexibility while learning the new things. Such people do not want to sacrifice their work hours and still want to have face-to-face interaction mixed with technological involvement while learning. So for, such distance learning courses technology can be very useful giving equal opportunity to everyone for education.

3. Cost Effective and Time Saving: Cost effectiveness and time saving abilities are the main measures of any program. Today e-learning software can cater to the needs of all the learners of a specific group all over the world at the same time. This makes it very cost effective and time saving. In less cost, more learners can be approached.
4. Interesting: Use of various methodologies in the same lecture makes it interesting. Students get bored with long and monotonous lecture-based classrooms. When use of audio visual aids with some activities is made, the lecture becomes interactive and students are involved in it.

2.3 Blended Learning Models

Depending on various factors like content, teacher, availability of time, technology and other resources, different models of blended learning are developed.

1. Station Rotation: This is the most common model of blended learning as it is followed in many of the educational institutes [2]. In this model, students rotate through different stations one of which necessarily has to be online station or computer laboratory. These stations have a fixed schedule.
2. Laboratory Rotation: This model is similar to the station rotation model as in this model as well as a fixed schedule of stations is followed. Here, one of the stations compulsorily has been the computer laboratory where teachers guide students to use resources. This is characterized by the dedicated use of the institute's computer laboratory.
3. Individual Rotation: This model allows students to rotate through different stations but individually. Here, every student does not have to rotate through every station. He rotates only through the stations scheduled for him. This individual schedule is prepared to cater the needs of every student.
4. Enriched Virtual: This model is used by many students as the number of virtual classrooms is increasing. In this model, students complete the course online and on their own. During this process, they do not have to meet the teacher on daily basis. They can meet the teacher intermittently. So, face-to-face teaching and learning do not take place regularly.
5. Flipped Classroom: In this model, role of a teacher is flipped or inversed. Here, students are introduced with the content at home and they complete the related activities and studies in the school in the presence of teacher. Here, face-to-face interaction with teachers is on daily basis.
6. Flex Blended Learning: This model is very popular in the alternate school setting where students are enrolled in part-time school program. In this model, students have to rely mostly on the online courseware and teacher acts merely as a facilitator. Online courseware forms the backbone of the students learning.
7. Project Based: In this model, students' project-based classroom learning is supported by online courseware. Here, students learn online as well as offline but they collaborate to complete activities and projects.
8. Self-Directed Blended Learning: In this model, students use online material in self-directed way. They use online courseware to solve their doubts and problems. They

can connect with the mentors physically or digitally. As they are self-directed, role of teachers changes here. Only when required, teachers have to guide them.

9. Inside Out Blended Learning: This is very well-planned model in which students move in different spaces like physical classroom, digital classroom and outside the physical classroom. This model plans to end up students activities outside the classroom. Here, role of digital classroom is critical and emphasis is given on outward projects and assignments.
10. Outside-in Blended Learning: In outside-in blended learning, experiences are planned to 'start' in the non-academic physical and digital environments students use on a daily basis, but finish inside a classroom. This could mean traditional letter grades and assessments forms, or less traditional teaching and learning that simply uses the classroom as a 'closed-circuit' publishing 'platform'—a safe space to share, be creative, collaborate, and give and receive feedback that grows student work.
11. Supplemental Blended Learning: In this model, students completely work online and get supplementary guidance in face-to-face classroom teaching or they work completely in face-to-face classroom supplemented by online material and courses.
12. Mastery-Based Blended Learning: Students rotate between online and face-to-face learning (activities, assessments, projects, etc.) based on the completion of mastery-based learning objectives [6].

3 Access to ICT in Institutes

3.1 Government of India Initiatives

In a developing country like India, use of technology or ICT-based methodology of teaching can bring a significant change in the teaching–learning process. Information and Communication Technology is the new and emerging technology in the field of education and has captured the attention of many educators and policy makers. Use of ICT makes learning more effective on individual level as a student is provided access to many educational resources and learning is made very easy. Government of India also has taken initiative to make ICT as important part of the education system of India. Since citizens of India are its most valuable resource, our billion-strong nation needs the nurture and care in the form of basic education to achieve a better quality of life. Through its very important department of Ministry of Human Resource Department (MHRD), government has started many ICT-based applications such as Shala Siddhi, e-Pathshala, ShalaDarshan and Saransh Portal. ICT-based videos and ICT mobile Apps. Since independence, different educational policies have recommended the importance and use of ICT in education. In 1986, the National Policy on Education (NPE) promoted the use of computers in schools. As a result, computer was introduced in Indian schools in 1986 [3–6]. The NPE states that professional education is to be imparted through exposure to computers, training and computer literacy. Hence, the 'Computer Literacy Studies in Schools' (CLASS) project was launched in 1984–85 to support computer literacy in schools at that time. However, the NPE (1986), as modified in 1992, stressed the need to employ educational technology to improve the quality of education. In 2004, two major centrally sponsored schemes, namely, Educational Technology (ET) and Computer Literacy and Studies in Schools (CLASS) were merged for a more comprehensive centrally sponsored scheme—ICT @ Schools. The ICT @ school scheme has been subsumed with Rashtriya Madhyamik Shiksha Abhiyan (RMSA). Recently, the Govt. of India has launched an integrated scheme for pre-nursery to Class 12, i.e. Samagra Shiksha under the Union Budget of 2018–19. It is to be noted that Samagra Shiksha subsumes the three flagship schemes of Sarva Shiksha Abhiyan (SSA), RMSA and Teacher Education (TE). Consequently, the above-mentioned ICT @ school scheme is covered under new Samagra Shiksha for improving school effectiveness in terms of equal opportunities for schooling and equitable learning outcomes. ICT @ school scheme aims to provide for schooling and equitable learning outcomes. Additionally, the National Policy on ICT (2012) was framed specifically for school education. MHRD has already presented different drafts of this policy, one in 2009, and the revised draft dated 24 February 2011 and the final revised one in 2012. At the same time, Universal Secondary Education report (2005) by the Central Advisory Board of Education (CABE) has figured ICT as an important norm of schooling. This policy initiative focuses on ICT use in School Education to devise, catalyze, support and sustain ICT.

3.2 Access to ICT in Urban Area

Government has realized blending of ICT in regular teaching. Whether it is urban or rural area, the government has tried to provide the facility. It is found that in comparison with the rural area, usage of ICT in urban area is more successful. In urban area, trained teachers are available and they are provided with a periodic training to update and upgrade their knowledge. As the service providers are at the doorstep, all the computers and other equipment are maintained. Urban area has electricity and Internet availability which makes it easier for the teachers and students to use the technology easily. Apart from this, teachers', students' and parents' attitude towards technology is very positive as they know its importance. Students also have a lot of exposure to technology which makes them interested in it.

The same scenario is seen in many private schools and colleges. As they use ICT-based teaching as their Unique Selling Product (USP), they promote this teaching methodology. Such schools have smart classrooms and use some educational software which makes learning fun filled and easy. They have trained staff in ICT and all the subject teachers are trained in using this software while teaching. Parents are also supportive and there are no budgetary constraints. So, the hardware and software are always in working condition. Students get the real benefit from technology-based learning.

3.3 Access to ICT in Rural Area

Though implementation of ICT in education sector brings many positive changes among students and teachers and entire teaching–learning process, all the students in rural area cannot avail the benefit of ICT in a developing country like India for some obvious reasons. [7] There are multiple issues and challenges confronting the implementation of ICT education in schools and educational institutions in these countries and the problems are much more magnified in case of schools located in remote villages and rural areas. Though the government is trying to provide these facilities to the students in rural area, it is not completely successful as some factors act as a barrier. Some of these factors are

1. Maintenance and Upgradation of the Equipments—[8] Once the equipment is installed, they require periodic upgradation as technology changes continuously. Hardware and software may require maintenance and repairing. All this requires budget, and budgetary constraints may create an issue. So, though equipments are seen, they may not be in working condition.
2. Lack of Trained Teachers—[9] The major problem is the lack of quality and trained teachers in the field of ICT. Even if the teachers are trained, they may not get periodic training to update their knowledge.
3. Attitude—Many traditional teachers are not able to understand the importance of the use of technology while teaching. They still want to follow the age old traditional teaching methodologies.
4. Language Constraint—To use technology and ICT-based equipment, one needs to have good command over English language. In rural area, English proficiency of the teachers is not very good. So, they avoid using these equipments.
5. Unavailability of Electricity and Internet—This is a major issue faced by the rural area. There is never uninterrupted supply of electricity and Internet. This makes using technology even more difficult. So, people avoid using technology completely [10, 11].

4 Results and Discussions

4.1 Objectives of the Study

Various studies have shown that many schools and colleges try to use ICT in the teaching–learning process but they are not completely successful in adopting technology-based educational system. There are a number of factors affecting the use of technology in education. To investigate and understand these factors, a survey was undertaken in schools and colleges. The survey aimed at seeking the contribution of ICT in teaching–learning process with special reference to English language and what are the factors that affect the use of ICT in education. A questionnaire was framed for students and teachers and primary data was collected.

The present study has been undertaken with the following specific objectives.

1. To identify students' openness to learn using Information and Communication Technology (ICT) rather than traditional lecture-based methodology
2. To know students' comfort level in using ICT
3. To identify the teachers' level of knowledge and comfort level in using ICT in classroom

4. To identify whether the students and teachers have access to ICT at home or in classroom
5. To identify the overall factors that affect the use of ICT in classroom.

4.2 Methodology

Success of any survey lies in its exact and useful analysis and conclusion. To get a broader idea of the factors affecting the use of ICT in education, data from students as well as teachers was collected.

4.3 Questionnaires

The main instrument of the survey was the questionnaire. Two separate questionnaires were prepared for the students and teachers. As the questionnaires were specially designed for this survey, they covered all the important questions. Both the questionnaires contained ten multiple choice questions and one descriptive question about suggestions. The questionnaires covered main areas—(a) Comfort level of the students and teachers in using Information and Communication Technology, (b) Amount to which students and teachers have access to ICT, (c) Attitude of students and teachers towards the use of ICT and their motivation to do it and (d) Overall factors affecting the use of ICT in education or classroom.

4.4 The Analysis

4.4.1 Population and Sample

Population for this survey consisted of students and teachers. A sample of 100 students and 100 teachers was used. The questionnaires were handed over to these students and teachers who were selected from different schools and colleges. Students and teachers were helped time to time while answering the questions to avoid misunderstanding of the questions by them (Tables 1, 2, 3, 4, 5, 6 and 7).

The problem of understanding and communication in English language is a big issue with engineering graduates in India. It is observed that the students with urban background are comfortable in English conversation and its understanding as compared to the rural background due to variety of reasons. However, it will be a difficult task to teach engineering graduates English as a language due to limitations of the curricula. Hence, it is decided to go to the grass root level of the problem. The language study is best carried out in secondary school level. Owing to this fact, the present analysis is carried out with a sample consist of secondary school teachers and students.

4.4.2 Analysis of Data Collected from Teachers

Hundred teachers from different secondary schools from Aurangabad region were taken as a sample for the present study. All the teachers possessed graduation degree with English language as a major and essential teacher training course. Further, all the teachers are active in English language teaching from last minimum 5 years. Before the collection of the questionnaire, they were informed properly about the experimentation and written consent was taken from the individuals. Table 1 shows the results of the

Table 1. Analysis of the QUESTIONNAIRE for teachers

S. No.	Questions	Answers given by the number of teachers		
		Yes	No	Cannot say
1	Do you think teachers should have knowledge about how ICT can support learning?	97	0	3
2	Do you think ICT in education prepares students for future career?	100	0	0
3	Do you think use of ICT in education facilitates student learning?	94	0	6
4	Do you use computers to prepare educational reports/PPTs?	81	19	0
5	Do you think use of technology in teaching will develop you as a teacher?	100	0	0
6	Do you think new technology is beyond your abilities?	27	64	9
7	Are you afraid of coping up with the unfriendly technological jargon?	45	43	12
8	Do you think use of technology in classroom takes away time from other school/college related activities?	27	70	3
9	Do you have availability of educational software/projector in your school?	91	9	0
10	Do you have a fear of being replaced by a computer one day?	88	3	9
11	Do you have knowledge of interactive white board facility for teaching–learning activities?	57	43	0
12	Do you know the blooms taxonomy for teaching/learning process?	23	77	00
13	Do you think proper training sessions are required for creating awareness of ICT in the teacher community?	83	17	00

Table 2. Analysis of the questionnaire for students

S. No.	Questions	Answers given by the number of teachers		
		Yes	No	Cannot say
1	Do you think use of ICT helps in learning English in a better way?	86	9	5
2	Do you play interactive games to learn English language?	46	48	6
3	Is the interactive white board more beneficial than a blackboard and a chalk piece?	87	5	8
4	Do you think availability of ICT in classroom has changed your way of learning?	84	6	10
5	Are you able to operate computer to prepare presentations/online information surfing/MS-Office/Excel Sheet?	69	31	00

Table 3. Classroom analysis

S. No.	Questions	Trends	Listen only to the lecture by the teacher	Listen to the views of a student	Listen to the lecture along with a video clip on the same subject	Use interactive white board	Search online information related to the topic	Perform a group or individual activity
1	How often do you do these activities in the classroom/home?	Never	2	16	11	24	13	10
		Sometimes	50	18	49	16	49	70
		Always	48	16	40	9	38	20

Table 4. Use of resources

S. No.	Issue under question	PC/laptop without internet access	PC/laptop with internet access	Digital reader (e.g. kindle)	Video gaming system	Mobile without internet connection	Mobile with internet connection	Music/video player	Data projector	Interactive white board	Educational software
1	Which of these are available for you to use at home/school/college?	19	55	6	18	12	82	46	31	32	40

Table 5. Use of computers in day-to-day life

S. No.	Questions	Answers given by the number of students					
1	How long have you been using a computer at home/school/college?	1 year	2 years	3 years	4 years	5 years	More than 5 years
		Data	8	5	19	7	54

Table 6. Use of ICT in class room

S. No.	Issue under question	Only lecture	ICT-based activity along with lecture
1	Would you like to have English classroom with only a lecture by the teacher or a video clip or interactive game or any ICT-based activity along with the lecture by the teacher?	2	98

Table 7. Behaviour analysis

S. No.	Do you agree with following statement	Response		
		Yes	No	Cannot say
1	It is important for me to learn using ICT for better understanding	81	5	14
2	Using ICT while learning is a real fun	85	11	4
3	Use of ICT makes me remember the concept forever	66	12	22
4	Learning computer will make me perform better in my future job	90	3	7
5	I lose track of time while using computer	26	45	29

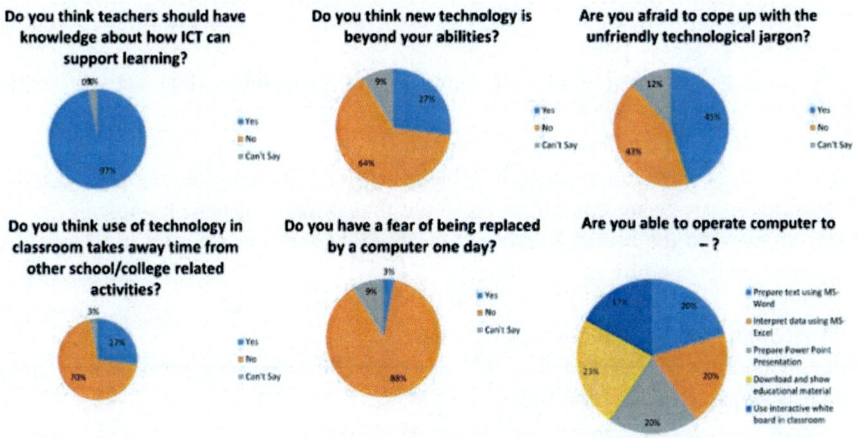

Fig. 1. Analysis of teachers response

data collected from the teachers. Figure 1 depicts the data analysis of the data collected from teachers. Following points are observed from the data collection:

1. 86% of the teachers agreed that ICT may help to improve the teaching–learning processes.
2. 87% teachers emphasised the use of interactive white boards in the teaching process of English.
3. 84% teachers agreed that ICT-based teaching–learning process may change the class room scenario positively.
4. 27% teachers have adaptability issue with the ICT-based technologies.
5. 88% teachers are afraid of being replaced by the growing use of technology in teaching–learning process.
6. Only 57% teachers have knowledge of interactive white board technology/facility.
7. Only 23% teachers have awareness about Blooms Taxonomy.
8. 83% of the teachers require a proper training for the use of ICT-based modules for classroom teaching–learning process.

It is been observed that most of the teachers are acquainted with the computers and Internet. However, only few of them are aware of various ICT technologies to be used in the classroom. Most of them use the traditional classroom teaching methodologies. Interestingly, a large group of teachers fear that the adoptability of ICT-based teaching–learning process may be a threat to their jobs. To cope with the situation, continuous training sessions are required for teachers regarding ICT technology and different pedagogical techniques.

4.4.3 Analysis of Data Collected from Students

Students were analysed on a variety of the scales such as their habits to use smart phones, reading, integrative learning, use of computers, class room environment, and traditional teaching–learning methods. Following observations were recorded (Fig. 2).

1. Only 46% students have the knowledge of interactive games as a learning methodology for English
2. 87% students agree that interactive white board and other ICT-based modules may change the class room scenario.
3. 69% students are aware of MS-Office utility and use of Internet for educational purpose.
4. 20% students had participated in some sort of group/individual activity while learning English as a language.
5. Almost 90% of the students have access to one or other gadget for ICT use. This scenario can be used to introduce ICT-based teaching–learning among the students.
6. 54% students are using computers for more than 5 years at their home.
7. 98% emphasised the use of ICT-based teaching–learning process in the classroom for better understanding of the subject.
8. 66% students emphasised that ICT-based techniques helped them to understand/remember the concept with ease.
9. 90% of the students agreed that ICT-based teaching–learning process may help them in future to secure good jobs.

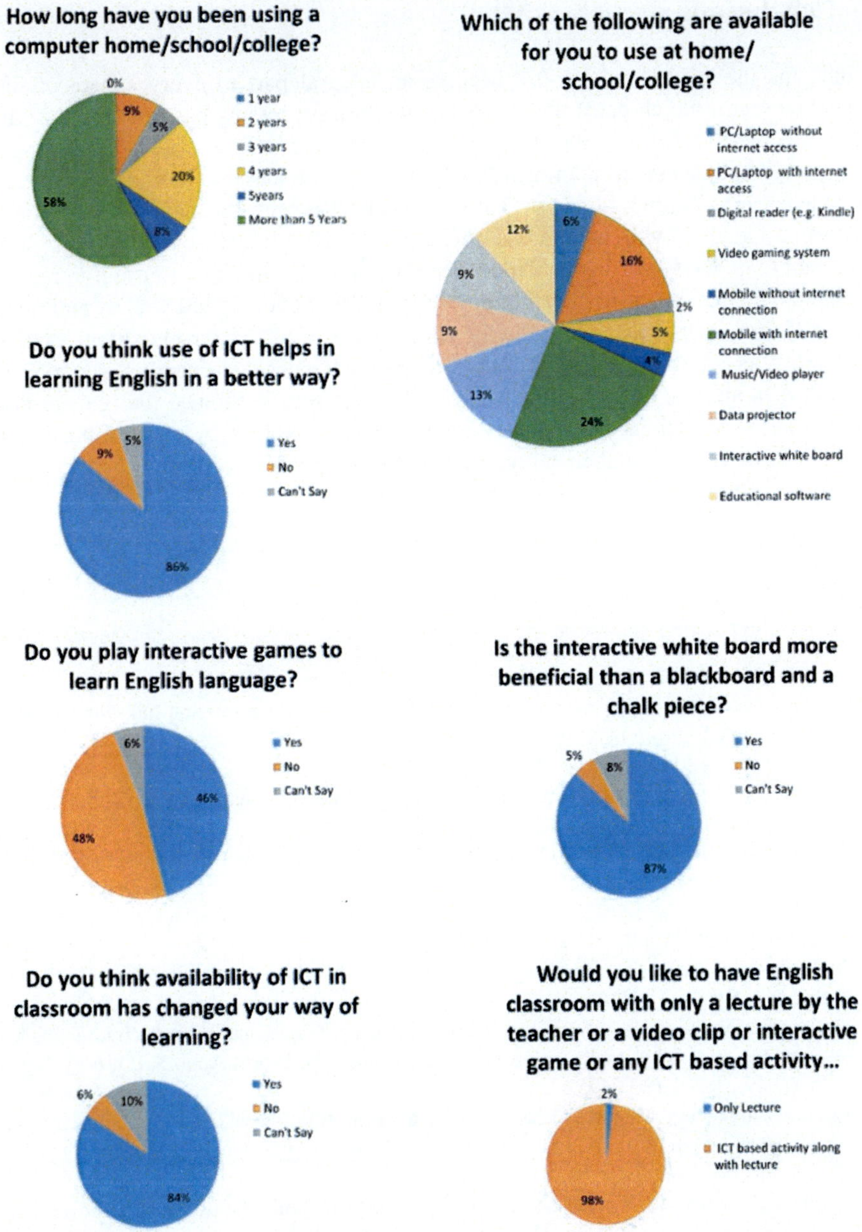

Fig. 2. Analysis of students response

5 Conclusions

Today, the use of technology has become an integral part in every sphere of life. Education sector which demands a huge information processing has to accept ICT for faster and better results. Blended learning is in demand. ICT is widely used for the support and enhancement of information delivery in the classroom. It definitely enriches the teaching–learning process. The traditional classroom has taken a new shape of classroom blended with computers, Internet and educational software in it. It is important to understand what is blended learning and why and how to use it in regular teaching. Teachers are seen showing power point presentations, video clips and using other ICT tools in the classroom. But the effective use of ICT in the classroom requires availability, accessibility and knowledge to be operated by teachers. A survey was conducted to analyse these factors. The result of the survey showed that government and private organizations are taking efforts to blend technology in teaching but unless all the stakeholders show their willingness and readiness to accept this change, and it can be an unavoidable part of our education system.

References

1. Bonk, C.J., Graham, C.R. (Eds.): Handbook of blended learning: Global perspectives, local designs. Pfeiffer Publishing, San Francisco, CA (in press)
2. Find the model that works for you: 12 types of blended learning, internet link: https://www.teachthought.com/learning/12-types-of-blended-learning/
3. Use of ICT in education (Available at): https://mhrd.gov.in/ICT/index.html
4. Mamur Ali: Professional development of teachers with ICT. Indian J. Educ. Technol. 1(1), 30–37 (2019)
5. National policy in ICT in school education: Department of School Education and Literacy Ministry of Human Resource Development Government of India 2012. (Available at)]: https:// mhrd.gov.in/ sites/upload_files/mhrd/files/ upload_document/ revised_policy%20-document%20ofICT.pdf
6. Budhedev, S.H.: Issues and challenges in bringing ICT enabled education to rural India. Int. J. Sci. Res. Edu. 4(1), (2016)
7. Husseini, S.M., Safa, L.: Factors affecting the use of information and communication technologies (ICTs) by Iranian agricultural faculty member. World Appl. Sci. J. 6(8), 1123–1127 (2009). ISSN 1818-4952
8. Factors influencing teachers' use of ICT in education. (Available at): https://www.researchgate.net/publication/284773965_Factors_Influencing_Teachers'_Use_of_ICT_in_Education
9. Chigona, A., Chigona, W.: An investigation of factors affecting the use of ICT for teaching in the western cape schools. In: 18th European Conference on Information Systems, pp. 1–12 (2014)
10. Deshpande, S., Shesh, A., Iyer, B.: English Language adoptability in engineering graduates: A case study. Adv. Intell. Syst. Comput. 810, 535–545 (2019)
11. Deshpande, S., Shesh, A.: Performance assessment and remedies using blended learning for professional students. Adv. Intell. Syst. Comput. **1025**, 783–790 (2020)

Automation in Hydroponics Farming Ecosystem

Jagruti Kishor Wagh[1], Rajendra V. Patil[2], Anil D. Vishwakarma[2], and Vijay D. Chaudhari[2](✉)

[1] Dr. Babasaheb, Ambedkar Technological University (DBATU), Lonere, Raigad, M.S, India
[2] Godavari Foundation's Godavari College of Engineering, Jalgaon, India
vinuda.chaudhari@gmail.com

Abstract. Hydroponics means growing plants without soil. As the quality of production in farming is decreases day by day, everyone demands nutrient rich food, but this demand cannot be fulfilled by using our traditional farming method. This method undergoes many problems. Some of them are availability of land and labor; another one is increased use of fertilizers which can affect quality of crop that in turn can affect the ability of soil fertilization. With poor soil fertility, the farmer can face many problems. He did not get production in huge quantity also the crop we will get is not nutrient rich. Other problems in traditional farming are frequent weather changes, rise in temperature, water pollution, etc. Under this condition, it will be very difficult in the future to grow a crop that will feed the entire population using traditional agriculture. We are using IOT technology which is very helpful to connect objects to the Internet for automation in farming. Hydroponics is an interesting new platform that requires less area for plantation and can produce more product than the conventional farming.

Keywords: Smart farming · Hydroponics · IOT

1 Introduction

Hydroponics is a method of farming in which we are using water to grow plants. It is a method of growing plants or vegetables in water mixed with mineral nutrient solutions instead of soil. Since, we are using water-based solution to grow plants, we have to control electrical conductivity of solution, pH, humidity, etc., for more production and nutrient rich food. Plants need air, water, sunlight, and nutrients. Soil cannot provide the nutrients required for proper growth of plants; it only works as structure to hold the plants. Hydroponics is ideal in any weather conditions because controller itself can control air, water, light, temperature, and humidity of plants. So the hydroponics

© Springer Nature Singapore Pte Ltd. 2021
P. Deshpande et al. (eds.), *Next Generation Information Processing System*, Advances in Intelligent Systems and Computing 1162,
https://doi.org/10.1007/978-981-15-4851-2_34

method is more relevant in such scenario, also the pants that are grown hydroponically grow faster and are healthier compared to soil-based plants. Our hydroponics system is made up of some sensors, i.e., pH sensor, EC sensor, air quality, humidity, and luminous sensor; Raspberry Pi 3 is used to monitor our system. Values from sensors are monitored by Raspberry Pi which regulates the system flow and a relay is used to manage water pump. At the end, system will inform user for any changes on computer screen via e-mail. As, our system is automatic; it is very useful for busy city people because controller will take care of most of the activities. Our system is IOT based which is great platform for software and hardware interactions.

2 Literature Review

Many researches have been done in designing and implementation of smart farming. All have different methods of farming. The author selected an arduino processor in his paper and output he is getting on android application so that notifications about the changes in HFE setup will be displayed on the user's mobile phone [1]. In this paper, electrode sensor is used for measuring water conductivity for the hydroponics system and is based on ARM processor that monitors the system [2]. In this paper, data collection, monitoring, and evaluation of the system are used to determine which approach is effective. Blynk App is used to control and monitor the smart farm [3]. In this paper, hydroponic system is modulated using supplementary LED lightening technology which is based on IOT implementation. Indoor environment is controlled by artificial lights. Butterhead lettuce was planted for observation purpose [4]. In [5], Google assistant is used. Whole system periphery is controlled by giving voice command and sensor values are displayed in Adafruit application. In [6], UBIDOTS cloud platform is used to connect a greenhouse with android application which is hosting a Web server. Here, user can control parameters from anywhere. In [7], Blynk App is used to control and monitor the smart farm. Node MCU is used for implementing the greenhouse monitoring. In [8], Nutrient Film Technique is used. Author developed a system using Arduino mega. In [9], hydroponic planter methods and its basic structure are explained, which is based on a vertical pipe method for suspended cultivation. In [10], author developed a smart hydroponic system with supplementary LED lights; the system was successful in collecting data real time for ease to user access. In [11], we developed a hydroponics system in which GSM is used and later, it will be removed. In [12], different techniques of hydroponics and their operations are explained. In [13], A Web application is developed that enables users to control the mechanisms for sprinkling, draining, and refilling the solution. In this system, user can monitor the pH level, temp, and humidity of plant manually. After studying the literature, the Table 1 shows the performance comparison of various systems in hydropnic farming.

Table 1. Performance comparison table

S. No.	Previous work	Recent work
1.	Most of the systems are using Arduino microcontroller	We are using Raspberry Pi which is minicomputer itself as shown in Fig. 1
2.	The system was not automated and needs human interface for observation, needs continuous observation of plant	Our system is fully automated, it does not require human interface every time to change the sensor parameters
3.	Developed an android application on mobile to send notification to users and monitor the system parameters	We developed an application on PC to turn ON and OFF the water pump automatically as shown in Figs. 7 and 8
4.	Some systems are using Raspberry Pi 2 model B but the system was complex	We are using Raspberry Pi 3 model B which makes system easy to understand and simple

3 Methodology

3.1 Block Diagram Description

The implementation of proposed system is shown in following Fig. 1. It consists of Raspberry Pi 3, power supply, and various sensors as shown. Figure 1 shows the block diagram of HFE. The proposed system is implemented to grow various plants onto single system. In this system, nutrient rich solution mixed with water is supplied to the crops. Various sensors are installed for monitoring the various parameters, i.e., pH level of the nutrient solution, humidity, light intensity, and the water level. The input from these sensors will enable the Raspberry Pi to regulate the water and nutrient flow mixture in correct proportion. The Raspberry Pi is programmed with suitable algorithm to regulate the flow systematically. It is treated as the most important part of this project. Raspberry Pi 3 model B is used as a controller. The program is written in the Python language. The server-side program constantly monitors the sensor parameters whether they are inside the edge extent. In the event that the qualities go down from the edge extend, telling message with the arrangement is sent to the user via e-mail that is an output module. Once the system is built, it can be tested according to the needs of individual crops and then all of which can be integrated. Raspberry Pi will continuously monitor several environmental conditions such as water pH, electrical conductivity of water, and luminosity to achieve the optimal growth of the plants.

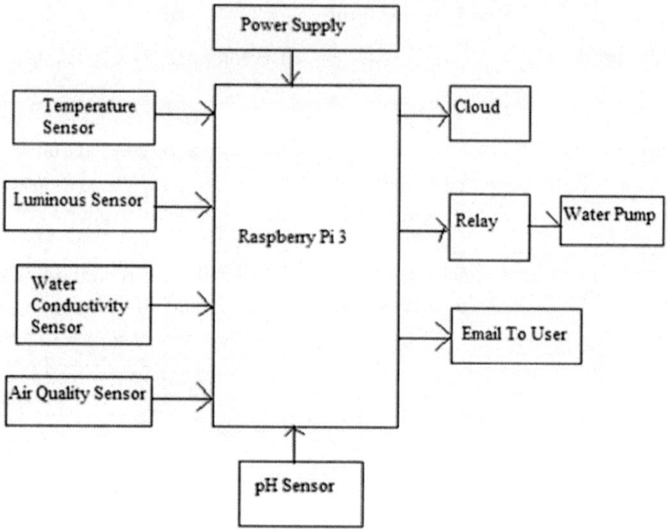

Fig. 1. Block diagram of the proposed system

3.2 Hardware Description

Raspberry Pi 3B+: It is a small microcontroller that can handle many functions like minicomputer. It is used where the system processing is huge. Raspberry Pi can perform all those advanced functions which Arm processor cannot able to do.

DHT11 Temperature Sensor: DHT 11 is a digital temperature and humidity sensor. It is low cost sensor. This sensor can be easily interfaced with Raspberry Pi, and it is smaller in size.

Luminous Sensor: We are using BH1750FVI luminous sensor; it is a digital light sensor which works on I2C interface. It connects to Raspberry Pi with only four wires. Light level can be read from it. To use this module, first you must have to enable the I2C interface on the Raspberry Pi.

Electrical Conductivity Sensor: The EC sensor measures the conductivity of a solution. In hydroponics farming, this sensor helps us to grow nutrient rich healthier plants by measuring nutrients, salt, and other contents of the solution. Accordingly, we can adjust the elements in solution.

Air Quality Sensor: Air quality sensors are used for detection of pollution in the surrounding area. These are suitable for indoor surrounding also. Here, we are using MQ135 sensor. It has high sensitivity to harmful gases, i.e., nitrogen, hydrogen, carbon dioxide, and benze steam. This sensor is used for detecting poisonous gases that are present in the air in homes and offices. It has long life.

pH Sensor: It is relatively stable against pressure fluctuations within the pH of system. It is commonly used for water Ph measurement; pH is a measure of acidity and alkanity of solution. It can measure up to 1–14. The value 7 is neutral.

Relay: A relay is a low power electromagnetic switch which is used to turn ON and turn OFF a circuit. The relay module is an electrically operated switch.

4 Software

The HFE will perform in the following steps (Fig. 2).

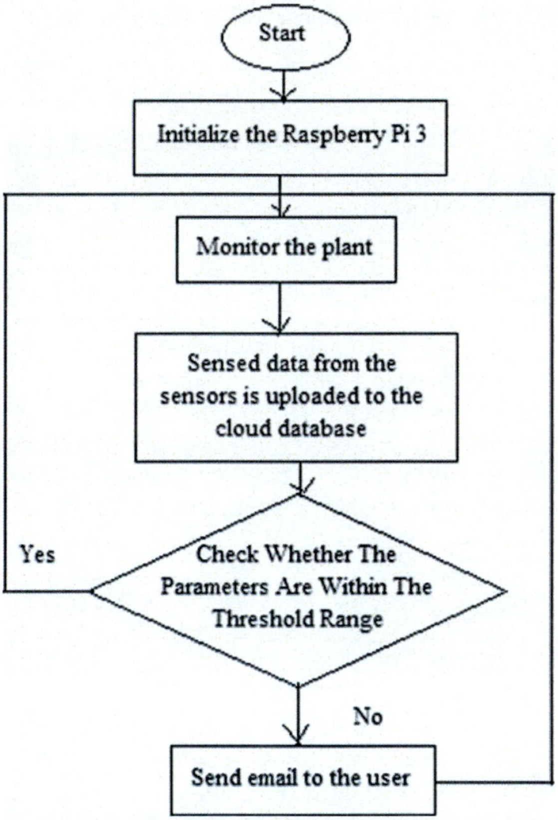

Fig. 2. Flowchart of the hydroponic system

Algorithm steps for flowchart
Step 1: Start the system
Step 2: Initialize the system on Raspberry Pi
Step 3: System will monitor the plant
Step 4: System will senses the data from sensors
Step 5: Data is sent to the cloud
Step 6: Check whether the parameters are within the threshold
Step 7: Sends an e-mail to user.

5 Results and Discussion

Figure 3 shows the setup when power supply is ON. First, insert microSD card into MicroSD card slot of Raspberry Pi. All the sensors are connected to Raspberry Pi. Then, we will get IP address of Raspberry Pi with the help of advance IP scanner. This IP address is in putty configuration. Now, we connect laptop with Raspberry Pi using VNC viewer and the information about sensor parameters is display on PC screen.

Fig. 3. Experimental setup when power supply is on

Figure 4 shows the process of plant monitoring. Our system will continuously monitor the plant and gives results as output of sensors. The information about DHT 11 sensor is displayed on PC screen and e-mail is sent to user via cloud server.

In our hydroponics system, results were obtained using different sensors for 24 h. Results were obtained with time for varying parameters of water quality. Figure 5 shows the plot of temperature varying with time, the temperature value crosses the edge value limit during that particular time interval, and the farmer will receive an alert message via e-mail to user. Similarly, Fig. 6 shows the plot of humidity versus time.

All the sensors are connected to Raspberry Pi and power supply is given to them. Then, Raspberry Pi reads the values from all sensors, respectively, and sends this collected information to the cloud server. If these values are less than the already set threshold values, then the relay gets ON and it switches ON the motor. The motor remains ON until the sensor values reach to already set threshold values. When it reaches to threshold value, the relay automatically switches off the motor. We also developed an application with the help of thingspeak.com on PC that helps to switch

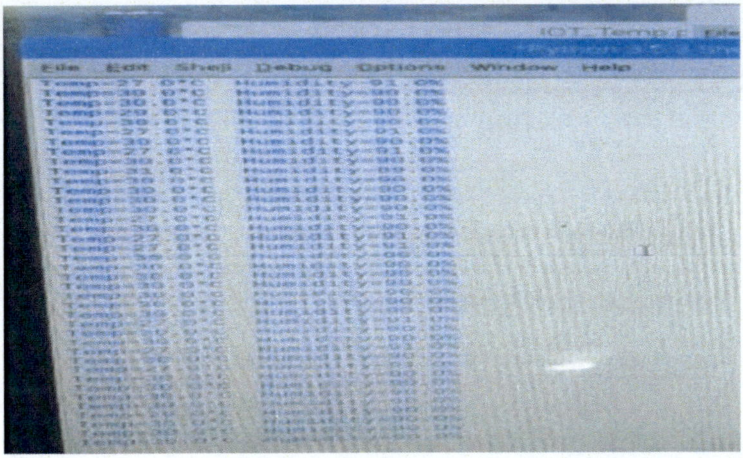

Fig. 4. Output of DHT11

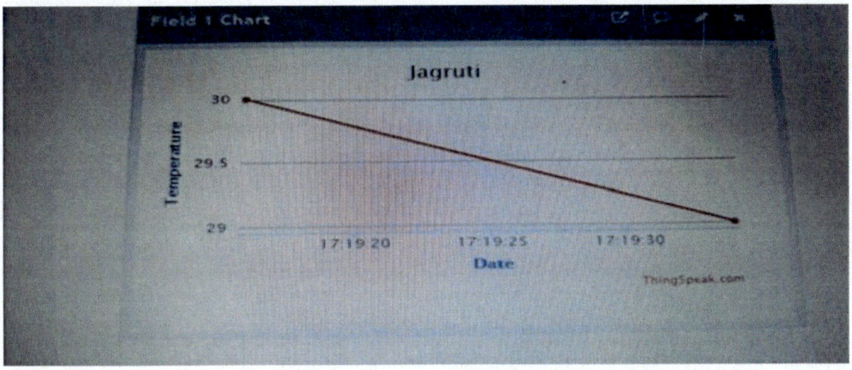

Fig. 5. Variation of temperature with time

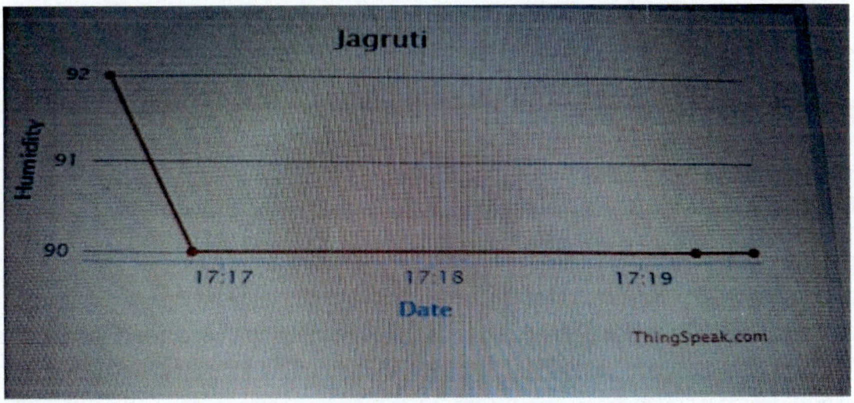

Fig. 6. Variation of humidity with time

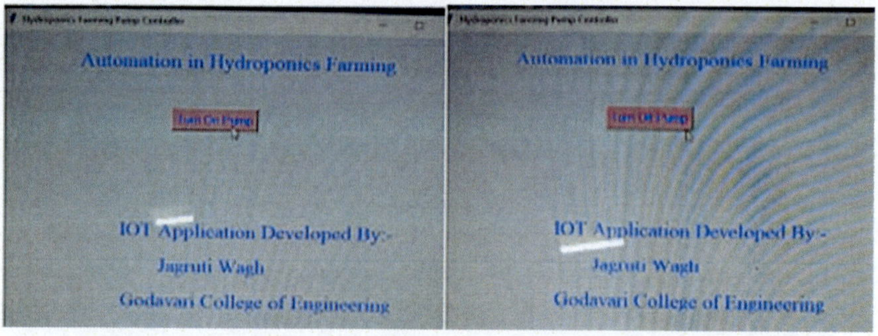

Fig. 7. Image of application window to turn ON and OFF the water pump

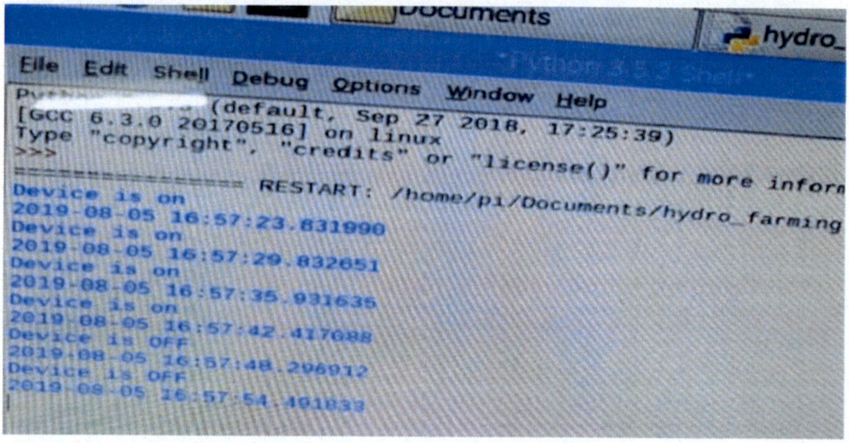

Fig. 8. Output of application window

ON and switch off the pump according to sensor parameters; with this application, we can decide when to turn ON and when to turn Off the pump manually. Figure 7 shows application window to turn ON the pump manually.

Figure 8 is the output of our application developed by thingspeak.com which shows the while device is ON or OFF.

The future work of our system is to collect environmental data from sensors and implement an artificial intelligence system that makes the hydroponics farming ecosystem to run automatically without human interference. Artificial intelligence is expected to improve efficiency and reduce costs associated with hydroponics farming, and also it will optimize the nutritional value of the crops. Recently, IoT-based systems are gaining popularity due to its non-contact remote operation, low cost, and compact size [14–17]. The present systems may be upgraded to a very precise and sophisticated IoT-based systems.

6 Conclusion

The purpose of implementing hydroponics farming ecosystem is to grow a nutrient rich and healthier plants in minimum space without soil. The plants that are grown in hydroponics do not need fertilizers so the maximum pest attacks are eliminated as compared to traditional farming. Farmer will get a production in quantity and quality can also be maintained.

References

1. Saaid, M.F. Sanuddin, A., Ali M., Yasin, M.S.A.I.M.: Automated pH controller system for hydroponic cultivation. In: IEEE International Conference on System Engineering and Technology (ICSET) (2015)
2. Ruengittinun, S., Phongsamsuan S, Sureeratanakorn, P.: Applied internet of thing for smart hydroponic farming ecosystem (HFE). In: 10th International Conference on Ubi-media Computing and Workshops (Ubi-Media), (2017)
3. Rajkumar, R.: A novel approach for smart hydroponic farming using IoT. Int. J. Eng. Res. Comput. Sci. Eng (IJERCSE) **5**(5), 18–23 (2018)
4. Namgyel, T., Siyang S., Khunarak, C., Pobkrut, T., Norbu, J., Chaiyasit, T. Kerdcharoen T.: IoT based hydroponic system with supplementary LED light for smart home farming of lettuce. In: 15th International Conference on Electrical Engineering/Electronics, Computer, Telecommunications and Information Technology, pp. 221–224 (2018)
5. Meenakshi, U., Lakshmaiah, M.V., Bilvika, B.: Design and implementation of Raspberry Pi 3 based embedded system for analysis of soil parameters. Int. J. Eng. Res. Appl. **7**(11), 34–39 (2017). ISSN: 2248-9622
6. Rakshitha, M., Shwetha, H.L.: Automation of hydroponics system using android application and ubidots platform. In: International Journal of Engineering Research and Technology (IJERT), NCESC Conference Proceedings, pp. 1–4 (2018). Published by, www.ijert.org. ISSN: 2278-0181
7. Satoh, A.: A hydroponic planter system to enable an urban agriculture service industry. In: 7th Global Conference on Consumer Electronics (GCCE 2018), pp. 281–284 (2018)
8. Harsha, A., Deekshith, K., Murali Krishna, B.K., Sachin, K.T., Sushanth.: Automated hydroponics greenhouse monitoring system using adafruit.io controlled by google assistant. Int. J. Eng. Res. Technol. (IJERT). In: Conference Proceedings ICRTT—2018, pp. 1–4. Published by, www.ijert.org. ISSN: 2278-0181
9. Namgyel T., Siyang, S., Khunarak, C., Pobkrut, T.: IoT based hydroponic system with supplementary LED light for smart home farming of lettuce. In: 15th International Conference on Electrical Engineering/Electronics, Computer, Telecommunications and Information Technology. pp. 221–224, (2018)
10. Vineela, T., NagaHarini, J., Kiranmai, Ch, Harshitha, G, AdiLakshmi, B.: IoT based agriculture monitoring and smart irrigation system using Raspberry Pi. Int. Res. J. Eng. Technol (IRJET) **05**(01), 1417–1420, (2018). e-ISSN: 2395-0056
11. Wagh, J.K., Patil, R.V., Ingale, H.T., Chaudhari, V.D.: Automation in hydroponics farming Eco-system: A review. Int. J. Innov. Eng. Sci. **4**(9), 53–56 (2019)
12. Sharma, N., Acharya, S., Kumar, K., Chaurasia, O.P.: Hydroponics as an advanced technique for vegetable production: An overview. J. Soil Water Conserv **17940**, 364–371 2018

13. Chris, J.G. Aliac, E.M.: IOT hydroponics management system. In: IEEE 10th International Conference On Humanoid, Nanotechnology, Information Technology (2018)
14. Patil, N., Iyer, B.: Health monitoring and tracking system for soldiers using internet of things (IoT). In: 2017 International Conference on Computing, Communication and Automation, pp. 1347–1352
15. Iyer, B., Patil, N.: IoT enabled tracking and monitoring sensor for military applications. Int. J. Syst. Assur. Eng. Manag. **9**, 1294–1301 (2018). https://doi.org/10.1007/s13198-018-0727-8
16. Deshpande, P., Iyer, B.: Research directions in the internet of every things (IoET). In: International Conference on Computing, Communication and Automation (ICCCA), pp. 1353–1357 (2017)
17. Deshpande, P.: Cloud of everything (CLeT): The next-generation computing paradigm. Adv. Intell. Syst. Comput. **1025**, 207–214 (2020)

Low Power, Low Voltage, Low Drop-Out On-chip Voltage Regulator

Gopal Agarwal[1(✉)] and Ved Vyas Dwivedi[2]

[1] Research Scholar, C U Shah University, Surendranagar, Gujarat, India
gopalgenius@gmail.com

[2] Professor, E and C Department, C U Shah University, Surendranagar, Gujarat, India

Abstract. Extremely low power, low voltage complete on-chip low drop-out voltage regulator (LDO) is presented. The LDO receives an unregulated supply of 0.5–1 V and regulates it to 0.4 V nominal output utilizing a 0.1 V on-chip reference voltage. A load capacitor of just 1pf is assumed to be present at LDO output. Designed voltage reference and complete LDO were simulated in 180 nm CMOS technology with variation in supply voltage and temperature. The maximum power dissipation of the complete LDO was found to be approximately 3nW and maximum line and load regulations were observed to be 6.62 mV/V and 8.08 mV/mA, respectively.

Keywords: LDO · On-chip · Regulation · PM · BW · UGB

1 Introduction

Modern system-on-chip designs integrate all the circuit functions (digital and analog) on a single substrate including power management of these functions. Each circuit function in such design has different power requirement from power source and the power-line isolation becomes critical in such mixed-signal environment. With growing market trend, where market is getting crowded with various hand-held application devices, power efficient operation of circuit functions is must for prolonged battery life. Plus, to achieve complete integration, and to save printed circuit board-space, power-management circuitry should be completely on-chip and the conventionally used large off-chip capacitor at the linear regulator outputs must be dropped from their design.

Digital circuit functions are expected to switch very fast at their clock frequency and hence would expect a sudden high current source/sink capability from the power supply. With the absence of large off-chip capacitor at linear regulator o/p powering the circuit function, this current must be supplied by the regulator itself. The regulator must be fast enough to respond to such load changes to restrict the drop in output voltage to acceptable limits. However, regulated output voltage accuracy is not that critical here. Analog circuit functions on other hand would expect a very accurate and less noisy voltage at their power-line while speed of regulator is not that critical.

Keeping above things in mind, power management of a system-on-chip is generally done in two steps. The battery voltage is given to an efficient switching regulator first which brings it down to low voltage domain for achieving power efficiency in

© Springer Nature Singapore Pte Ltd. 2021
P. Deshpande et al. (eds.), *Next Generation Information Processing System*,
Advances in Intelligent Systems and Computing 1162,
https://doi.org/10.1007/978-981-15-4851-2_35

subsequently connected linear regulators. Switching regulator output goes to various LDOs where each individual LDO would be supplying power to a different circuit function and each would be designed to meet the requirements of respective circuit function it is powering.

This work focuses on one such LDO which can be utilized to supply power to 0.5 V near-threshold logic circuits. Minimizing the LDO quiescent power and complete on-chip integration was the key aspects targeted in this work.

2 Basic LDO and the OFF-CHIP Capacitor

The basic block diagram of an LDO is shown in Fig. 1. It consists of resistive sampling network, an error amplifier, a reference voltage $V_{REF,}$ and a PMOS pass device.

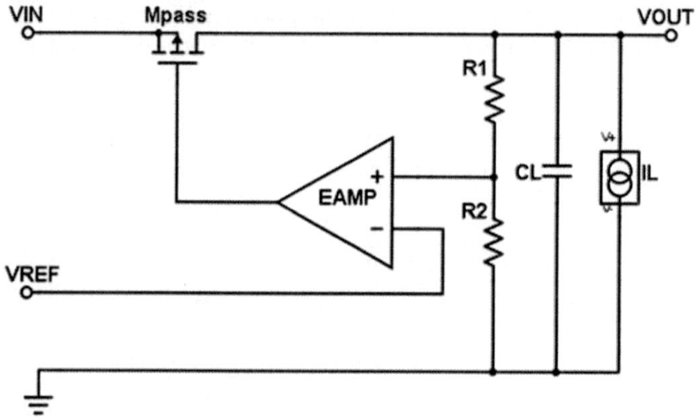

Fig. 1. Basic LDO block diagram

V_{OUT} tracks V_{REF} across slow variations of process or temperature and slow/fast variations in V_{IN} or load current, due to negative feedback loop which consists of error amplifier, pass device, and sampling network.

Tracking of V_{REF} for certain nominal conditions of process, temperature, V_{IN}, and load current happens by achieving V_{OUT} to be V_{REF} times the feedback gain/attenuation using the sufficient open-loop gain typically >40 dB of the control loop. Bandwidth of the loop is not that critical for this task.

In case parameters deviate from the nominal conditions, deviation of V_{OUT} from V_{REF} within the specified limit, (e.g., 10%) requires a much smaller gain, about 20 dB, but a very high bandwidth of control loop is essential.

Large off-chip capacitor affects both loop stability and transient response of LDO. With large off-chip capacitor present, dominant pole occurs at LDO output while first non-dominant pole occurs at gate of pass transistor. Stability is achieved either by placing the first non-dominant pole beyond UGF or by cancelling the first non-dominant pole with the help of zero created by the R_{ESR} of off-chip capacitor. Without this off-chip capacitor, the dominant pole occurs at gate of pass transistor and the first non-dominant pole occurs at the LDO output which is load current dependent. Since the load capacitance assumed here is just 1pf, and UGF is very low, the LDO output pole occurs beyond UGF and the system is stable with a phase margin of approximately 90°. Also, the large off-chip capacitor helps in supplying the sudden load current demand of the load, and it helps in improving the load transient response of the LDO.

With the unregulated input reaching to 0.5 V domain, designing of analog feedback loop with good gain has become difficult. Various techniques are being proposed having digital feedback loop to regulate the output voltage in a LDO. With the unregulated input of 0.5 V, the pass device is expected to work in subthreshold region of operation and will not have much driving capability. Hence, the size of pass transistor becomes extremely huge when the amount of load current to be driven is large.

Various LDO architectures without off-chip capacitors have been proposed in the literature. Besten and Nauta [1] designed an LDO using NMOS as pass transistor controlling its gate voltage using a slow control loop. NMOS in the output stage has inherent advantage of fast transient response, low output impedance and hence, better PSRR performance but at the expense of circuit complexity which results in higher quiescent current consumption.

Milliken et al. [2] explored the Miller effect due C_{GD} for improving stability and explored differentiating action of C_{GD} to output voltage change to improve transient response. The load transient performance was a 100 mV output voltage drop for a 50 mA load step, in 1 μs.

Okuma et al. [3] proposed digital LDO eliminating all analog circuits in the feedback loop to provide low noise and tunable power supply voltage to 0.5 V near-threshold logic circuits. The drop-out voltage was 50 mV and they have achieved 98.7% current efficiency with 2.7 μA quiescent current and 200 μA load current.

Yang et al. [4] proposed an LDO which is inverter based, achieves rail-to-rail regulation, and a PSR of over -20 dB at 0.2 V lowest supply voltage. The LDO consumes 410 nA quiescent current at 0.2V V_{dd}, and 32 μA at 0.6V V_{dd}.

Huang et al. [5] proposed a analog-assisted (AA) digital LDO with tri-loop control. A fine-tuning loop improves output accuracy while coarse-tuning loop helps to reduce recovery time. With a total load capacitance of 100 pF, in response to a 10 mA/1 ns load step, maximum measured voltage undershoot/overshoot was 105 mV.

3 Voltage Reference

A low power, low voltage reference using the peaking current mirror circuit with MOSFETs operated in the subthreshold region is designed as shown in Fig. 2.

Large lengths of the devices used here help in very low power operation of the circuit. The total current consumption of the circuit in nominal condition was just 3.4 pA. The nominal output of the circuit is 96 mV. The I_{bias} terminal will be going to LDO as an input to obtain LDO block's bias current.

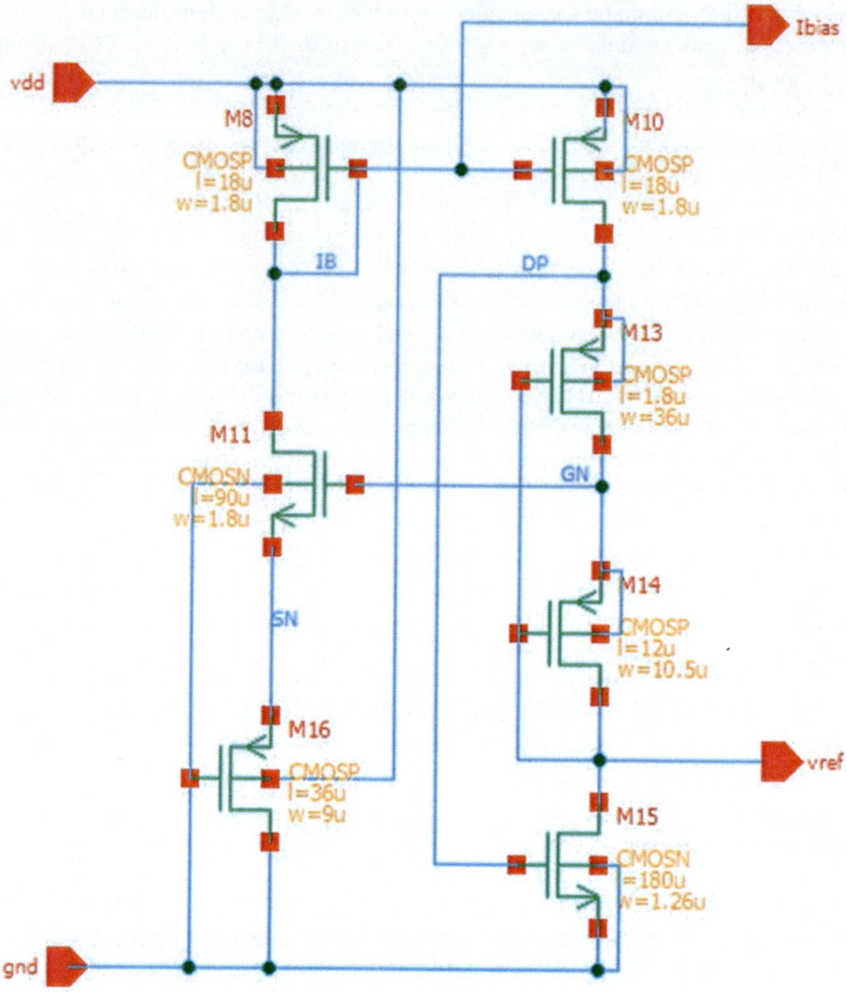

Fig. 2. Voltage reference

Figure 3 demonstrates the performance of the voltage reference across temperature and supply voltage.

V_{REF} varies from 87.36 to 112.61 mV across supply and temperature variation.

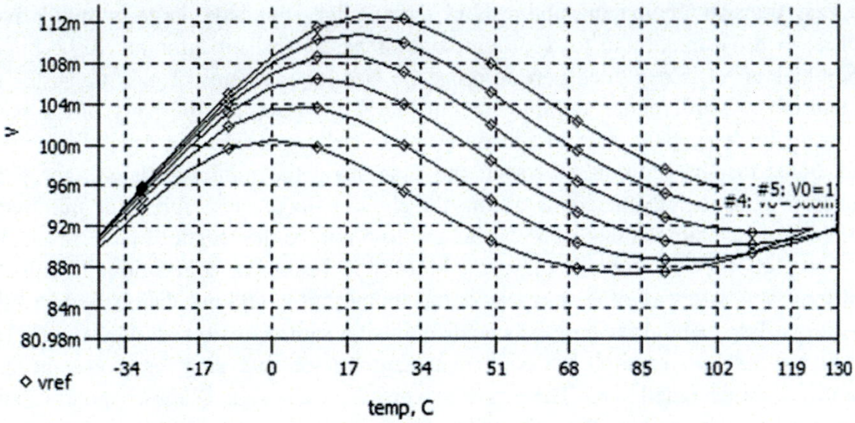

Fig. 3. V_{REF} variation with respect to temperature and V_{DD}

4 Proposed LDO

The proposed LDO is shown in Fig. 4.

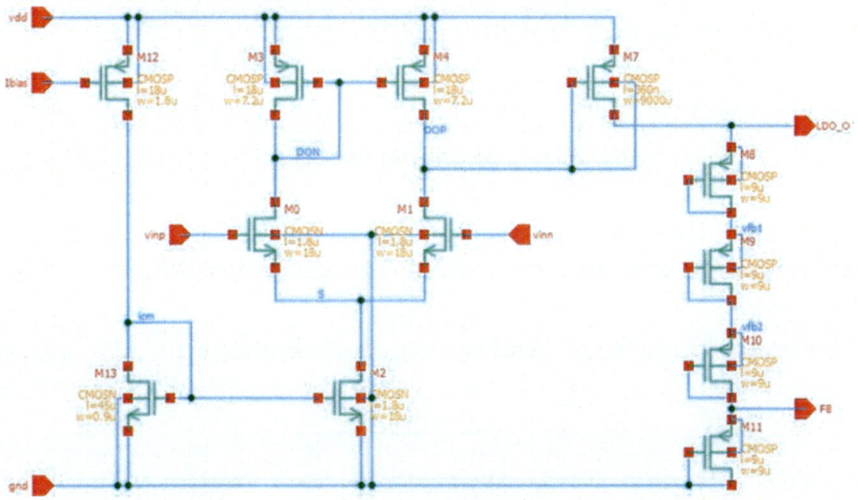

Fig. 4. Proposed low voltage, low power LDO

The designed LDO is biased using a mirrored current from voltage reference block to track the changes with respect to process, temperature, and voltage. The designed LDO is a two-stage design with first stage being the differential amplifier and in second stage, gain is achieved using pass transistor. All the transistors operate in subthreshold region of their operation.

Pass transistor being in subthreshold region, does not have large current drive. Hence, to drive maximum load current, size of pass transistor turns out to be huge. Body bias effect is exploited here to relax the size requirements of pass transistor to certain extent. Differential amplifier output biases the gate and the body of the pass transistor increasing the gain of second stage as well as the driving capability of the pass transistor. In addition, it was found that using the minimum length for pass transistor was not advantageous as the threshold voltage was higher for minimum length. Separate simulations were carried out to optimize the length of pass transistor.

Sampling network consists of purely MOSFETs instead of conventionally utilized resistors. This saves lot of area as low quiescent current would mandate large resistors and hence large area. Matching was achieved using same sizes for all the MOSFETs.

Figure 5 shows small-signal equivalent diagram for analyzing the loop gain and stability of the designed LDO. There exist two poles in the design. Dominant pole is at the gate of pass transistor which is due to high small-signal output resistance of differential amplifier stage and due to the equivalent large gate capacitance of pass transistor. First, non-dominant pole occurs at the LDO output which is due to the small-signal output resistance of the LDO and due to the expected load capacitance of about 1 pF.

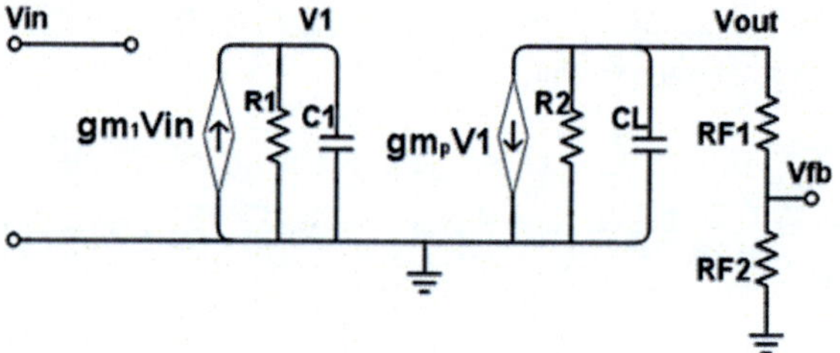

Fig. 5. Small signal analyses

gmp is the total transconductance of pass transistor, i.e., the summation of transconductance due to gate effect and due to body effect, $RF1 = \frac{3}{(gm||ro)}$ and $RF2 = \frac{1}{(gm||ro)}$.

The open-loop gain transfer function of the system is given by:

$$\frac{V_{fb}}{V_{in}} = -\frac{1}{4}\frac{(g_{m1}R_1)(g_{mp}R'_o)}{(1+sR_1C_1)(1+sR'_oC_L)}$$

where, g_{m1} = transconductance of i/p differential stage, R_1 = equivalent o/p resistance at first stage, C_1 = equivalent capacitance at first stage, and $R'_o = r_{op}||(R_{F1}+R_{F2})$.

As can be seen, there are two poles in loop transfer function. Here, the effect of RHP zero due to C_{GD} is neglected. The pole expressions can be approximated as:

$$P_1 = \frac{1}{R_1C_1} \text{ and } P_2 = \frac{1}{R'_oC_L}$$

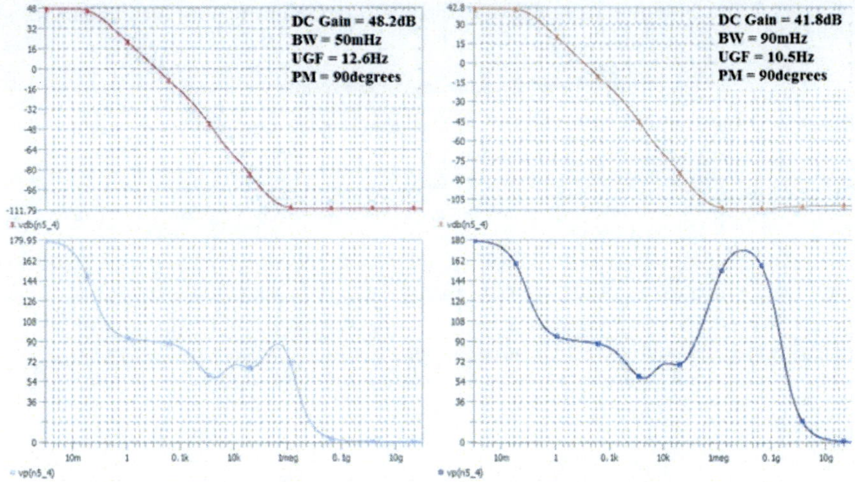

Fig. 6. Open-loop gain plots with V_{IN} = 0.5 V and R_L = 10 KΩ and 400Ω

5 Simulation Results

The circuit was designed and simulated using 180 nm CMOS technology. Figure 6 shows loop gain plot in nominal condition for min (40 μA) and max (1 mA) load current.

As can be seen, the loop gain is about 40 dB and loop is stable at approximate 90° PM. Table 1 summarizes loop gain results across voltage and temperature variation.

Loop is found to be all-time stable with good PM and enough loop gain of about 40 dB.

Table 1. Summary of loop gain results across voltage and temperature variation

	R_L = 10 KΩ				R_L = 400Ω			
	DC gain (dB)	BW (mHz)	UGF (Hz)	PM (°)	DC gain (dB)	BW (mHz)	UGF (Hz)	PM (°)
V_{IN} = 0.5 V								
T = −40 °C	50.67	9.8	3.55	89	47	14	3	89
T = 27 °C	48.2	50	12.6	90	41.8	90	10.5	90
T = 130 °C	43.81	355	56.2	90	35.17	800	50	90
V_{IN} = 1 V								
T = −40 °C	52.8	12.5	5.25	88	51.68	13	5.13	88
T = 27 °C	52.28	65	28.2	89	51.72	63	24.6	90
T = 130 °C	46.11	324	66	90	47.85	240	60.3	90

Tables 2 and 3 summarizes the line and load regulation results for the LDO.

Table 2. Line regulation results

Minimum load line regulation (R_L = 10 KΩ)					Maximum load line regulation (R_L = 10 KΩ)				
T	V_{IN}	0.5 V	1 V	Regulation	T	V_{IN}	0.5 V	1 V	Regulation
−40°C	V_{OUT}	0.356 V	0.363 V	6.94 mV	−40°C	V_{OUT}	0.351 V	0.359 V	8.07 mV
	4*V_{REF}	0.359 V	0.364 V	2.18 mV*		4*V_{REF}	0.359 V	0.364 V	3.31 mV*
27°C	V_{OUT}	0.383 V	0.450 V	66.8 mV	27°C	V_{OUT}	0.378 V	0.446 V	68.6 mV
	4*V_{REF}	0.384 V	0.450 V	1.3 mV*		4*V_{REF}	0.384 V	0.450 V	3.13 mV*
130°C	V_{OUT}	0.373 V	0.378 V	5.21 mV	130°C	V_{OUT}	0.365 V	0.374 V	8.24 mV
	4*V_{REF}	0.371 V	0.377 V	0.03 mV*		4*V_{REF}	0.371 V	0.377 V	3 mV*

Regulation (without asterisk) = Regulation including effect of supply variation on reference voltage
Regulation (with asterisk) = Regulation excluding effect of supply variation on reference voltage

Table 3. Load regulation results

V_{IN} = 0.5 V					V_{IN} = 1 V				
T	R_L	10 KΩ	400 Ω	Regulation	T	R_L	10 KΩ	400 Ω	Regulation
−40°C	V_{OUT}	0.356 V	0.351 V	4.73 mV	−40°C	V_{OUT}	0.363 V	0.359 V	3.60 mV
	4*V_{REF}	0.359 V	0.359 V			4*V_{REF}	0.364 V	0.364 V	
27°C	V_{OUT}	0.383 V	0.378 V	5.33 mV	27°C	V_{OUT}	0.450 V	0.446 V	3.50 mV
	4*V_{REF}	0.384 V	0.384 V			4*V_{REF}	0.450 V	0.450 V	
130°C	V_{OUT}	0.373 V	0.365 V	8.08 mV	130°C	V_{OUT}	0.379 V	0.374 V	4.88 mV
	4*V_{REF}	0.371 V	0.371 V			4*V_{REF}	0.371 V	0.377 V	

The worst-case line regulation is 6.62 mV/V and worst-case load regulation is 8.08 mV/mA.

Table 4 summarizes power dissipation results for the designed LDO.

Table 4. Power dissipation results

	I_{VREF} (pA)	I_{LDO} (pA)	P_{VREF} (pW)	P_{LDO} (pW)	P_{TOTAL} (pW)	I_{VREF} (pA)	I_{LDO} (pA)	P_{VREF} (pW)	P_{LDO} (pW)	P_{TOTAL} (pW)
T (°C)	V_{IN} = 0.5 V & R_L = 10 K Ω					V_{IN} = 0.5 V & R_L = 400 Ω				
−40	2.58	88.33	1.29	44.13	45.46	2.58	86.28	1.29	43.14	44.43
27	3.40	440.2	1.70	220.11	221.82	3.40	436.68	1.70	218.34	220.04
130	11.99	2850	5.99	1425	1431	11.99	2830	5.99	1415	1421
−40	5.97	128.15	5.97	128.15	134.12	5.97	125.51	5.97	125.51	131.48
27	6.73	844.42	6.73	844.42	851.15	6.73	839.76	6.73	839.76	846.49
130	11.68	2930	11.68	2930	2941.68	11.68	2920	11.68	2920	2931.68

Table 5. Performance comparison

Parameter	[6] 2008	[3] 2010	[7] 2016	[8] 2016	[9] 2017	[4] 2017	[5] 2018	This work
Process	350 nm	65 nm	130 nm	65 nm	65 nm	65 nm	65 nm	180 nm
Area (mm^2)	0.053	0.042	0.355	0.029	0.076	0.016	0.034	–
Type	Analog	Digital	Digital	Digital	Digital	Analog	Digital	Analog
Architecture	NA	SR based	SR based	ADC based	SAR/PD	NA	SR + AA	NA
Vin (V)	1.05 min	0.5 min	0.5–1.2	0.5–1	0.5–1	0.6	0.5–1	0.5–1
Vout (V)	0.9 typ	0.45 typ	0.45–1.14	0.45–0.95	0.3–0.45	0.3–0.55	0.45–0.95	0.4
Max Fs (MHz)	–	1	400	200	240	NA	10	–
Min IQ (μA)	4.02–164	2.7	24	12.5	14	32	3.2	**3nA**
Line regulation (mV/V)	1.1	3.1	–	–	–	–	–	6.62
Load regulation (mV/mA)	0.06	0.65	–	–	–	–	–	8.08
Resolution/SR bits	–	256	7 bit/128	NA	NA	NA	9 bit/28	–
Cout (nF)	1000	100	1	0.4	0.4	0.04	0.1	**0.001**
ΔIL (mA)	50	0.2	1.4	0.4	1.06	10	10	0.96
ΔVout (mV)	6.6	40	90	40	40	133.9	105	75
IL transient (ns)	10	–	NA	NA	1	150	1	2 s

The worst-case power dissipation for the complete LDO is just 3 nW which is lowest for an LDO reported so far in literature. Performance comparison in Table 5 demonstrates designed LDOs superiority as far as IQ and Cout are concerned.

6 Conclusion

A complete on-chip LDO with extremely low power dissipation is presented. The worst-case power dissipation is just 3 nW with a worst-case line regulation of 6.62 mV/V and worst-case load regulation of 8.08 mV/mA is achieved.

References

1. den Besten, G.W., Nauta, B.: Embedded 5 V-to-3.3 V voltage regulator for supplying Digital IC's in 3.3 V technology. IEEE J. Solid-State Circ. **33**(7), 956–962 (1998)
2. Milliken, R.J., Silva-Martinez, J., Sánchez-Sinencio, E.:. Full on-chip CMOS low-dropout voltage regulator. IEEE Trans. Circ. Syst. I Reg Pap **54**(9), 1879–1890 (2007)
3. Okuma, Y., et al.: 0.5-V input digital LDO with 98.7% current efficiency and 2.7-μA quiescent current in 65 nm CMOS. In: IEEE Custom Integrated Circuits Conference 2010, San Jose, CA, 2010, pp. 1–4. https://doi.org/10.1109/cicc.2010.5617586
4. Yang, F., Mok, P.K.T.: A 65 nm inverter-based low-dropout regulator with rail-to-rail regulation and over −20 dB PSR at 0.2 V lowest supply voltage. IEEE Int. Solid-State Circuits Conf. (ISSCC) Dig. Tech. Pap 106–107 (2017)
5. Huang, M., Lu, Y., Seng-Pan, U., Martins, R.P.: An analog-assisted tri-loop digital low-dropout regulator. IEEE J. Solid-State Circ. **53**(1), 20–34, (2018). https://doi.org/10.1109/jssc.2017.2751512
6. Lam, Y.H., Ki, W.H.: A 0.9 V 0.35 um adaptively biased CMOS LDO regulator with fast transient response. IEEE Int. Solid-State Circ. Conf. 442–443, (2008)
7. Nasir, S.B., Gangopadhyay, S., Raychowdhury, A.: All-digital low-dropout regulator with adaptive control and reduced dynamic stability for digital load circuits. IEEE Trans. Power Electron. **31**(12), 8293–8302 (2016)
8. Seok, M.: Fully integrated low-drop-out regulator based on event driven PI control. IEEE Int. Solid-State Circuits Conf. (ISSCC) Dig. Tech. Pap. 148–149 (2016)
9. Salem, L.G., Warchall, J., Mercier, P.P.: A 100 nA-to-2 mA successive-approximation digital LDO with PD compensation and sub-LSB duty control achieving a 15.1 ns response time at 0.5 V. IEEE Int. Solid-State Circ. Conf. (ISSCC) Dig. Tech. Pap. 340–341 (2017)